Springer
*Berlin*
*Heidelberg*
*New York*
*Hong Kong*
*London*
*Milan*
*Paris*
*Tokyo*

Keith D. Alverson • Raymond S. Bradley •
Thomas F. Pedersen (Eds.)

# Paleoclimate, Global Change and the Future

With 88 Figures, 51 in Colour, and 4 Tables

 Springer

**Editors**

*Keith D. Alverson*
Executive Director
PAGES International Project Office
Bärenplatz 2
3011 Bern, Switzerland
E-mail:*alverson@pages.unibe.ch*

*Raymond S. Bradley*
Distinguished Professor
and Head of Department
Department of Geosciences
Morrill Science Center
611 North Pleasant Street
Amherst, MA 01003-9297, USA
E-mail: rbradley@geo.umass.edu

*Thomas F. Pedersen*
Professor and Director
School of Earth and Ocean Sciences
University of Victoria
Petch 168, P.O. Box 3055 STN CSC
Victoria, BC, V8W 3P6, Canada
E-mail: *tfp@uvic.ca*

Cover photos:
1. The "Marion Dufresne" research vessel. Photo: L'Institut Français pour la Recherche et la Technologie Polaire, Plouzané, France
2. The Northern Icefield margin on Mount Kilimanjaro. Photo: D. Hardy, University of Massachusetts, Amherst, USA
3. The Rio Loa valley in the Atacama Desert of Northern Chile. Photo: C. Kull, PAGES IPO, Bern, Switzerland
4. The Lauterbrunnen valley in the Bernese Alps, Switzerland. Photo: C. Kull, PAGES IPO, Bern, Switzerland

ISSN 1619-2435
ISBN 3-540-42402-4  Springer-Verlag Berlin Heidelberg New York

Library of Congress Cataloging-in-Publication Data applied for

Bibliograhic information published by Die Deutsche Bibliothek
Die Deutsche Bibliothek lists this publication in the Deutsche Nationalbibliografie; detailed bibliographic data is available in the Internet at <http://dnb.ddb.de>.

Springer-Verlag Berlin Heidelberg New York
a member of BertelsmannSpringer Science+Business Media GmbH

http://www.springer.de

© Springer-Verlag Berlin Heidelberg 2003
Printed in Germany

The use of general descriptive names, registered names, trademarks, etc. in this publication does not imply, even in the absence of a specific statement, that such names are exempt from the relevant protective laws and regulations and therefore free for general use.

Product liability: The publishers cannot guarantee the accuracy of any information about the application of operative techniques and medications contained in this book. In every individual case the user must check such information by consulting the relevant literature.

Typesetting: Camera-ready by Isabelle Larocque
Cover design: E. Kirchner, Heidelberg

Printed on acid-free paper      32/3111/as  5 4 3 2 1

# Preface

WE LIVE IN UNUSUAL TIMES. Greenhouse gas concentrations are increasing rapidly and are now much higher than they have been for at least 420,000 years. Global average temperatures exceed anything seen in the last thousand years. The evidence is now overwhelming that such changes are a consequence of human activities, but these are superimposed on underlying natural variations. Climate on Earth naturally undergoes changes driven by external factors such as variations in solar output and internal factors like volcanic eruptions. How can we distinguish the human from the natural impacts? And what might the changes herald for the future of human societies as population pressure grows, as fossil fuel consumption increases and as land cover is altered?

Such questions are compelling, and the need for answers urgent. But the search for answers will only be successful when we have developed insight into the full range of natural variability of the climate system. That range is illustrated by the events of the past, and it is only by unravelling those events that we will be able to predict the future, and our place in it, with confidence.

This book stands as a progress report in the search for the past. It highlights a number of the extraordinary discoveries about the operation of the Earth System through time that have been made by natural scientists around the world over the last few decades. The great gains described in these pages have been wrought through exploration across the face of the planet and beyond: on land, sea, lakes, ice caps, via satellite observations and through simulations run on silicon chips. But that is only one dimension of the search, for critical to the future of human society is an improved understanding of the sensitivity of civilizations to climate change. Increasingly, paleoclimatologists are working with social scientists to disentangle the impacts of evolving social pressures and cultural practices from those induced by past climate change.

The scientific findings in these pages give cause for both exhilaration and concern. The exhilaration lies in appreciating the remarkable increase in our understanding of the complexity and elegance of the Earth System. The concern is rooted in recognizing that we are now pushing the planet beyond anything experienced naturally for many thousands of years. The records of the past show that climate shifts can appear abruptly and be global in extent, while archaeological and other data emphasize that such shifts have had devastating consequences for human societies. In the past, therefore, lies a lesson. And as this book illustrates, we should heed it.

Keith Alverson, Ray Bradley, Tom Pedersen, September 2002.

# Contents

**The Late Quaternary History of Biogeochemical Cycling of Carbon**

**Terrestrial Biosphere Dynamics in the Climate System: Past and Future**

# Contributors

*J. Allen*
Department of Biological Sciences Environmental Research Centre, University of Durham
South Road, Durham DH1 3LE, Great Britain

*K. Alverson*
PAGES IPO, Bärenplatz 2, 3011 Bern, Switzerland

*E. Balbon*
Laboratoire des Sciences du Climat et de l'Environnement, Laboratoire mixte du CNRS-CEA, 4
Avenue de la Terrasse, 91198 Gif-sur-Yvette, France

*J-M. Barnola*
UPR 5151-Laboratoire de Glaciologie et de Géophysique de l'Environnement, LGGE Centre National
de la Recherche Scientifique, 54 Rue Molière, FR-38402 Saint-Martin-d'Hères, France

*P.J. Bartlein*
Department of Geography, University of Oregon, 13th & Kincaid, 107 Condon Hall, Eugene, OR
97403-1251, United States of America

*T. Blunier*
Climate & Environmental Physics, Physics Institute, University of Bern, Sidlerstrasse 5, 3012 Bern,
Switzerland

*R.S. Bradley*
Climate System Research Center, Department of Geosciences, University of Massachusetts, Amherst,
MA, 01003, United States of America

*K.R. Briffa*
Climatic Research Unit, University of East Anglia, Norwich, NR4 7TJ, United Kingdom

*J. Cole*
Department of Geosciences, University of Arizona, Gould-Simpson, room 208, Tuscon, AZ 85721-
0077, United States of America

*Y.C. Collingham*
Institute of Ecosystem Science and Environmental Research Centre, University of Durham, School of
Biological and Biomedical Sciences, South Road, Durham, DH1 3LE, United Kingdom

*E. Cook*
Tree-Ring Lab/Lamont Doherty Earth Observatory, Columbia University, Rt. 9W, Palisades, NY
10964, United States of America

*E. Cortijo*
Laboratoire des Sciences du Climat et de l'Environment, CNRS/CEA Domaine du CNRS
Avenue de la Terrasse, FR-91198 Gif-Sur-Yvette. France

*R. D'Arrigo*
Tree-Ring Lab/Lamont Doherty Earth Observatory, Columbia University, Rt. 9W, Palisades, NY
10964, United States of America

*J. Dearing*
Department of Geography, University of Liverpool, PO Box 147, Liverpool L69 3BX, United
Kingdom

*R. J. Delmas*
Laboratoire de Glaciologie et Géophysique de l'Environnement, Domaine Universitaire B.P. 96, Saint Martin d'Hères Cedex, 38402 France

*M. Diepenbroek*
Alfred Wegener Institute for Polar and Marine Research, Am Alten Hafen 26, DE-27568 Bremerhaven, Germany

*C. M. Eakin*
NOAA/National Geophysical Data Center, World Data Center for Paleoclimatology, 325 Broadway E/GC, DSRC 1B139, Boulder, CO 80305-3328, United States of America

*L. François*
CICT, 118, route de Narbonne, FR-31062 Toulouse CEDEX 4, France

*R. François*
Woods Hole Oceanographic Institution, Dept. of Marine Chemistry and Geochemistry, Woods Hole, MA 2543, United States of America

*Z. Gedalov*
Climate Impacts Group, Joint Institute for the Study of the Atmosphere and Ocean (JISAO), Box 354235, University of Washington, Seattle, WA 98195-4235, United States of America

*E.C. Grimm*
Illinois State Museum, Research & Collections, 1011 East Ash Street, Springfield, IL 62703, United States of America

*M. Hoepffner*
CNES, MEDIAS- FRANCE, BPI 2102 18 avenue E. Belin, FR-31401 Toulouse-Cedex 4, France

*M.K. Hughes*
Lab. of Tree-Ring Research, University of Arizona, 105 W. Stadium, Bldg. 58, Tucson, AZ 85721, United States of America

*B. Huntley*
Environmental Research Centre, University of Durham, Durham, DH1 3LE, United Kingdom

*F. Joos*
Physikalisches Institut - Klima- und Umweltphysik, Universität Bern, Sidlerstr. 5, CH-3012 Bern, Switzerland

*C. Kull*
PAGES IPO, Bärenplatz 2, CH 3011 Bern, Switzerland

*L. Labeyrie*
Unité Mixte CEA-CNRS, Laboratoire des Sciences du Climat et de l'Environnement, Domaine du CNRS Bat 12, av. de la Terrasse, FR-91198 Gif sur Yvette Cedex, France

*K. Lambeck*
Research School of Earth Sciences, Australian National University, Canberra, ACT 0200, Australia

*I. Larocque*
PAGES IPO, Bärenplatz 2, CH 3011 Bern, Switzerland

*J. McManus*
Woods Hole Oceanographic Institution, 121 Clark Laboratory, MS#23, Woods Hole, MA 2543, United
States of America

*F. Oldfield*
Department of Geography, University of Liverpool, PO Box 147, Liverpool L69 3BX, United
Kingdom

*Y. Ono*
Laboratory of Geoecology, Graduate School for Environmental Earth Science, Hokkaido University,
Kita10 Nishi5, Kita-ku, Sapporo, Hokkaido 60, Japan

*T.J. Osborn*
Climatic Research Unit, University of East Anglia, Norwich, NR4 7TJ, United Kingdom

*J. Overpeck*
Department of Geosciences and Institute for the Study of Planet Earth, University of Arizona, Tucson,
AZ 85721, United States of America

*D. Paillard*
Laboratoire de Modélisation du Climate et de l'Environnement, CEA/DSM, Centre d'Etudes de
Saclay, 91191 Gif-sur-Yvette, France

*T.F. Pedersen*
School of Earth and Ocean Sciences, University of Victoria, PO Box 3055, STN SCC, Victoria, BC,
Canada, V8W 3P6

*J-R. Petit*
Laboratoire de Glaciologie et de Géophysique de l'Environnement, LGGE Centre National de la
Recherche Scientifique, 54 Rue Molière, FR-38402 Saint-Martin-d'Hères, France

*D. Raynaud*
UPR 5151-Laboratoire de Glaciologie et de Géophysique de l'Environnement, LGGE Centre National
de la Recherche Scientifique, 54 Rue Molière, FR-38402 Saint-Martin-d'Hères, France

*R. Spahni*
Climate and Environmental Physics, Physics Institute, University of Bern, Sidlerstrasse 5, 3012 Bern,
Switzerland

*T. Stocker*
Climate and Environmental Physics, University of Bern Sidlerstrasse 5, 3012 Bern, Switzerland

*J.L. Turon*
Dept. de Géologie et Océanographie, Université Bordeaux, 1 Av. des Facultés
FR-33405 Talence, Cédex, France

*C. Waelbroeck*
Domaine du CNRS, LSCE, labo mixte CEA-CNRS, bat. 12, FR-91198 Gif-sur-Yvette
France

*T. Webb III*
Dept of Geological Sciences, Brown University, 324 Brook St., Room 101. Providence, RI 02912-
1846, United States of America

*C. Whitlock*
Department of Geography, University of Oregon, Eugene, OR 97403, United States of America

*J.W. Williams*
National Center for Ecological Analysis and Synthesis (NCEAS), 735 State St., Suite 300, Santa Barbara, CA 93101-3351, United States of America

*S.G. Willis*
Institute of Ecosystem Science and Environmental Research Centre, University of Durham, School of Biological and Biomedical Sciences, South Road, Durham, DH1 3LE, United Kingdom

*Y. Yokohama*
Space Sciences Laboratory, University of California at Berkeley, CA 94720-7450, United States of America

# The Societal Relevance of Paleoenvironmental Research

F. Oldfield[1], K. Alverson

PAGES IPO, Bärenplatz 2, CH 3011 Bern, Switzerland

[1]current address: Department of Geography, University of Liverpool, PO Box 147, Liverpool L69 3BX, U.K.

## 1.1 Introduction

As the third millennium opens, it is clear that human beings are having a discernible impact on global climate. Profound changes are underway, but their attribution to specific causes poses a problem. What fraction can be assigned to human activities? Can we be sure that human impacts are not subordinate to natural variability? How can we gauge the severity, likely long-term effects and possible consequences of changes that we are inducing? Part of the answer lies in the exploration of the past. By understanding how climate has varied naturally in geologically recent times, we enhance our ability to peer into the future. This objective, simply stated, belies a remarkable complexity in the climate system and its linkages with other environmental systems. This book addresses the challenge posed by this complexity with a view to shedding light on current and future changes.

The meaning of the term 'global change' has become somewhat narrowed in recent literature. Increasingly, it has become linked exclusively to the major changes to the earth system that are currently underway. More often than not, the term is used to denote the inferred consequences of human actions. Here, we give the term its literal meaning without restriction to the most recent times, and without prejudgment with regard to underlying causation. Thus the shifts in climate that took place at the end of the last glacial period, as well as the biospheric responses to these shifts, are just as much examples of global change as are contemporary anthropogenic changes such as greenhouse gas concentration increases in the atmosphere, water and land degradation, and declining biodiversity.

The dominant theme of this book is past climate change and its links to other environmental systems at both global and regional scales. The book presents a synthesis of research in a broad, interdisciplinary field, the scope of which is defined by the major goals of the IGBP Past Global Changes (PAGES) project (Oldfield 1998, Alverson et al. 2000). The overriding concern of the research community contributing to the PAGES project has been to provide a quantitative understanding of the climate and environment in the geologically recent past and to define the envelope of natural variability alongside which anthropogenic impacts on the earth system may be assessed. Within this almost boundless remit, focus has been achieved by concentrating on those aspects of past environmental change that most affect our ability to understand and, wherever possible, predict and respond to future changes.

The 'raw material' available for this task includes:

- a wide variety of archives, both natural (ice and sediment cores, tree-rings, corals for example) and documentary;
- an even greater range of 'proxy' signatures decipherable within these archives using techniques such as microfossil, sedimentological, geochemical and stable isotope analysis;
- a range of dating techniques involving methods as diverse as layer counting and radioisotope decay series; and
- numerical models used to reconstruct past climates within the bounds of known dynamical constraints.

The main focus of the research summarized in this volume is on continuous paleoenvironmental records, with annual to decadal time resolution over the last few millennia, and mainly with decadal to century scale resolution, spanning the last several hundred thousand years. In order to document both lower frequency variability and the full range of transient extreme events recorded within the present interglacial, it is necessary to study the variability of the entire Holocene, which began with the dramatic transition out of the Younger Dryas cold period around 11.5k BP. Figure 1.1 shows how rapidly ice accumulation rates and oxygen isotopes in ice (a complex proxy of temperature) changed in Central Greenland at this time (Dansgaard et al. 1993). This remarkable climatic shift, recorded in many high resolution archives, was a rapid warming event felt over much of the earth's surface and is a striking example of global change by any definition. As is clear from the preceding part of the record, also in figure 1.1., it was by no means unique.

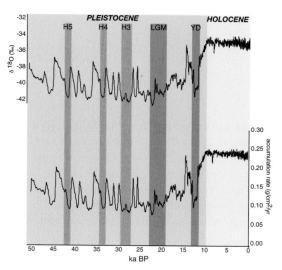

**Fig. 1.1.** Accumulation and isotopically inferred temperature over the past 50,000 years as measured in Greenland ice cores (Dansgaard et al. 1993).

Paleo-reconstructions of lower temporal resolution may be equally valuable in terms of their location and quantitative implications. Temperatures during glacial times reconstructed from noble gas ratios in ground water archives (Stute et al. 1995, Stute and Talma 1998, Weyhenmeyer et al. 2000) and temperatures of the last millennium reconstructed from bore-hole temperature inversions (Huang et al. 1996, Huang et al. 2000) are but two such examples.

The rationale for emphasizing past climate change rests on the following propositions:

- Climate has varied continuously on all timescales. Irrespective of the extent to which human activities lead to changes in the global climate system, it will continue to vary in the future. This is unquestioned; doubts surround only the nature of future variability and the degree to which the consequences of human activities will influence or perhaps even dominate it. Documenting past variability therefore has a vital role to play in understanding the present climate and predicting future change.

- Climate change affects human societies both directly and indirectly. Although the nature of the interaction between climate and human society is mediated by a wide range of cultural processes and varies greatly for different types of biophysical environment and social organization, the socio-economic impact of climate variability is substantial.

- Increasingly, policy development in virtually every sphere of life, from sustainable subsistence agriculture to infrastructure insurance in technological societies, rests, in part, on scenarios of future climate based on models that require empirical refinement and validation. An essential component of this validation consists of providing accurate reconstructions of past conditions.

- Part of the basis for understanding and predicting the course of future climate change lies in increasing our knowledge of the spatial and temporal patterns, causes and consequences of past variability.

Figures 1.2 and 1.3 show highly condensed impressions of temperature and hydrological variability respectively over a range of timescales. Figure 1.2 shows, from top to bottom, temperature variability associated with glacial cycles as inferred from $\delta^{18}O$ measurements in the Vostok ice core (Petit et al. 1999), millennial scale temperature variability during the last glacial period and the Holocene as inferred from $\delta^{18}O$ in Greenland ice cores (Dansgaard et al. 1993), and Northern Hemisphere average temperature over the last millennium as estimated from a network of multiproxy reconstructions and instrumental data (Mann et al. 1999). Figure 1.3 presents examples of past hydrological variability on three different timescales as derived from paleoproxy measurements. Sea level variability associated with glacial cycles (Waelbroek et al. 2002), lake level changes in Lake Abhé, Ethiopia during the Holocene (Gasse 2000) and lake level changes in Lake Naivasha, Kenya over the last millennium (Verschuren et al. 2000).

These figures illustrate several noteworthy points:

- climate variability is not just reflected in temperature – extreme hydrological variability, which is often of much greater importance to human populations, has been documented on all timescales in the past.

- underlying *global* climate change are very diverse *regional* variations which reflect the mechanisms responsible for the global change.

- although major, abrupt transitions, reflecting reorganization of the global system are most evident during glacial periods, they are not absent in the Holocene, especially in regional hydrological variability at lower latitudes.

- during the late Holocene, although natural forcings and boundary conditions were similar to those operating today, climate variability

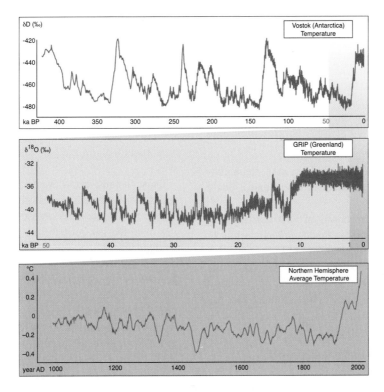

**Fig. 1.2.** Examples of past temperature variability on three different timescales as derived from paleoproxy measurements. Temperature variability associated with glacial cycles as inferred from δD measurements in the Vostok ice core (Petit et al. 1999), millennial scale temperature variability during the last glacial period and the Holocene as inferred from $\delta^{18}O$ in Greenland ice cores (Dansgaard et al. 1993), and Northern Hemisphere average temperature over the last millennium as estimated from a network of multiproxy reconstructions and instrumental data (Mann et al. 1999).

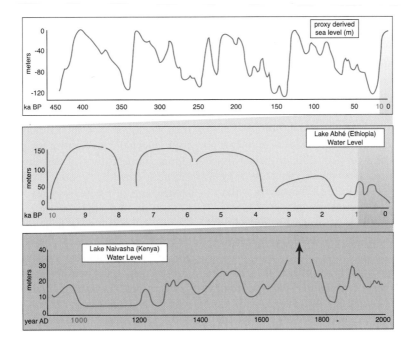

**Fig. 1.3.** Examples of past hydrological variability on three different timescales as derived from paleoproxy measurements. Sea level variability associated with glacial cycles (Waelbroek et al. 2002), lake level changes in Lake Abhé, Ethiopia during the Holocene (Gasse 2000) and lake level changes in Lake Naivasha, Kenya over the last millennium (Verschuren et al. 2000).

often greatly exceeded anything that is seen in modern instrumental records.

Irrespective of any evaluation of the effects of human activities, natural climate variability alone is thus of vital concern for the future. When we set this observation alongside the projected impact of anthropogenic forcing, and consider natural and anthropogenic factors together, several important questions arise:

- How does the paleo-record improve our understanding of earth system function and climate variability?
- What can reconstructions of past climate contribute to the detection and quantification of human impacts on climate?
- What does the past record tell us about the global and regional implications of potential future climate change for ecosystems and human societies?

## 1.2 A paleo-perspective on earth system function

Although past changes in climate could be described without reference to other aspects of earth system history, this would give no sense of the processes and functional linkages that are crucial for understanding and prediction. Climate variability reflects complex interactions between external forcing, ocean-atmosphere-biosphere-cryosphere dynamics and a range of environmental feedbacks. Among the relevant external forcing mechanisms, those operating on the longest timescales reflect the astronomical cycles that modulate the pattern of solar energy impinging on the earth and provide the chronometer for the onset of successive major glacial episodes in the geologically recent past. At the other extreme are the transient effects of major volcanic eruptions that impact climate primarily through the effects of atmospheric aerosols they produce. Between the two are changes in solar irradiance operating on decadal to millennial timescales. A wide range of internal modes of variability arises from feedbacks and system responses to external forcing. Among these are components of the earth system that operate on millennial timescales with a potentially high degree of hysteresis such as the deep ocean and the Antarctic ice sheet. Other parts of the earth system have more rapid characteristic response times. Many processes within the terrestrial biosphere and upper ocean for example operate over years, decades or centuries.

One effect of the interactions between external

forcings and internal system dynamics is that at any point in time the state of the earth system reflects both contemporary processes and those that are inherited from the past. This highlights the need for an understanding of earth system function that is firmly rooted in knowledge of the past. Furthermore some the processes that crucially modulate current modes of climate variability have undergone major rearrangements in the past and we do not yet know why.

These issues are addressed in the chapters that follow. Chapter 2 focuses on the history of atmospheric trace gases and aerosols on timescales ranging from the last millennium to the last four glacial cycles. Special attention is devoted to the role of changes in trace gas concentrations associated with periods of rapid climate change both within and at the end of glacial periods. Chapter 3 explores dynamical processes operating on timescales from decades to hundreds of millennia. Both data and model examples are used to explore past millennial scale climate variability, rapid changes, and glacial cycles. The carbon cycle is the main theme of chapter 4, though other chemical species critical to the functioning of the earth system and its living components are also considered. An overview of the history of marine and terrestrial sources and sinks of carbon and changes in the fluxes between various reservoirs is provided. Further consideration is given to the terrestrial biosphere in chapter 5, which deals with both its responses to climate change and the role it has played in modulating climate. The chapter also provides a paleo-perspective on future management and conservation issues. Chapter 6 highlights and analyzes the changes in climate that have occurred during the last thousand years, with a view to evaluating the relative importance of different forcing and feedback mechanisms, outlining patterns of natural climate variability over this time interval, and providing a dynamic baseline against which to assess anthropogenic greenhouse gas forcing. Reconstructions of the changing patterns of forcing and climate response through time help to identify the causal mechanisms for recent climate variability. They also provide strong evidence for a significant anthropogenic influence on global climate over the last few decades (Mann et al. 1998, Crowley and Kim 1999, Mann et al. 1999, Crowley and Lowery 2000, Stott et al. 2000, Stott et al. 2001). This approach complements studies that seek to detect anthropogenic climate change through a 'fingerprinting' approach (Forest et al. 2002). Paleoclimatic reconstructions therefore make a distinctive contribution to the often highly politicized debate about the causes of the recent warming trend and the extent to which increased greenhouse gas concentrations are contributing to it. Chapter 7 is concerned with the role humans have played as past drivers of change in terrestrial and aquatic ecosystems. Timescales of transformation resulting from human impact range from millennia to the last few decades. Human activities must be considered alongside natural environmental changes and the relation between the two is shown to be an interactive one. The final chapter aims to draw these many themes together and provides an integrated account of the main insights set out in the book, to identify the main implications for human societies and to propose research priorities for the future.

In seeking to gain a preliminary view of future climate change impacts, one of the most striking perspectives arises from comparing the range of predicted future changes in global temperature with estimates of the changes that occurred during the last millennium (Figure 1.4, see also Chapter 6, Section 6.3). At the global level, the most dramatic of these past changes were far below even the lowest predictions of future climate change. Yet this same period included major regional climate changes that were of great human significance in the past and would undoubtedly have dramatic consequences for present day societies. This brings us to a fuller consideration of the interactions between climate variability and human societies.

## 1.3 Past climate variability, human societies and human impacts

Studies of the relationship between past climate variability and human societies have often been marked by antithetical perceptions within the social science and physical science communities. Archaeological and anthropological research encompasses interpretations of socio-economic and cultural change, resource use and subsistence practices (Pringle 1997, Redman 1999), but the direct evidence for potentially damaging climate change is usually derived independently from different archives and lines of evidence (Cullen et al. 2000, Hodell et al. 2001). As a result, even where temporal correlations can be proposed between major societal changes and shifts in climate, they could be viewed as little more than coincidences. Taken to its extreme, the 'cultural' view attributes major changes in past societies, even the collapse of ancient civilizations, entirely to human actions. Although human actions are clearly important, too

one-sided an interpretation is not supported by research which ascribes the collapse of civilizations as diverse as the classic Akkadian (Cullen et al. 2000, DeMenocal et al. 2000), Mayan (Hodell et al. 1995, Hodell et al. 2001) and Anasazi (Dean et al. 1985, Dean et al. 1999) cultures to abrupt and persistent climatic changes (Figure 1.5, see also Chapter 7, Section 7.8). Such studies do not discount the role of societal factors, but assert that, at times, climate variability has been a critical factor influencing societal stability.

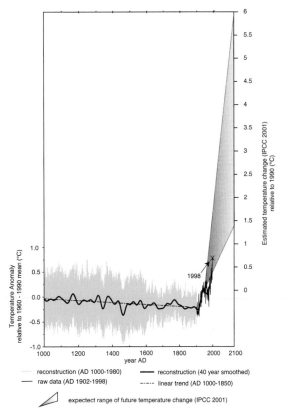

**Fig. 1.4.** A multiproxy reconstruction of mean annual Northern Hemisphere temperature (Mann et al. 1999) extrapolated to the range of IPCC estimates for the year 2100 (Houghton et al. 2001). A statistical confidence interval for the reconstruction is also shown.

### 1.3.1 The Anthropocene

Acknowledging the strong dependence of human-environmental interactions on socio-economic variables raises a further issue. The interaction between natural processes and human activities is complex. Moreover, the balance between the two has shifted dramatically (Vitousek et al. 1997, Meybeck et al. 2001), for the effects of human activities, especially during the last two centuries, have led to transformations much more significant than those resulting from climate change over the same period.

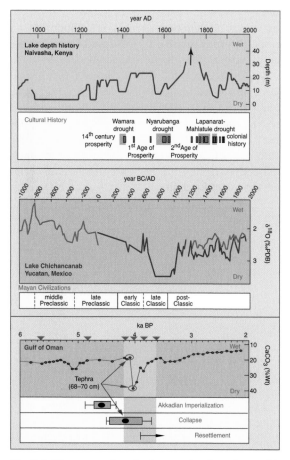

**Fig. 1.5.** Three examples of paleorecords where the combination of environmental and cultural history, coupled with rigorous chronological constraints, points to a strong link between the incidence of drought and the collapse of human cultures. In the upper panel, changing lake level in Lake Naivasha, Kenya as inferred from the sediment record is superimposed with a rough reconstruction, based on oral histories, of societal prosperity in the region (Verschuren et al. 2000). In the middle panel, drought in Yucatan, Mexico inferred from changes in the stable isotope ratios in two species of ostracod (red and blue lines) is shown to coincide with the collapse of Mayan Civilization (Hodell et al. 2001). In the lower panel a steep fall in carbonate percentage in a marine sediment record from the Gulf of Oman, representing a major episode of dust deposition, is directly linked to drought conditions associated with the demise of the Akkadian civilization (Cullen et al. 2000). This figure is reproduced from Alverson et al. (2001).

They have brought us into what may realistically be termed the "Anthropocene" (Crutzen 2002) and in so doing, endowed us with a 'no-analogue' biosphere as the canvas upon which future climate changes and human activities will interact. Moreover, the contemporary, no-analogue Anthropocene biosphere is now the initial condition on which changes related to increasing greenhouse gas concentrations will be superimposed. The Anthropocene is also marked by greenhouse gas levels well outside the range of at least the last 400,000

years (Figure 1.6, see also Chapter 2, Section 2.6) and global average temperatures that are the warmest for at least the past millennium (Figure 1.4, see also Chapter 6, section 6.6). At the same time, the pace of population growth, the level of technology and the degree of globalization of the world economy have endowed us with 'no-analogue' patterns of global and regional social organization. Does this imply that the insights to be gained from a deeper understanding of past human-environment relations are irrelevant for the future? Certainly they must not be overstated, but there are at least three important considerations that merit attention (1.3.2 to 1.3.4).

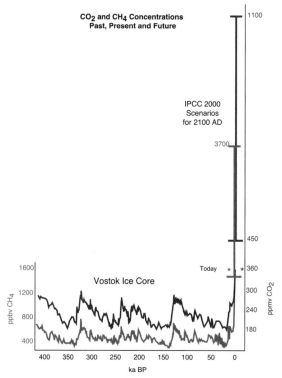

**Fig. 1.6.** Greenhouse trace gas ($CO_2$ and $CH_4$) changes over the last four glacial as recorded in the Vostok ice core (Petit et al. 1999) extrapolated to present day values and compared with the range of IPCC scenarios for the year 2100 (Houghton et al. 2001). This figure is reproduced from Alverson et al. (2001).

## 1.3.2 Societal responses to past climatic change

Over the last few years, paleoenvironmental, documentary, archaeological and anthropological approaches have begun to overcome the antagonism between dogmatic or over-mechanistic interpretations. The results reveal the complex nature of the relationship between environmental change and socioeconomic structures over longer time spans and during large or rapid environmental changes

(Chapter 7, Section 7.8). The insights obtained are particularly valuable with regard to contemporary societies whose patterns of social organization and resource use are most comparable to the earlier societies, and whose potential adaptation or mitigation strategies are most similar to those employed in the past.

### 1.3.3 Decadal-centennial modulation of modes of climate variability

One of the most striking successes in climate prediction has stemmed from the growing ability to anticipate the onset of El Niño. Although ENSO is one of the best known modes of interannual climate variability, with major implications for human activities, it is not the only one. The economic costs of El Niño events can be large (US$10$^{10}$) (Cane et al. 2000), as can ecosystem responses (Mantua et al. 1997, Hare and Mantua 2000). Its predictability is recognized to be dependent on understanding, among other things, the interplay between ENSO and longer, decadal to century scale modes of variability such as the Pacific Decadal Oscillation. One of the essential roles of paleoclimate research is to explore the variability of ENSO and especially changes in its teleconnections with a view to understanding their physical basis (Moore et al. 2001, Villalba et al. 2001). In this way, ENSO predictability under the changed conditions implied by future climate scenarios may be improved. The same applies to other modes of climate variability.

### 1.3.4 Vulnerability to extreme events

The impacts of extreme climatic events on even the most technologically advanced modern human societies are often severe. The period of instrumental records is too short to indicate the full range of hydrological and ecological stresses that will occur in the future. Even where relatively well informed planning is possible, decisions are necessarily based on assumptions about magnitude frequency relationships. As climatic boundary conditions change and as patterns of variability shift into different modes, these assumptions are compromised (Schrott and Pasuto 1999, Brown et al. 2000, Knox 2000, Messerli et al. 2000).

Figure 1.7, for example, shows the frequency of floods along the Colorado River drainage in the Southwest United States reconstructed for the past several thousand years (Knox 2000). The dramatic increase in flood frequency in the last few centuries is certainly not a simple expression of natural

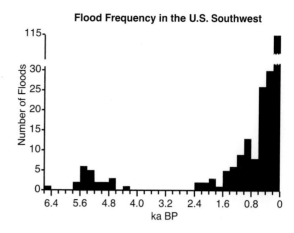

**Fig. 1.7.** The frequency of floods (Colorado River, Southwest United States, Knox 2000). The dramatic increase in flood frequency in the last few centuries is certainly not a simple expression of natural variability. However, looking further back to times well before substantial human influence could have occurred, it is clear that flood magnitude and frequency statistics can change substantially on centennial timescales.

variability. However, looking further back to times well before substantial human influence could have occurred, it is clear that flood magnitude and frequency statistics can change substantially on centennial timescales.

Figure 1.8 shows one example of an attempt to quantify the record of extreme events, in this case typhoons in the northern South China Sea (Huang and Yim 2001). Here typhoon frequency is reconstructed over three different timeframes using instrumental, documentary, and sedimentary records. Unfortunately, in this case there is little overlap between the time periods that these records cover, making cross validation difficult. Nonetheless, the records indicate that there is substantial natural variability in the frequency and magnitude of typhoons.

## 1.4 Hydrological variability

The most severe impacts of climate variability on human populations are often due to extreme drought, storm and flood events. Many environmental archives, as well as early documentary records provide clear evidence of such extremes. Persistent droughts, well beyond the range of those recently experienced, have been common in the past (Swetnam and Betancourt 1990, Gasse 2000, Verschuren et al. 2000) suggesting a high probability of occurrence in the future. Responsible planning must recognize this and allow for future environmental impacts associated with the wider range of hydrological variability documented in the past.

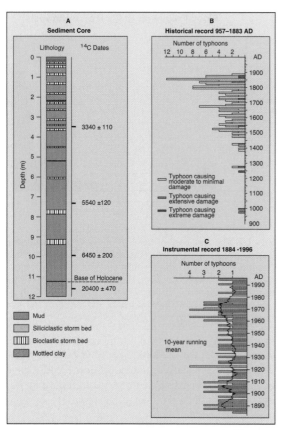

**Fig. 1.8.** An example of an attempt to quantify the record of extreme events, in this case Typhoons in the Northern South China Sea (Huang and Yim 2001). Here typhoon frequency is reconstructed over three different timeframes using instrumental, documentary, and sedimentary records. Unfortunately, in this case there is little overlap between the time periods that these records cover, making cross validation difficult. Nonetheless, the records indicate that there is substantial natural variability in the frequency of typhoons.

A common theme implicitly linking many environmental and human concerns is that of the quality and availability of fresh water. The paleorecord highlights enormous variability in water resources in many parts of the world where population pressure and demands for water are currently rapidly increasing. Availability of clean fresh water is already a dominant concern in many areas of the world and one likely to be greatly exacerbated in the future. In this regard, the course of environmental change in mountain regions is of special significance, since they provide a high percentage of the water used in many densely populated parts of the world (Loaiciga et al. 1996, Beniston 2000, Meybeck et al. 2001).

Groundwater is a major human resource. Under favorable conditions, it also acts as an archive of climatic and environmental change, through information obtained from conservative chemical species, noble gas ratios and isotopic signatures (Gasse

et al. 1987, Fontes et al. 1993, Stute and Talma 1998, Edmunds et al. 1999). The groundwater paleoarchive not only helps to quantify source contributions, ground water age, recharge rates and the future consequences of extraction, but also the effects of diffuse and point source contamination. Such considerations are of vital human importance especially in semi-arid areas where finite aquifers are being 'mined' much faster than they can be replenished. Rapidly falling water tables in such regions presage severe societal stress. The exhaustion of groundwater in some regions, for example Gujarat State, NW India, has already eliminated the principal buffer that could have been used to blunt the impact of potential future drought. In other regions, such as the mid-western United States similar depletion of groundwater is underway.

Documenting and better understanding the basis for the full range of potential climate driven hydrological variability is thus one of the most urgent and significant tasks for paleoscience. It is important to recall in this context that even the most sophisticated numerical climate models do not yet provide a realistic representation of precipitation patterns or amounts at regional scales.

## 1.5 Ecosystem processes

As the foregoing discourse shows, the human rationale for PAGES science extends into the area of past changes in the entire biosphere, whether as a result of climate change, human activities or the interaction between the two. Any consideration of human impacts on the biosphere entails an involvement with those aspects of global change that are, in terms of the distinctions made by Turner et al. (1990) 'cumulative' rather than 'systemic' (Chapter 7, Table 7.1). Such changes have, so far, had a much greater impact on hydrology, soils, vegetation and renewable resource systems over the last 150 years than have changes in climate. This is true despite the fact that over roughly the same time period, the change from the end of the Little Ice Age (ca. AD 1850) through to the last decade of the 20th century has, in many areas, already been one of the greatest climate changes recorded during the late Holocene (Bradley 2000).

Two main challenges arise from the above. The first is to reconstruct changes in ecosystems and hydrological regimes during the period of human impact. The second is to understand the consequences of these changes, especially those affecting gas and energy exchanges between the terrestrial biosphere and the atmosphere. For example, there is growing evidence from models, present day observations and paleodata/model comparisons (Ganopolski et al. 1998, Brovkin et al. 1999, Kleidon et al. 2000) that feedbacks to the atmosphere from land cover changes play an important role in modifying regional climate.

Even within the limited time-frame of the last two centuries, paleoscience has played a major role in documenting and improving our understanding of many environmental systems and processes (Battarbee 1990, Appleby and Oldfield 1992). This reflects, in part, the importance of decadal and century timescales for understanding the interaction of the many environmental processes upon which ecosystem function depends. Significant human impact on the environment and especially on terrestrial ecosystems did not begin two centuries ago; in many parts of the world, it began many millennia before present. On these longer timescales, the paleoperspective also provides an important complement to studies based on present day observations and modeling.

## 1.6 Landcover change

The present day landscape that forms the canvas upon which human endeavors and climate play out their counterpoint is itself the product of that counterpoint. Consider regions where soils are degraded, nutrient status diminished, organic matter content depleted or moisture retaining capacity is reduced. It is important to know, before modeling such systems or adopting the present day conditions as some kind of baseline, the extent to which current conditions are the product of current practices, or an inheritance from the past. Equally, it is important to explore the extent to which prevailing conditions reflect processes that are easily reversible or remediable, or are the product of past, non-linear shifts in ecosystem function that are much less tractable. The historical record is rich in examples of persistently reduced productivity arising from the transgression of thresholds in, for example, nutrient cycling or moisture retention (Bahn and Flenley 1992). Such switches have occurred in different parts of the world over at least the last six thousand years. It is often impossible to establish the relative importance of climate variability and human activity in driving systems over a given threshold since the two are interactive and can be mutually reinforcing in their effects.

Land cover changes brought about by human activity have a major influence on the impacts of climate variability, especially extreme events. Land cover modulates the expression of floods, their geomorphological impact (Starkel 1987, Eden and Page 1998) and the human hazards to which they give rise. There are many examples of how landcover changes, such as urbanization, deforestation and agriculture, have totally transformed downstream hydrological and sediment regimes. In many parts of the world, the period of direct observation and monitoring is not long enough to capture the full interaction of these processes. It is thus essential to build knowledge of past land cover change into the any evaluation of flood hazard based on magnitude/frequency statistics. The Yangtze River floods of 1998 that displaced in excess of 200 million people are a classic example of extreme climatic conditions exacerbated by land cover change, in this case deforestation in the watershed and loss of flood storage capacity downstream as a result of wetland reclamation for agriculture.

The past provides essential evidence for understanding the nature of human-environment interactions and optimizing them for the future. To consider the complex and rapidly changing scheme of human-environment interaction without regard to its antecedents is not a realistic enterprise. Human activities are as much drivers of contemporary and future environmental change as are anticipated changes in climate. The interplay between the two types of forcing is of vital concern and the history of their interaction is an important component in assessing future impacts.

The cumulative effects of land cover changes have global significance; they respond to and interact with the systemic changes currently under way. Even so, it is often most profitable to examine them within a more limited spatial framework since they are strongly differentiated regionally and it is at the regional level that their human implications are most evident. Global syntheses of such processes thus embrace representative case studies and recognize their diversity, rather than agglomerate them into global generalizations. It is through their effects on environmentally dependent resource systems that future changes in climate will most often impact human activities.

## 1.7 Biodiversity

The biodiversity that we now value is as much a product of environmental history as is any other aspect of the biosphere. Past interactions between climate variability and human activities (Chapter 7, Section 7.6) have served both to create and to jeopardize much of the biodiversity that is of vital concern at the present day. One of the key elements in any appraisal of biodiversity must be an understanding of the processes that have allowed for its development in the first place, its persistence, and finally those processes that threaten the survival of key taxa and habitats (Chapter 5, Section 5.4). This applies at all spatial scales from localized hotspots of high endemic diversity to major biomes such as the dwindling areas of equatorial rainforest. These historical aspects of biodiversity, though often neglected, pose a vital challenge to paleoenvironmental scientists at the present day.

## 1.8 Testing climate models with paleodata

Scenarios of how climate will vary in the future are dependent on model simulations. In order to be accepted as credible indicators of future conditions, such models must also be shown to be capable of reproducing conditions known to have occurred in the past. Models thus form one of the major links between past global change and decision making, which is inherently oriented towards the future. Subsequent chapters in this book include numerous examples of the interdependence of modeling and paleoreconstruction for model validation and process understanding. Paleostudies, for example, have played a major role in demonstrating the importance of terrestrial biosphere feedbacks on climate. Paleostudies also provide a basis for evaluating the dynamics of known past examples of abrupt climate change and exploring their potential relevance to the future (Claussen et al. 1999, De Menocal et al. 2000). One example is the possibility that future warming could lead to a decrease or shut down of North Atlantic Deep Water formation. The human consequences of any 'surprise' such as this are not well known but some of the physical and environmental consequences can be estimated by analogy with past periods when they are known to have occurred.

The emphasis in paleoclimate modeling is shifting towards transient experiments and regional simulations. Another technique, relatively new to paleo studies but long used in operational weather forecasting is inverse modeling, which allows known uncertainties in both data and model dynamics to be explicitly accounted for when addressing questions of model-data comparison (LeGrand and Wunsch 1995, LeGrand and Alverson 2001). These new modeling directions point towards a change in the interplay between models and data, with more emphasis on process understanding. Many concerns

about future environmental change are rooted at the local level and this is precisely where paleodata can make the strongest contribution to placing contemporary or model predicted variability in a longer term perspective. The trend towards regional model simulations will further reinforce the relevance of paleoresearch in decision making, provided quantitative regional reconstructions are available. At present regional model output based on various downscaling techniques remains highly uncertain. With this degree of uncertainty likely to remain within the near future, it is worth tapping the potential of paleorecords for climate change impact assessment studies. It is possible to generate scenarios for future change based on a range of variability reconstructed from proxy data. Of course this approach has limitations, since future forcing may generate variability outside the range of past extremes.

Moreover, a strong case can be made for using paleodata qualitatively to assess system predictability. In the case of well buffered systems that have not exhibited catastrophic responses to extreme events in the past, predictability may be relatively high. By contrast, there are also systems that have clearly shown repeated non-linear responses to high magnitude events. For these systems future behavior will remain extremely difficult to model or predict.

## 1.9 A paleoperspective on future global change

The scope of this book is wide-ranging. Drawing out the main conclusions from themes that are as diverse as ocean biogeochemistry, atmospheric physics, terrestrial paleoecology and environmental archaeology is a formidable task. The final chapter seeks to do this with an eye on those processes and interactions that are most significant for the future of human societies. The past does not provide a prescriptive guide to the future, but analyzed and interpreted with sufficient insight, it does inform evaluation of present day trends, future probabilities and likely human consequences.

The following are some of the key points to emerge from the present introduction:

- ongoing natural climate variability has affected every part of the earth system on all timescales
- the past range of climate variability exceeds that captured by the short period of instrumental observations
- atmospheric greenhouse gas concentrations

have varied in parallel with temperature throughout the last four glacial cycles

- current levels of greenhouse gas concentrations in the atmosphere are well above the upper limit of natural variability for at least the past 400,000 years
- understanding of the nature of the link between greenhouse gas concentrations and temperature in the past is a prerequisite for evaluating the consequences of recent increases
- understanding past ecosystem processes improves our ability to predict the likely effects of future global change
- in the past, areas that are today densely populated have experienced droughts and floods that would have disastrous consequences in the future
- the paleorecord is rich in examples of climate change coinciding with major changes in human societies
- humans have long had impacts on environmental processes with a range of consequences for ecosystems and sustainability
- the paleorecord can help to disentangle the natural climate variability and human activities as drivers of environmental change and to evaluate their interactions.
- studies of the past are an essential component of global change research which is otherwise limited by the short period of direct observations

Kenneth Boulding (1973) aptly describes the challenge we face:

"...whereas all experiences are of the past, all decisions are about the future... it is the great task of human knowledge to bridge this gap and find those patterns in the past which can be projected into the future as realistic images..."

Clearly, the paleorecord provides essential insight into past earth system variability. Without this perspective a full understanding of how the system works and why it has changed (an elusive goal even wearing the rosiest of glasses) would be unattainable. Attempts to anticipate future changes will inevitably fail without considering these past experiences. By synthesizing the evidence from the past and tying it to human concerns, this volume stands as a PAGES response to Boulding's challenge.

# The Late Quaternary History of Atmospheric Trace Gases and Aerosols: Interactions Between Climate and Biogeochemical Cycles

D. Raynaud
Laboratoire de Glaciologie et de Géophysique de l'Environnement, LGGE Centre National de la Recherche Scientifique, 54 Rue Molière, FR-38402 Saint-Martin-d'Hères, France

T. Blunier
Climate & Environmental Physics, Physics Institute, University of Bern, Sidlerstrasse 5, 3012 Bern, Switzerland

Y. Ono
Laboratory of Geoecology, Graduate School for Environmental Earth Science, Hokkaido University Kita10 Nishi5, Kita-ku, Sapporo, Hokkaido 60, Japan

R. J. Delmas
Laboratoire de Glaciologie et de Géophysique de l'Environnement, Domaine Universitaire B.P. 96, Saint Martin d'Hères Cedex, 38402 France

Contributors: J-M. Barnola, F. Joos, J-R. Petit, R. Spahni

## 2.1 Introduction: anthropogenic and natural changes

Human activities perturb the atmosphere and thereby influence the global climate. Prominent examples are greenhouse trace gases and sulfate aerosols which both affect the radiative balance at the surface of the Earth. This is also true for black carbon and organic carbon aerosols emitted by burning of biomass and fossil fuel, as well as eolian mineral dust originating from changes in land use and land cover.

Given these ongoing anthropogenic changes, understanding the past record of atmospheric composition is important for several reasons:

- Over the past hundred years it is not always an easy task to separate atmospheric changes induced by human activities from those related to natural variability. Only the longer term past record provides the context of natural variability within which recent anthropogenic perturbation has taken place.
- The rate of anthropogenic perturbation is very high. Since 1750 the atmospheric $CO_2$ concentration has increased by 30% and $CH_4$ by 150%. Did similar rates occur in the past and, if so, how did they affect climate and environment?
- Complex feedback mechanisms exist between the climate system and the trace gas content of the atmosphere. Past changes can be used to study such atmosphere-climate feedbacks and the processes involved.
- The paleorecord offers a unique opportunity to test the ability of climate models to simulate interactions between atmospheric composition and climate, under different climatic conditions.

The Earth has experienced major changes in its atmospheric composition over the past several hundred thousand years. During this time, atmospheric composition and climate have interacted alongside external influences (or forcings), such as changes in the pattern of solar radiation incident on the Earth's surface. The aim of this chapter is to review the history of atmospheric composition as it interacted with climate during the last four major climatic cycles (roughly the last 400,000 years). We concentrate on processes operating on time scales covering the interval between 100,000 years and a century. There are two major reasons for choosing this time period: (1) the atmospheric record further back in time is generally not well documented and (2) the last four glacial cycles encompass a wide spectrum of climatic conditions.

In this chapter, we begin with a brief introduction on past variations of greenhouse gases and aerosols during several different, climatically significant periods of the past: glacial-interglacial cycles (section 2.3), abrupt climatic changes (section 2.4), the Holocene (section 2.5), and the last millennium (section 2.6). In the last section (2.7) we highlight significant conclusions that can be drawn in terms of climate and biogeochemical cycles. The following section (2.1) is devoted to explaining the significance of the ice core archive of atmospheric gases and aerosols.

### 2.1.1 Greenhouse gases

Greenhouse gases are transparent to incoming solar radiation but opaque to the infrared radiation emitted by the earth's surface. The main greenhouse gas is water vapor whose atmospheric content is dominantly influenced by climatic conditions, primarily temperature. Water vapor concentration in the atmosphere varies widely in space and time, though its long term temporal variability is not well constrained by data. It is worth noting that the direct emission of water vapor by human activities is negligible compared to natural fluxes. Other greenhouse gases exist in much smaller concentrations in the atmosphere, but their increase due to anthropogenic activities may drastically affect our future climate. Most of them, such as $CO_2$, $CH_4$, $N_2O$ and tropospheric ozone were present in the atmosphere prior to the changes brought about by anthropogenic activities. Some trace-gases, like CFC's, exist solely due to humans.

Some trace gases, which are not themselves radiatively active, nonetheless interact through atmospheric chemistry processes with greenhouse gases. CO, for example, although not itself a greenhouse gas is involved in setting the oxidizing capacity (OH cycle) of the atmosphere, and therefore affects the atmospheric sink of $CH_4$.

### 2.1.2 Aerosols

An aerosol is a suspended liquid or solid particle in a gas. Aerosols constitute a minor (~ one ppb by mass) but important component of the atmosphere. Indeed, aerosols play an active role in atmospheric chemistry. Natural aerosols may be divided into two classes, primary and secondary aerosols, arising from two different basic processes:

Primary aerosols derive from the dispersal of fine materials from the earth's surface. There are two major categories of natural primary aerosols: sea salt and soil dust. Most sea salt particles are produced by evaporation of spray from breaking waves at the ocean surface whereas mineral dust particles are mostly generated by winds in arid continental regions and, sporadically, by explosive volcanic eruptions which emit huge amounts of ash particles. The continental sediments formed during glacial periods by the deposition of wind blown dust are called loess.

Secondary aerosols are formed by chemical reactions and condensation of atmospheric gases and vapors. The sulfur cycle dominates the tropospheric secondary aerosol budget. In pre-industrial conditions, it is mainly linked to marine biogenic activity, which produces large amounts of gaseous dimethylsulfide (DMS). Once in the atmosphere, DMS oxidizes primarily into sulfuric acid, ultimately present in the atmosphere in the form of fine aerosol droplets. This fine aerosol is also partly of volcanic origin. Presently, anthropogenic $SO_2$ emission linked to fossil fuel, biomass burning and industrial processes dominates over all natural sources of atmospheric sulfur. In addition, it has been suggested recently that organic matter, frequently associated with sulfur species, may constitute a major fraction of secondary aerosol particles. Organic carbon and "black carbon" (called soot) are the main organic aerosol particles. They are primarily produced by the emission of smoke from biomass burning.

Primary and secondary particles may interact strongly in the atmosphere, turning atmospheric aerosols into a very complex mixture. They are removed from the air by both dry and wet deposition.

Atmospheric aerosols influence climate in two ways: directly through reflection and absorption of solar radiation, and indirectly by modifying the optical properties and lifetimes of clouds. In addition to ash, large explosive volcanic eruptions sporadically inject (a few times per century) huge amounts of $SO_2$ into the stratosphere. The sulfuric acid veil formed after such eruptions may persist for several years at an elevation of about 20 km, markedly cooling the global climate.

Dust, be it desert, volcanic or soot particles, when deposited on the surface of snow and ice may decrease its albedo and enhance surface melting. This process, along with high altitude temperature increases, is contributing to the retreat of glaciers and snowfields in high mountain areas worldwide. It has even been suggested that bacteria living on the surface of glaciers could be nourished by these aerosol deposits, further changing the albedo and melting rate.

Large amounts of mineral substances (e.g. iron, nitrate, phosphorous) are transported as aerosols. In some cases, these substances can act as nutrients, enhancing marine biogenic activity and the rate of the atmospheric $CO_2$ sequestration in the ocean. Moreover, continental carbonate aerosol inputs may potentially change the alkalinity of shallow oceanic layers, modifying surface ocean chemistry and consequently air/sea exchanges of $CO_2$. Since atmospheric dust concentrations were strongly enhanced during glacial periods, these processes must be considered in paleoclimatic studies.

As previously mentioned, dimethylsulfide (DMS) is produced in the ocean by marine biogenic activity. This gas is the primary natural source of atmospheric sulfate. Non-sea-salt sulfate (or nss-$SO_4$) and methanesulfonic acid (MSA) are the two most important compounds formed as DMS is oxidized in the atmosphere. MSA is of particular interest in that it is uniquely attributable to marine biogenic activity. When oxidized in the atmosphere, DMS forms acidic gases or fine aerosols which are a major source of cloud condensation nuclei. It has been proposed that a higher emission rate of DMS may lead to an albedo increase due to enhanced cloud cover which, consequently, decreases absorption of solar radiation and lowers surface temperature (Liss et al. 1997, Charlson et al. 1998). This climatic process can in turn influence the rate of DMS production. If a sea-surface temperature (SST) decrease were to enhance production of DMS, it would provide a positive feedback to SSTs. However, both the direction and magnitude of this feedback is currently unknown (Watson and Liss 1998).

The determination of MSA in polar ice is relatively recent (Saigne and Legrand 1987) and there are indications that MSA ice records may suffer post-deposition modifications (Mulvaney et al. 1992). Measurements at Vostok Station suggest that 80 to 90% of this compound may have escaped the ice matrix in the first few meters (Wagnon et al. 1999). This finding casts doubts on the actual significance of MSA concentrations measured in polar ice cores. Despite these difficulties, MSA has become a species of primary importance in ice core studies: it has been proposed as a tool for reconstructing the past history of major El Niño events (Legrand and Feniet-Saigne 1991) and documenting the co-variations of marine biogenic activity with known past climate variability (Legrand et al. 1991, Legrand et al. 1992). Furthermore, it has been proposed that marine biogenic activity might be regulated by atmospheric deposition of continental dust, containing Fe and nitrate onto the ocean surface, particularly in high nutrient low productivity areas (chapter 4). However, it has yet to be demonstrated that marine biogenic activity was actually enhanced during ice ages due to higher dust input to the ocean.

Anthropogenic activities increase the amount of secondary particles in the atmosphere, as well as soot. The majority of anthropogenic aerosols exist in the form of sulfate and carbon, but also of nitrogen compounds (e.g. nitrate and ammonium). Presently, anthropogenic activities contribute approximately 20% to the global aerosol mass burden, but up to 50% to the global mean aerosol optical depth. It is generally accepted that the net global radiative forcing due to anthropogenic aerosols is significant and negative (i.e. it tends to cool the average global temperature) (Sato et al. 1993). However, quantification of their climatic effects remains difficult, in particular due to large uncertainties associated with the indirect impact of aerosols on clouds.

## 2.2 The significance of past atmospheric records

### 2.2.1 Aerosol incorporation and gas occlusion in ice

Gaseous and particulate impurities are mostly advected from low and mid latitudes to the polar areas by wind, generally in the troposphere, though sometimes by a stratospheric pathway. Local sources are generally negligible.

Aerosol particles are incorporated in snowflakes by nucleation scavenging at cloud level, by below-cloud scavenging (this process is, however, rather unimportant in clean air conditions) and by dry deposition onto the snow surface. It is generally accepted that wet deposition is dominant when the accumulation rate is high (> 20 g cm$^{-2}$ yr$^{-1}$). On the other hand, at low accumulation sites, such as Dome C and Vostok stations in central Antarctica, dry deposition dominates. At Vostok, where accumulation is currently 2.3 g cm$^{-2}$ yr$^1$, dry deposition contributes probably up to 80%, except for acid species such as $HNO_3$, HCl, MSA and carboxylic acids, for which the dry deposition process seems to be negligible. Moreover, the stability of these species in firn is questionable, in particular at very low accumulation sites (De Angelis 1995, Wagnon et al. 1999, Wolff 1996).

The relationships linking snow chemistry to atmospheric concentrations (the so-called "transfer functions") are still not fully elucidated for aerosol species and reactive gases. In many cases, glaciochemists have used empirical formulas to interpret the ice records in terms of past atmospheric composition, assuming that the processes involved in the transfer function did not change under different climate conditions and are therefore well represented by the empirical relation. Lack of understanding of the transfer processes is a critical point in our interpretation of the past aerosol records.

Regarding air composition, glaciers and ice sheets are unique as they are the only paleoarchives which directly record atmospheric composition in the form of trapped air bubbles. The top layer (roughly the first 50 to 130m) is formed by porous firn, which is not yet consolidated ice. Atmospheric air exchanges with the air in this layer. Air in the porous firn thus consists of a mixture of air that was last in equilibrium with the atmosphere at different times (Schwander 1996). As a consequence, at a given time, the concentrations of gas species in the firn are not the same as their atmospheric concentrations due to physical, and (in case of reactive gases) chemical, processes. Air occluded in ice at the bottom of the firn column does not have an exact age reflecting its last contact with the atmosphere, but an age distribution. Thus the temporal resolution of any trace gas record from ice cores is inherently limited. This age distribution is a function of both temperature and accumulation, which are highly variable from site to site (Table 2.1). Generally, it increases with decreasing temperature and accumulation. The width of the age distribution is as low as 7 years at high accumulation/ high temperature sites but can be several centuries for Antarctic low accumulation/ low temperature sites (Schwander and Stauffer 1984).

In the top few meters of firn, air is well mixed by convection and so has essentially the same composition as the atmosphere. Below this zone the air in the firn is static and mixes mainly by molecular diffusion. An equilibrium between molecular diffusion and gravitational settling is reached for each gas component (Craig et al. 1988, Schwander and Stauffer 1989). As a consequence, two gas components with different molecular weights fractionate with depth relative to their initial relationship in the atmosphere. The magnitude of this fractionation is well known, allowing an accurate, corrected ice core record to be constructed. The process is most important for isotope records but can be neglected when dealing with concentration records.

Another consequence of the air occlusion at the bottom of the firn column is that the age of the air in an occluded air bubble is less than the age of the surrounding ice. This age difference ($\Delta$age hereafter) is the difference between the age of the ice and the mean age of the air at the depth of occlusion. The age of the ice at close off is the dominant term of $\Delta$age. $\Delta$age can be calculated using a model for the firn densification process. Although the process is well understood, the accuracy of the calculation is limited due to uncertainty in past accumulation and temperature. However, $\Delta$age calculated with the densification model is in excellent agreement with independent measurements of $\Delta$age based on temperature diffusion processes in the firn for central Greenland ice cores (Severinghaus et al. 1998, Leuenberger et al. 1999, Severinghaus and Brook 1999). $\Delta$age can be substantial. Under present day conditions $\Delta$age is on the order of a few centuries for high accumulation/ high temperature sites like

Central Greenland (Schwander et al. 1993). At Vostok, where accumulation and temperature are low, $\Delta$age is about 3000 years (Schwander and Stauffer 1984, Barnola et al. 1991). Under colder climate conditions, which is paralleled by lower accumulation, $\Delta$age increases significantly.

Below the close off zone the air bubbles shrink in size as the ice flows to deeper strata under the increasing pressure of overlying layers. When the pressure gets high enough the gas is transformed from air bubbles into air hydrates (Miller 1969). For example, first hydrate formation has been observed at 500 m depth at Vostok and 1022 m at Summit Greenland, respectively (Shoji et al. 2000). Hydrates decompose into bubbles again once the core is recovered and relaxed. However, a full reformation of air bubbles takes several decades.

### 2.2.2 How reliable are the climate records obtained from ice cores?

How sure are we that the measured gas composition represents the atmospheric concentrations at trapping time? As pointed out above, the record of atmospheric gas composition is smoothed in the ice. This smoothing may be significant and averages over centuries for Vostok station (Schwander and Stauffer 1984, Barnola et al. 1991, Rommelaere et al. 1997), where the accumulation rate is very low and the temperature among the coldest. Would a concentration increase like the one we observe over the last few centuries be visible in the Vostok record? In Figure 2.1 we show how the present anthropogenic $CO_2$ increase would be recorded at Vostok. For simulating the changes in atmospheric $CO_2$ concentrations after the year 2000 we use the Bern model (Joos et al. 1996). As a lower limit for the future $CO_2$ concentration we set the anthropogenic $CO_2$ emission to zero after the year 2000. The smoothed "Vostok record" is obtained using the Schwander model for gas occlusion (Schwander et al. 1993) with an extension for gradual gas occlusion over the close off interval (Spahni, in preparation). The resulting concentration propagation with a maximum of about 315 ppmv and a very slow decrease to the pre-industrial background clearly stands out. At no place in the Vostok record has such a high concentration with such a concentration trend been measured. Strictly speaking the Vostok record does not exclude a pulse-like atmospheric $CO_2$ signal of a few decades duration with concentrations as high as today. However, this would require both a large carbon release (order 200 GtC) and an equally large uptake within a few decades only. Such an oscillation is not compatible with our present view of the global carbon cycle.

**Table 2.1.** Characteristics of selected ice cores

| PLACE | NAME | LOCATION | ELEVATION (m.a.s.l.) | ACCUMULATION (cm ice a$^{-1}$) | TEMP. (°C) |
|---|---|---|---|---|---|
| Greenland | GISP | 72°36'N, 38°30'W | 3214 | 25 | -31.4 |
| Greenland | GRIP | 72°35'N, 37°38'W | 3238 | 23 | -31.7 |
| Peru | Huascáran | 9°07'S, 77°37'W | 6048 | 140 | -9.8 |
| Antarctica | Law Dome (DSS) | 66°46'S, 112°48'E | 1370 | 60 | -22 |
| Antarctica | D47 | 67°23'S, 137°33'E | 1550 | 30 | -25 |
| Antarctica | H15 | 69°05'S, 40°47'E | 1057 | 26 | -20.5 |
| Antarctica | Dome C | 75°06'S, 123°24'E | 3233 | 2.7 | -54 |
| Antarctica | Vostok | 78°28'S, 106°48'E | 3488 | 2 | -55 |
| Antarctica | Byrd | 80°01'S, 119°31'W | 1530 | 11.2 | -28 |
| Antarctica | South Pole | 90°S | 2835 | 8 | -49.4 |

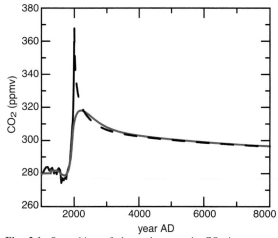

**Fig. 2.1.** Smoothing of the anthropogenic $CO_2$ increase at Vostok. Solid line: Atmospheric $CO_2$ concentration from ice cores and atmospheric measurements up to the year 2000. Dotted line: Propagation of the atmospheric $CO_2$ concentration calculated with the Bern model (Joos et al. 1996) setting the hypothetical anthropogenic carbon source to zero after the year 2000. Red solid line: prediction of how the atmospheric signal would be recorded in the Vostok ice core, calculated with the Schwander model for gas occlusion (Schwander et al. 1993) with an extension for gradual gas occlusion over the close off interval (Spahni in preparation).

Not all glaciers and ice sheets are equally useful regarding the past atmospheric composition. Where melting occurs, gas content and gas composition may be altered by chemical reactions taking place in aquatic systems or by physical gas exchange between the gaseous and the aquatic sections. In Greenland and Antarctica surface melting is sporadic. At these sites the gas occlusion occurs by dry sintering of the firn described above. As soon as there is melting the trapping gets more complex and erratic. Melt layers may block the exchange to the overlying atmosphere and due to the solubility of gases the gas composition in the melt layer may change tremendously. Such "wet trapping" is mostly observed in Alpine and mid latitude glaciers. Under dry, cold conditions of polar areas no chemical reactions were observed during the trapping of trace gases. This is demonstrated by the generally good match of atmospheric and ice core data. However, very slow chemical reactions are able to alter gas concentrations in ice cores due to the long time available for the reaction to proceed. This has been observed in Greenland where high impurities in the ice leads to significant *in situ* $CO_2$ production via acid-carbonate interactions and oxidation of organic material (Delmas 1993, Anklin et al. 1997, Haan and Raynaud 1998). It has been demonstrated that the Antarctic records provide the most reliable data of changes in global atmospheric $CO_2$ (Raynaud et al. 1993), probably within a few ppmv. Antarctic results are consistent despite the fact that the coring sites have different accumulation, temperature and impurities. $CO_2$ measurements made several years apart on the same core show no significant changes. Bacterial activity has the potential to alter the trace gas composition and its isotopic signature. Viable bacteria have been found in several ice cores to significant depths (Abyzov et al. 1993, Christner et al. 2000). Bacterial formation of trace gases is probably important in temperate glaciers, where much organic matter is trapped in the ice. In polar regions the bacterial concentration is lower and bacterially induced alteration of the trace gas composition has yet to be confirmed. However, $N_2O$ measurements from the Vostok ice core for the penultimate glacial period show values that are suspected to originate from *in situ* bacterial production (Sowers 2001). This hypothesis is supported by the fact that the layers with unusual $N_2O$ values contain high dust content, which may correlate with high bacterial content (Christner et al. 2000). In another ice core from Greenland, unreasonably high values have been found sporadically. However, in general, $N_2O$ results from old and new cores from the northern and the southern hemisphere agree (Flückiger et al. 1999). The significance of bacterial alteration for trace gas records (especially $N_2O$) has yet to be demonstrated.

A further consideration is post-coring effects from storage and contamination during the extraction of the ice. These effects are important when analyzing trace metals or the isotopic composition of gases, but are generally not important with respect to the concentration measurements discussed here.

## 2.3 Glacial-interglacial cycles

Figure 2.2 provides a longer-term perspective on atmospheric trace gas changes over 4 glacial-interglacial cycles covering roughly the last 400,000 years.

### 2.3.1 Greenhouse gases

The Vostok ice core shows a remarkable correlation between temperature and greenhouse gas concentrations on glacial-interglacial scales (Figure 2.2; Petit et al. 1999). Concentrations are generally higher during interglacials than during glacials. Although there is no continuous, quantitative proxy record for water vapor, the most important greenhouse gas, it was certainly lower during cold and dry glacial periods. $CO_2$ concentrations oscillate between 180-200 ppmv during the coldest glacial periods and 280-300 ppmv during full interglacials. The $CH_4$ concentration is about 700 ppbv during interstadials and only half of that during cold periods. Model results indicate that changes in the oxidizing capacity of the atmosphere can explain only 10-30% of this variability (Thompson 1992). Therefore the changes must have been caused by changes in the sources of this gas. A complete record of the $N_2O$ concentration over glacial-interglacial cycles does not exist. However, measurements over the last termination and during millennial scale Dansgaard-Oeschger events indicate variations with the same general pattern as for $CH_4$, with lower concentrations during cold periods (Flückiger et al. 1999). The Vostok $N_2O$ record indicates values for the last interglacial similar to those recorded during the pre-industrial period in other cores (Sowers 2001); as already mentioned in section 2.2, the $N_2O$ record for the penultimate glacial period is potentially altered, possibly by bacterial activity (Sowers 2001).

The average rates of concentration changes between glacial-interglacial conditions are much smaller than those observed for the period covering the last 250 years (industrial era): about 35, 120 and 25 times slower for $CO_2$, $CH_4$ and $N_2O$, respectively.

One outstanding question on glacial-interglacial time scales is why atmospheric $CO_2$ rose by 80 to 100 ppmv during glacial to interglacial warmings. There is no doubt that the ocean must have played a major role since it represents the largest pool of available $CO_2$ that could have been delivered directly to the atmosphere, and because the rapidly expanding continental biosphere probably acted as a $CO_2$ sink during these major warming periods. There is a third player, namely changing rates of weathering of silicate and carbonate rocks. Although it certainly played a role, this process is too slow to explain the rapidity of transitions seen in the Vostok record. We can therefore assert that the 170-190 Gt of carbon, which accumulated in the atmosphere during the several millennia corresponding to glacial to interglacial transitions, must have been the result of a net flux out of the oceans. This is a minimum estimate, since during the same time the uptake of atmospheric $CO_2$ by the continental biosphere, although not well constrained, was almost certainly positive. Estimates for this continental uptake range between 0 and 1200 GtC with a best estimate of around 400 GtC. In conclusion, there must have been a transfer of several hundred to about one thousand gigatons of carbon from the ocean to the atmosphere and terrestrial ecosystems during glacial-interglacial transitions. The role of the different processes in explaining glacial-interglacial variability in $CO_2$ is discussed in Chapter 4.

The sequence of events between $CO_2$ and temperature in the glacial-interglacial record of ice cores indicates that the warming at high southern latitudes slightly preceded the start of the $CO_2$ increase (Petit et al. 1999, Pépin et al. in press, Monnin et al. 2001). This highlights the fact that changes of $CO_2$, as well as of other greenhouse trace-gases, at the scale of the glacial-interglacial cycles have been forced by climatic mechanisms acting on their different reservoirs. Then, the real question is: how large has the role of changing greenhouse trace-gases been, in forcing the observed climate changes?

The initial forcing due to the direct radiative effect of increasing greenhouse trace-gases ($CO_2$ + $CH_4$ + $N_2O$) during the glacial-interglacial transitions is estimated to have produced a global warming of about 0.95°C (Petit et al. 1999). This initial forcing would have been amplified by rapid feedbacks due to associated water vapor and albedo modifications (sea ice, clouds, etc.), as is also the

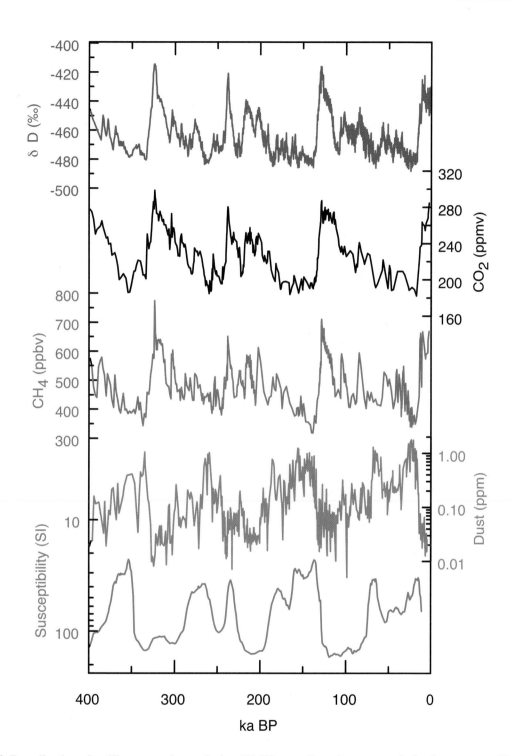

**Fig. 2.2.** Four climatic cycles. Climate records over the last 400,000 years. Deuterium, a proxy for local temperature, $CO_2$, $CH_4$ and dust (logarithmic scale) from the Vostok ice core (Petit et al. 1999, Pépin et al. in press). The Vostok dating is based on the glaciological time scale described in Petit et al. (1999). Below the Vostok records, the stacked low field magnetic susceptibility record (logarithmic scale) of the Chinese loess (Kukla et al. 1988) is plotted in SI units. The Chinese loess record has been obtained by averaging data from 2 different localities. The dating is established by considering the SPECMAP age for the last interglacial soil and the Bruhnes/Matuyama reversal taken at 780 kyr. The dating in-between has been interpolated.

case with the increasing load of anthropogenic greenhouse trace gases. Results from different climate simulations make it reasonable to assume that these trace gases and their associated rapid feedbacks have contributed significantly (possibly about half, that is 2-3°C) to the globally averaged glacial-interglacial temperature change (Berger et al. 1998, Weaver et al. 1998).

It is interesting to note that the $CO_2$ increases associated with the four glacial-interglacial transitions start several thousand years before any intense deglaciation (Petit et al. 1999, Pépin et al. 2001). This observation leads to a sequence of climate forcings as follows: changing the orbital parameters initiated the glacial-interglacial climatic changes, then the greenhouse gases amplified the weak orbital signal, accompanied several thousand years later by the effect of decreasing albedo during the retreat of the Northern hemisphere ice sheets. The lag of continental ice volume changes to temperature and $CO_2$ is also identified at the scale of the 100,000-year cycle (Shackleton 2000).

### 2.3.2 Aerosols and DMS

Atmospheric aerosol is an important component of the climatic system, but its role has not been investigated as much as that of greenhouse gases. In addition to temperature and gas records, ice core studies have provided information about the variations of atmospheric aerosol content under different climatic conditions in the past. Tropospheric and stratospheric primary and secondary aerosols (Section 2.1.2) have been documented.

Soil dust and sea salt emissions are one to two orders of magnitude higher than those of secondary aerosol. However, due to their very small size (generally <1 micron), gas-derived aerosols have a longer residence time in the air and are therefore able to be transported over larger distances than continental dust or sea spray. In central polar regions and under interglacial conditions, glaciochemical data indicate that sulfate aerosol generally dominates over primary aerosol (sea salt and continental dust) in the troposphere. In glacial periods, the situation is quite different.

Both Greenland and Antarctic ice cores show a remarkable increase in eolian dust supply during glacial periods (Thompson and Mosley-Thompson 1981, De Angelis et al. 1987, Petit et al. 1990, Mayewski et al. 1994, Mayewski et al. 1997). However, this effect is much more marked in Greenland than in Antarctic ice cores. In Greenland, dust deposited during glacial periods was alkaline and is associated with $Ca^{++}$ concentrations in the Last Glacial Maximum (LGM) ice that are about 50 times higher than the Holocene average (Fuhrer et

al. 1993). Higher dust content in Greenland ice cores during glacial times can be explained by increased desert area in central Asia and a strengthening of the Asian winter monsoon. Biscaye et al. (1997) employed natural radioisotope measurements to show that this dust originates from Central Asia. Figure 2.2 illustrates the changing dust concentrations over 4 glacial cycles in Antarctica (Vostok) and the Chinese Loess Plateau. The fluctuations in these two records are similar. Note that the Vostok and Chinese time scales in Figure 2.2 are not synchronized and that dating uncertainties do not allow the establishment of the degree of synchroneity.

Kohfeld and Harrison (2000) have mapped the distribution of eolian dust accumulation rates at the LGM and mid-Holocene. Their study indicates that the dust deposition rate at LGM was approximately 2-5 times higher than at present, downwind of modern dust source areas. Their map illustrates the fact that most of the eolian dust in the world is concentrated in Asian and African monsoon areas.

The Asian Monsoon plays an especially important role in the transportation of eolian dust. The large eolian dust accumulation rates in Japan (Xiao et al. 1995, Ono and Naruse 1997) and the Japan Sea (Tada et al. 1999), downstream of both winter monsoon winds and the westerlies, support this idea. Electron spin resonance measurements on quartz grains suggest that the intensification of westerlies played an important role in the transportation of eolian dust from Central Asia to the Western Pacific during glacial periods (Ono et al. 1997). The reasons why eolian dust over Antarctica (at least in the Vostok and Dome C regions) was enhanced during glacial times is still uncertain. However, enlargement of desert area and outwash plains in Patagonia, strengthening of the southern westerlies, and the emergence of the continental shelf of Patagonia due to lower sea-level, as well as the dryness of the atmosphere (responsible for a less efficient removal of the aerosols), are all likely to have been involved (Delmonte et al. in press). Oba and Pederson (1999) estimated an eolian input of carbonates from the Asian continent to the Japan Sea at LGM, based on the age difference between bulk carbonates and mono-specific *Foraminifera*. Their measurements revealed a huge input of old carbonate by eolian transportation on the order of 100 µg cm$^{-2}$ yr$^{-1}$, which is about 2 orders of magnitude higher than that estimated for the NW Pacific.

The timing of the decline in dust accumulation in the Huascáran ice core, at the end of LGM (Figure 2.3), also correlates well with the decrease of organic carbon accumulation in the central Panama Basin (Pedersen and Bertrand 2000).

Chinese loess provides one of the most complete records of the interplay between fluctuations in dust supply and climatic changes on glacial-interglacial time scales (e.g. Kukla et al. 1988, An and Porter 1997, An 2000). Loess deposition occurred more rapidly during glacials than interglacials (Ding et al. 1995, Porter and An 1995) and the grain size of loess deposited during glacials is coarser than in warmer periods (e.g. Ding et al. 1995, Porter and An 1995, Xiao et al. 1995, Lu et al. 1999, Xiao et al. 1999).

In glacial times, because of the weakening of the summer monsoon, the climate of Central and East Asia became dry. Pollen and fauna data indicate that the landscape changed to a more steppic or desertic one (Li et al. 1995). Not only the expanded desert, but also an emerged continental shelf due to lower sea level, enhanced the dust source area. The increased flux of finer sand and silt into dust source areas such as the Taklamakan desert, by fluvioglacial processes associated with the expansion of glaciers, provided an additional source of dust. The coarsening of Chinese loess grain size suggests a glacial increase in winter monsoon wind speeds as well (Xiao et al. 1995, Xiao et al. 1999). During the interglacials, a strengthened summer monsoon enhanced soil formation (pedogenesis), although the accumulation of loess continued (Fang 1995, Chen et al. 1997). This paleosol development in loess regions suggests vegetation cover (forest or steppe), more stable ground surface conditions and a much wetter climate due to enhanced summer monsoon rains. These landscape changes explain the decreased eolian dust supply from Central and East Asia during the interglacial periods.

Sulfur and nitrogen biogeochemical cycles, which are closely linked to environmental conditions, changed significantly during the ice age periods. Glaciochemical data from deep Greenland and Antarctic ice cores reflect these changes, in different ways, due to the different environmental conditions that prevailed in the northern and southern hemispheres during ice ages.

A major feature of the glacial ice at Vostok is its high content of continental dust and sea-salt aerosol as shown by the concentration patterns of insoluble dust particles (Petit et al. 1990), sodium and calcium (Legrand et al. 1988). This effect had been already pointed out some 10 years before at other polar sites (Cragin et al. 1977). The changes measured in ice depend on several factors (emission rates, meridional transport, atmospheric transformation and deposition mechanisms) which all may have changed in the past in relation to climate.

In the southern hemisphere, the atmospheric circulation has not been so dramatically modified as in

the northern hemisphere. Most recent investigation (Basile et al. 1997) shows that over the last glacial and interglacial periods, the dominant source of dust for East Antarctica was always the southern part of South America. The geographical extent of this source was strongly modulated by sea-level changes. Despite the high inputs of crustal dust in glacial conditions, Vostok ice remained acidic during the study time periods (Legrand et al. 1988).

Concentrations of ions related to the major global sulfur and nitrogen cycles have been determined. Nitrate deposition in Vostok snow increased during glacial periods in parallel with crustal dust inputs (Legrand et al. 1988). Regarding the sulfur cycle, the authors concluded from ice core data that marine biogenic DMS emissions around Antarctica were enhanced during cold climatic stages (Legrand et al. 1991, Legrand et al. 1992), an assertion which needs further confirmation due to the post-depositional issues associated with MSA (see section 2.2.1).

## 2.4 Abrupt climatic changes during the last ice age

Superimposed on the large scale glacial-interglacial changes are climatic events known as Dansgaard-Oeschger cycles, which occur with a frequency of 2-3 kyr and with amplitudes close to the glacial-interglacial changes themselves (see Chapter 3). These cycles affect large portions of the world and it is not surprising that they also have an effect on major trace gases (Figure 2.4).

### 2.4.1 CH$_4$ variations

For the most part, CH$_4$ concentrations change in concert with northern hemisphere temperature (Figure 2.4). Due to the age difference between reconstructed temperature and the gas signal from the same depth, it is not a priori clear which signal leads (Schwander et al. 1993). However, an innovative method based on thermal fractionation of nitrogen isotopes is able to answer this question for periods of fast temperature change (Lang et al. 1999, Severinghaus and Brook 1999). This method indicates that CH$_4$ changes lag Greenland warming by up to a few decades at the last glacial termination and at the end of the Younger Dryas. This shows that CH$_4$ changes are primarily a response to these abrupt northern climate changes rather than a driver. The CH$_4$ signal recorded in polar ice cores is a global integration of sources and sinks. The CH$_4$ sink is generally assumed to be fairly stable (Thompson 1992), suggesting that a major change of the CH$_4$ source must be responsible for CH$_4$

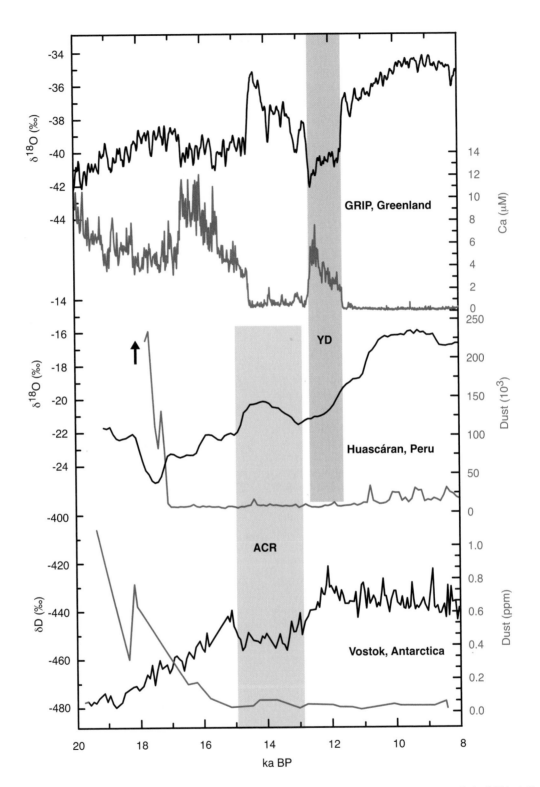

**Fig. 2.3.** Dust records over the last glacial-interglacial transition. Dust (solid red lines) and isotope records (solid black lines) from Greenland, Peru and Antarctica. Top: δ¹⁸O temperature (Dansgaard et al. 1993) and Ca (Fuhrer et al. 1993) record from the GRIP ice core. No continuous dust record exists for the central Greenland cores. However, Ca is considered to be representative of terrestrial dust (Fuhrer et al. 1993). Middle: δ¹⁸O and dust from Huascáran, Peru (Thompson et al. 1995) is plotted here on the time scale presented in Thompson (2000). Bottom: δD and dust record from Vostok (Petit et al. 1999). The gray areas indicate the location of the Younger Dryas (YD) and the Antarctic Cold Reversal (ACR) for Greenland and Antarctica, respectively.

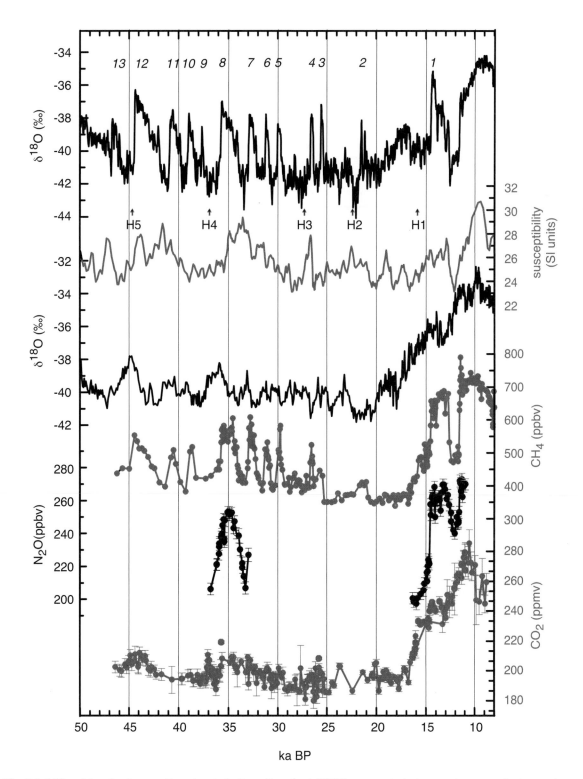

**Fig. 2.4.** Millennial scale changes. From top to bottom: Greenland (GRIP) oxygen isotopic record, a proxy for temperature (Dansgaard et al. 1993). Magnetic susceptibility from a Chinese loess record, a sensitive indicator of Asian summer monsoon (Fang et al. 1999). Antarctic (Byrd station) isotope record, a proxy for temperature (Johnsen et al. 1972). Methane record from Greenland (Blunier et al. 1993, Chappellaz et al. 1993, Dällenbach et al. 2000). N$_2$O record from Greenland (Flückiger et al. 1999). CO$_2$ record from Antarctica (Byrd station) (Stauffer et al. 1998, Marchal et al. 1999). Greenland and Antarctic records have been synchronized by using the global methane record (Blunier et al. 1998, Stauffer et al. 1998). Numbers 1-13 indicate Dansgaard-Oeschger events in the Greenland record. H1-H5 give the location of Heinrich events.

concentration changes associated with rapid climate changes in the past. The major $CH_4$ source under natural conditions is methanogenesis in wetlands. Large portions of high latitude present day $CH_4$ wetland sources were covered under several kilometers of ice during the glacial. Thus, it is tempting to attribute rapid $CH_4$ variations during the last glacial period and the transition following the Younger Dryas to variations in the low latitude wetland production. Records linked to monsoon activity in the Arabian region (Schulz et al. 1998) support this conjecture. However, recent reconstructions of the $CH_4$ source distribution during D-O events and the last transition has given new insight into the cause of $CH_4$ change (Brook et al. 2000, Dällenbach et al. 2000). This reconstruction is based on the bipolar $CH_4$ difference between Greenland and Antarctica. It shows that $CH_4$ changes associated with the transition from glacial to interglacial and the YD oscillation were indeed dominated by variability in tropical sources. However, mid to high latitude sources were already active during the Bølling-Allerød and did not decrease significantly during the YD cold period. According to Dällenbach et al. (2000), except during the coldest part of the last glacial period, high to mid latitude sources contributed significantly to the total $CH_4$ budget, and even dominated the concentration changes during D-O events. One conclusion of this study is that atmospheric $CH_4$ concentration during the last glacial is not an indicator of tropical precipitation, as has been suggested based on the concentration data from the Greenland site alone (Chappellaz et al. 1993). This implies that wetlands situated just south of the ice sheets were still active even under glacial climatic conditions.

In contrast to the fluctuations during the glacial, which seem to be primarily responding to changing northern hemispheric sources, the YD changes are associated with an emission change in the tropics (Brook et al. 2000, Dällenbach et al. 2000). This inference is supported by precipitation changes during the YD seen in lake level changes in Africa (Gasse 2000).

Recently, wetland emissions for the LGM and present day were modeled (Kaplan 2002). Kaplan's approach confirms in principle the results from the ice core approach but suggests that changes in wetland emissions are not fully responsible for the concentration change from glacial to interglacial.

An alternative explanation for rapid $CH_4$ changes focuses on the enormous amount of $CH_4$ stored in permafrost and continental margins in the form of hydrates (Nisbet 1990). Recently, Kennett et al. (2000) have shown that rapid climate change may be associated with the release of $CH_4$ from hydrates

in the Santa Barbara Basin (Kennett et al. 2000). However, it is likely that this source would not be able to maintain high concentration levels and that another source (likely wetlands) must be responsible for sustaining high atmospheric concentrations. Due to the diversity of the clathrate source it is difficult to estimate its potential influence. An answer as to whether or not clathrates release is important during rapid climate change events may come from measurements of the $\delta^{13}C$ and $\delta D$ signature of $CH_4$.

## 2.4.2 N₂O variations

The $N_2O$ record is not as complete as the $CH_4$ record (Figure 2.4). Currently available $N_2O$ records span one D-O event, the transition from the last glacial to the Holocene, the Holocene (Figure 2.5) (Flückiger et al. 1999, Flückiger et al. 2001) and the penultimate deglaciation (Sowers 2001). Some particularly high concentrations were found over very short core sections in a Greenland core. Elevated values were also found over the penultimate deglaciation in the Vostok core (Sowers 2001). Besides bacterial alteration, a chemical reaction in the ice was suggested for these higher values (Flückiger et al. 1999, Sowers 2001).

The undisturbed part of the $N_2O$ record covaries, like $CH_4$, largely with Greenland temperature and thus seems to be related to climate changes in the northern hemisphere. Within this general pattern of agreement, differences between the $\delta^{18}O$ and the $CH_4$ records do exist. During the Bølling-Allerød and D-O event 8 the $CH_4$ concentration reaches its high interstadial level simultaneously with the $\delta^{18}O$ inferred temperature maximum. The $N_2O$ concentration maximum, on the other hand, lags by several hundred years. The decrease in $N_2O$ concentration at the end of warmer epochs starts at the same time or even before decreases in $\delta^{18}O$ and $CH_4$, but the lowest $N_2O$ concentrations occur about 500 years after the minimum $\delta^{18}O$ values and about 300 years after minimum $CH_4$ concentrations. This delay of the $N_2O$ decrease is not explained by the longer lifetime of $N_2O$ as compared to $CH_4$. Assuming that the sink of $N_2O$, photodissociation in the stratosphere, did not vary significantly in the past, $N_2O$ sources must have decreased slowlier after a climatic cooling than did those of $CH_4$. Unlike $CH_4$, $N_2O$ does not decrease to near glacial levels during the YD. This has been attributed to different changes in the strength and nature of their respective sources. While $CH_4$ sources were reduced to close to late glacial levels during the YD, $N_2O$ sources declined by less than 40% of the full glacial-interglacial difference (Flückiger et al. 1999).

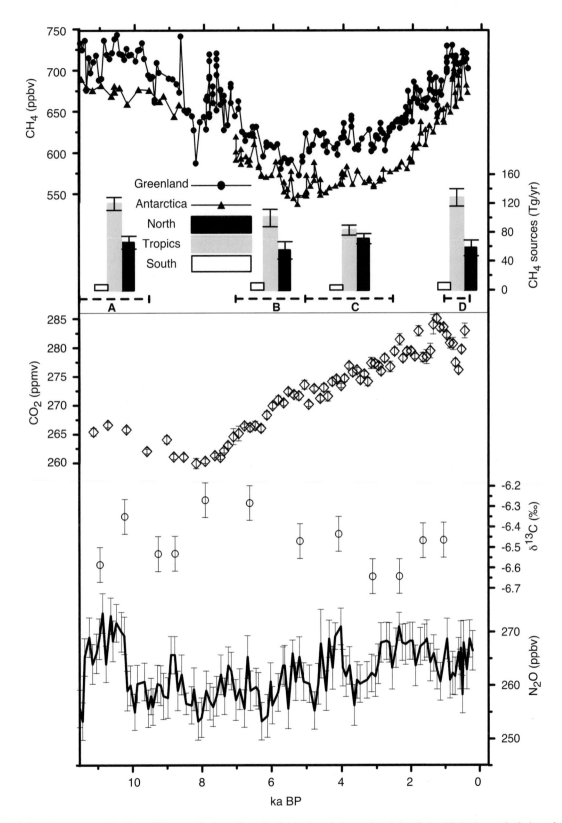

**Fig. 2.5.** Holocene records. Top: $CH_4$ records from Greenland (dots) and Antarctica (triangles) with budget calculations for latitudinal bands (Chappellaz et al. 1997). Calculations have been made with a box model. Boxes span from 90°N-30°N, 30°N-30°S and 30°S-90°N. Middle: $CO_2$ and $\delta^{13}C$ record from Taylor Dome Antarctica (Indermühle et al. 1999). Bottom: $N_2O$ record from Dome C (EPICA project) (Flückiger et al. 2001).

Other than $CH_4$, whose sources are almost entirely on land, $N_2O$ has both oceanic and terrestrial sources. On land, $N_2O$ is produced in soils. The process depends mainly on the input of nutrients, moisture status, temperature and level of oxygenation. The size of the oceanic $N_2O$ source depends not only on the rates of $N_2O$ production *in situ*, but also on gas transfer rate from the depth of production to the surface. Climate changes can influence both sources but their individual contributions, and even the sign of their responses, are difficult to estimate. As both oceanic and terrestrial sources are able to change rapidly, even well constrained estimates of the rate of $N_2O$ increase (e.g. from the last glacial to the Bølling-Allerød) are not sufficient to differentiate the sources. Thus, currently available data do not yet allow us to infer the sources responsible for $N_2O$ concentration changes seen in the paleorecord.

### 2.4.3 $CO_2$ variations

Measurements of the $CO_2$ concentrations in the air bubbles entrapped in ice cores from Greenland reveal large $CO_2$ variations associated with D-O events. Levels of 200 and 250 ppmv predominate during the cold and warm phases of the last glacial, respectively (Stauffer et al. 1984). However, measurements on Antarctic ice cores do not show variations of this magnitude (Neftel et al. 1988, Oeschger et al. 1989, Indermühle et al. 2000). It is now believed that the high $CO_2$ concentrations during warm phases in the Greenland ice cores do not represent the atmospheric concentration but are a spurious signal caused by an acid-carbonate reaction or by oxidation of organic material in the ice (Delmas 1993, Anklin et al. 1995, Tschumi and Stauffer 2000). *In situ* production of $CO_2$ in Antarctica can not completely be ruled out, but it is thought to be small.

The high-resolution record of the atmospheric $CO_2$ concentration from the Taylor Dome ice core contains four distinct millennial scale peaks of 20 ppmv during the period from 60 to 20 kyr BP (Figure 2.4). These long-term $CO_2$ variations correlate with Antarctic temperature reconstructions based on stable isotope measurements on the Vostok ice core (Indermühle et al. 2000). Strong D-O events in the northern hemisphere are preceded by a warming in Antarctica (Blunier and Brook 2001). The southern hemisphere warming, and apparently also the atmospheric $CO_2$ concentration increase, occur when Greenland temperatures are lowest (Indermühle et al. 2000). Possible mechanisms for this asynchronous behavior are discussed in Chapter 3.

Recently, a high resolution record from Dome C has confirmed and provided further details that expand on earlier measurements from Antarctica over the last glacial termination (Monnin et al. 2001). It is now clear that the Antarctic temperature rise started at least 200 years before the beginning of the $CO_2$ increase. The $CO_2$ concentration increases in stages from glacial to interglacial level over the termination. These stages can be compared best to the $CH_4$ signal which is also a first order approximation for northern hemispheric temperature changes over this period. Surprisingly, the steps in $CO_2$ and $CH_4$ concentration changes are identical, pointing to a northern hemispheric influence on the $CO_2$ increase. A first $CO_2$ increase of about 35 ppmv happens during a first moderate increase of the $CH_4$ concentration from about 17-15.5 kyr BP. Over a section of ~15.5 to 14 where the $CH_4$ concentration is roughly constant $CO_2$ increases slowly. At the sharp $CH_4$ increase, which parallels the sharp Greenland temperature increase into the Bølling-Allerød period, $CO_2$ rises almost instantaneously by about 10 ppmv. $CO_2$ slightly decreases over the Bølling-Allerød period, which roughly corresponds to the Antarctic Cold Reversal, to finally increase by 30 ppmv over the YD period to its Holocene level. The general shape of the $CO_2$ increase corresponds to the Antarctic isotope record, pointing to a southern ocean explanation for the increase. However, the details of the $CO_2$ increase have some characteristics of the northern hemispheric climate change, pointing to a non-negligible northern hemispheric influence on the global carbon cycle at least over the last termination. The northern hemispheric influence presumably reflects changes in the north Atlantic deep water formation (Marchal et al. 1999).

### 2.4.4 Dust

Eolian dust shows a remarkable response to the abrupt climatic changes during the last glacial-interglacial transition and during the last glacial period (Figure 2.3). A rapid increase of eolian dust in Greenland cores during the Younger Dryas, for example, is interpreted as the result of increased wind speed (Alley 2000). Further back in the last ice age, the chemical records in Greenland ice cores show that dust content is clearly connected to D-O rapid events (Fuhrer et al. 1993).

Changes in the rates of deposition of eolian dust during D-O cycles and Heinrich Events have also been reported in Chinese loess (Figure 2.4) (Porter and An 1995, Chen et al. 1997, Fang et al. 1999) and in marine records from the Arabian (Leuschner and Sirocko 2000), South China (Wang et al. 1998) and Japan Seas (Tada et al. 1999).

The highest resolution records in Chinese loess

(100-200 year resolution) capture millennial-scale variability which may reflect the rapid alternation of summer and winter monsoon systems (Figure 2.3), perhaps linked to changes in the westerlies (Guo et al. 1996, Guo et al. 1998). During warmer intersta-dial periods, the westerlies were situated in the north of the Tibetan Plateau, as in the modern summer, but during colder periods they were lo-cated south of the Himalaya for the entire year, as in the present winter (Fang et al. 1999). Marine re-cords from the Japan Sea and the South China Sea suggest that interactions between the westerlies and monsoons are correlated with sudden warming and cooling in the North Atlantic (Wang and Oba 1998, Tada et al. 1999, Wang et al. 2001). A satisfactory dynamical explanation has yet to be proposed for the apparent correlation of continental and marine Asian dust records with climate in the North Atlan-tic. A fuller understanding of these connections is likely to require higher resolution records.

## 2.5 The Holocene

Studying the Holocene is of special interest for documenting the interactions between climate and biogeochemical cycles under climatic conditions close to those prevailing today. These interactions provide insight into feedback processes relevant to understanding changes that are expected to occur in the near future. Thus, the study of the Holocene provides information about the sensitivity of atmos-pheric trace gas concentrations to latitudinal changes in different climatic forcing such as insola-tion.

Figure 2.5 shows trace gas concentrations of $CH_4$, $CO_2$ and $N_2O$ over the Holocene period. Late glacial and early Holocene dust variations from Greenland, tropical South America and Antarctica are shown in Figure 2.3.

### 2.5.1 $CH_4$ variation over the Holocene

High resolution $CH_4$ data are available for the entire Holocene (Blunier et al. 1995, Chappellaz et al. 1997). Results have been obtained from two Ant-arctic ice cores (D47 and Byrd) and a Greenland core (GRIP). The concentrations obtained from the southern cores are systematically lower than those from the northern cores (Figure 2.5). This latitu-dinal difference in concentration reflects the hetero-geneous latitudinal distribution of $CH_4$ sources and sinks and the fact that the atmospheric residence-time is only one order of magnitude longer than the inter-hemispheric exchange time.

The main atmospheric sink of $CH_4$ is oxidation by the OH radical. The only means currently available

to investigate the past strength of this sink is through photochemical model studies (see for in-stance Thompson 1992, Crutzen and Brühl 1993, Thompson et al. 1993, Martinerie et al. 1995). All models suggest that OH concentrations were 10-30% higher just prior to industrialization and also only slightly higher in the glacial period. Thus, the primary reason for past $CH_4$ concentration changes is probably variation in the sources. The largest pre-industrial source was wetlands, which constituted over 70% of the total source (Houghton et al. 1995). Other sources, such as termites, wild animals, wild-fires, oceans and $CH_4$ hydrates in permafrost and continental shelves are a relatively minor source.

An average difference of 44±4 ppbv occurred during the early Holocene (11.5 to 9.5 kyr BP). The minimum bipolar gradient, a difference of 33±7 ppbv, occurred from 7 to 5 kyr BP whereas the maximum gradient (50±3 ppbv) took place from 5 to 2.5 kyr BP (Chappellaz et al. 1997).

Using the $CH_4$ mixing ratio distribution inferred from the polar ice cores together with a box model, Chappellaz et al. 1997 inferred the time-varying strength of Holocene $CH_4$ sources in the tropics (30°S-30°N) and the mid to high latitudes of the northern hemisphere (30-90°N). The model inferred source histories are shown in Figure 2.5. The model indicates that Holocene $CH_4$ variations were pre-dominantly driven by changes in tropical sources, together with some minor modulations from boreal sources. The large early Holocene tropical source was most likely due to increased wetland extent accompanying what is reconstructed as the wettest period of the Holocene in the Tropics (Petit-Maire et al. 1991, Street-Perrott 1992). During the periods from 7-5 and 5-2.5 kyr BP, continual drying of the tropical regions is thought to have taken place. This is consistent with the inferred reduction in terrestrial $CH_4$ sources during this period. In boreal regions massive peat growth began in the mid-Holocene (Gorham 1991, Vardy et al. 2000). Surprisingly, however, the high latitude methane source strength does not appear to have increased substantially during the Holocene. It is possible that the increase in boreal wetlands comprised a northward shift of pre-existing wetland regions rather than the creation of net wetland area.

The period from 1-0.25 kyr BP is marked by a small $CH_4$ gradient and a high absolute level. The model infers from these observations a major tropi-cal source over this period. As pointed out above, the wetland areas of North Africa experienced con-tinuous drying from the mid Holocene until today. Therefore the inferred low latitude $CH_4$ source increase of the last millennium may reflect either local increase in wetness or early anthropogenic contributions to the $CH_4$ budget (Kammen and Marino 1993, Subak 1994).

### 2.5.2 $CO_2$ increase over the Holocene

The Antarctic Taylor Dome $CO_2$ and $\delta^{13}C$ records (Indermühle et al. 1999) (Figure 2.5) have recently revealed that the global carbon cycle has not been in steady state during the Holocene (Figure 2.5). The records show an 8 ppmv decrease in the $CO_2$ mixing ratio and a 0.3‰ increase in $\delta^{13}C$ between 10.5 and 8.2 kyr BP, and then over the following 7 kyr a fairly linear 25 ppmv $CO_2$ increase accompanied by a ~0.2‰ $\delta^{13}C$ decrease (Figure 2.5). Inverse methods based on a one-dimensional carbon cycle model applied to the Taylor Dome record (Indermühle et al. 1999) suggest that changes in terrestrial biomass and sea surface temperature are primarily responsible for the observed $CO_2$ changes. In particular, the $CO_2$ increase from 7 to 1 kyr BP could correspond to a cumulative continental biospheric release of about 195 GtC, in connection with a change from warmer and wetter mid-Holocene climate to colder and drier pre-industrial conditions. This model result is not sensitive to surface-to-deep ocean mixing and air-sea exchange coefficients because of their faster time scales. Changes in the biological carbon isotope fractionation factor, which varies according to the contribution from C3 and C4 plants due to environmental changes, may contribute up to 30-50% of the observed and modeled changes. The resulting additional uncertainty in the cumulative biospheric carbon release is ± 30 GtC at 7 kyr BP and ± 70 GtC at 1 kyr BP.

### 2.5.3 The Holocene $N_2O$ level

Nitrous oxide concentrations are relatively stable in the Holocene with a mean value of 265 ppbv and variations in the order of ±10 ppbv (Figure 2.5) (Flückiger et al. 2001). These small variations can be explained by source variations of less than 5%. Whether these variations were due to oceanic or terrestrial sources and in what way they may have been related to climatic changes remains an open question that might be answered with the help of isotope analyses of $N_2O$.

### 2.5.4 Dust

Natural aerosol supply decreased drastically after the LGM due to shrinking desert area, vegetation growth, and lower wind speeds (Figure 2.4). Holocene dust concentrations, as recorded in the Huascáran ice core, were more than two orders of magnitude lower than glacial levels. The variability in Holocene dust concentration was also small compared to glacial times (Thompson et al. 1995) and is

characterized by much more subtle century and millennial scale variability. The $CH_4$ and $CO_2$ variations observed during the Holocene confirm that the latitudinal distribution of continental ecosystems and the hydrological cycle experienced significant changes during this period, which is also revealed by continental proxies. It should be noted that anthropogenic land use may have already begun to influence dust concentrations prior to the industrial revolution.

More recently, millennial-scale oscillations have been observed in Antarctic ice (Delmonte et al. in press). The results indicate that the important property, in this case, is the variation in dust size distribution, which reflects the intensity of the atmospheric circulation over Antarctica.

## 2.6 The last millennium

### 2.6.1 Greenhouse gases

Trace gas concentrations varied little over the pre-industrial period of the last millennium (Figure 2.6). For $N_2O$, the precision of the data is not sufficient to interpret small natural changes during the last millennium. The $CO_2$ record suggest slightly increased concentrations from ~AD 1200-1400, during Medieval times, and slightly reduced concentrations from ~AD1550-1800, during the Little Ice Age. However, caution should be used in interpreting these variations as natural. Although all ice core records show generally the same picture, it is possible that a small contribution from chemical reactions in the ice has caused these variations (Barnola 1999).

The $CH_4$ variations appear to parallel the main climatic features during this period. High $CH_4$ concentration at the end of the 12[th] century corresponds to warmer conditions in many regions (see Chapter 6, section 6.5) and a decrease in $CH_4$ concentration correlates with the subsequent Little Ice Age. Variations in the oxidizing capacity of the atmosphere can only explain a minor part of pre-industrial fluctuations; the main contribution must stem from changes in methane emissions (Blunier et al. 1993). Climatic fluctuations on this time scale probably influenced the largest natural sources, i.e. wetlands, but the lack of climate proxy data on a global scale precludes a quantitative estimate of this effect. The role of pre-industrial anthropogenic sources may have been significant before the major population growth (Kammen and Marino 1993, Subak 1994) but it cannot account for the full amplitude of $CH_4$ variations.

Ice-based gas records do not reach to the present, due to the occlusion process. Continuous instrumental records of the atmospheric concentration of

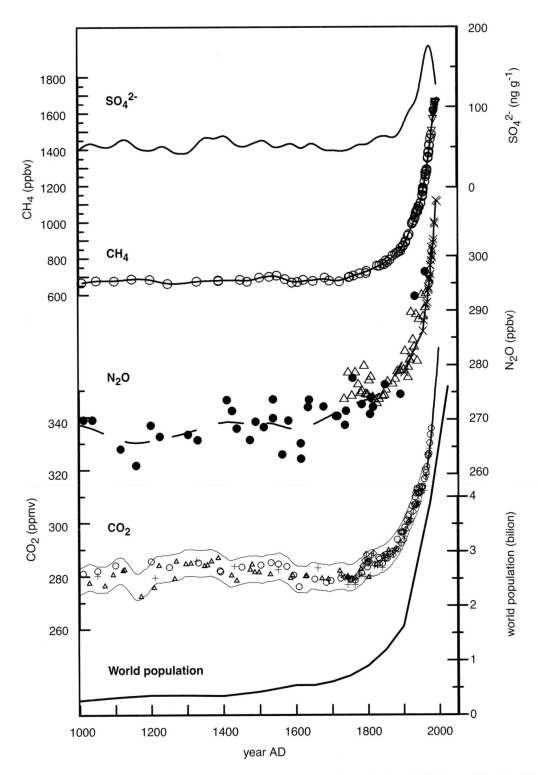

**Fig. 2.6.** Records over the last millennium. Sulfate record from Greenland (Bigler et al. in press). $CH_4$ record from Law Dome, Antarctica, including firn measurements (open circles) and direct atmospheric measurements (open triangles) from Cape Grim, Tasmania (Etheridge et al. 1998). $N_2O$ measurements from ice cores (Machida et al. 1995, Flückiger et al. 1999) and firn air (Battle et al. 1996). $CO_2$ data from Antarctic ice cores compiled by Barnola (1999). The ice core measurements overlap with direct atmospheric measurements (solid line) (Keeling and Whorf 2000). World population (McEvedy and Jones 1979).

the most important trace gases start in 1958 for $CO_2$ (Keeling et al. 1989), and 1978 for $CH_4$ and $N_2O$ (Blake and Rowland 1987, Prinn et al. 1990). Subsequently, a relatively dense monitoring network has been established which allows observation not only of the ongoing anthropogenic increase in these gas concentrations, but also of annual variations of the atmospheric mixing ratio as a function of latitude. The link between atmospheric data and data from ice cores can be made by concentration measurements on the firn air. The evolution of $CO_2$, $CH_4$ and $N_2O$ from ice, firn and atmospheric records over the last millennium is shown alongside world population growth in Figure 2.6.

A continuous record of $CO_2$ over the past millennium which overlaps with the period of instrumental data is contained in the Law Dome ice core (Etheridge et al. 1996). It is beyond doubt that the atmospheric $CO_2$ concentration has been increasing since the industrial revolution, reaching a level unprecedented over more than 400,000 years. The pre-anthropogenic level of $CO_2$ during the last millennium is approximately 200 ± 10 ppmv (Etheridge et al. 1996).

$CH_4$ concentrations have been measured in several ice cores by several laboratories for changes during the last millennium (Craig and Chou 1982, Khalil and Rasmussen 1982, Rasmussen and Khalil 1984, Stauffer et al. 1985, Pearman et al. 1986, Etheridge and Pearman 1988, Etheridge et al. 1992, Dibb et al. 1993, Nakazawa et al. 1993, Etheridge et al. 1998). All these records show an increase during the last 200 years from a global mean value of about 700 ppbv to about 1700 ppbv today. This increase parallels remarkably well the human population increase (Figure 2.6). The main reason for the $CH_4$ concentration increase is the increase from sources associated with population growth, such as domestic ruminants, rice paddies, biomass burning, landfills, and fossil fuel exploitation (Fung et al. 1991). Models suggest that about 20% of the $CH_4$ increase over the last millennium reflects a relative decline in oxidation by hydroxyl ion concentration (Thompson 1992).

$N_2O$ measurements in ice started later than those of $CO_2$ and $CH_4$. Although early measurements had high scatter, measurement and extraction techniques have been improved in recent years. Ice core data indicate a pre-industrial mean value of 270 ±5ppbv (Flückiger et al. 1999) compared to an atmospheric concentration of 314 ppbv in 1998 (Houghton et al. 2001). This $N_2O$ increase is again closely correlated to the world population increase (Figure 2.6). Both natural and anthropogenic sources are primarily biogenic. $N_2O$ increases are due to the increase in agricultural area and changing agricultural cultiva-

tion processes such as the introduction of legumes in crop rotation and the use of animal manure (Kroeze et al. 1999). A fraction of the increase may be derived from a progressive increase in denitrification in coastal waters in some regions (Naqvi and Jayakumar 2000).

### 2.6.2 Aerosols

#### *Pre-industrial*

Changes in the atmospheric aerosol load during the last millennium are documented by historical and ice core records. Prior to industrialization, soot was generated by forest fires and agricultural biomass burning. Land use changes involving extensive deforestation and development of dry field agriculture led to increases in aerosol input from the land surface. One of the largest recorded historical agriculture and drought-induced aerosol supply events was the dust bowl of the 1930s in the western USA. A rapid aerosol increase during the 16th century recorded in the Quelccaya Ice Cap, Peru, is interpreted to be the result of agricultural development by the Incas in the Andes (Thompson et al. 1988). The Little Ice Age is also characterized by elevated dust concentrations in several ice cores (Mosley-Thompson et al. 1993, Thompson 1995). The Chinese historical record of dust fall also indicates increased dust supply during the Little Ice Age, in the intervals from 1621 to 1700, and 1811 to 1900 (Zhang 1984). This historical record is interpreted as a proxy for dust storm frequency in the desert areas of western China.

Large volcanic eruptions also provide abrupt and intense periods of increased primary (ash) and secondary (sulfuric acid) aerosol concentrations in the atmosphere. The deposits from past cataclysmic eruptions, including many precisely dated by documentary records, are well preserved in polar ice cores and easily detected by ice electrical conductivity measurements (ECM). The climatic effect of these volcanic aerosols is hemispheric to global scale cooling lasting for one to several years. Temperature reconstructions from tree rings indicate a northern hemispheric cooling of 0.3 to 0.8°C correlating with volcanic eruptions (Briffa et al. 1998).

#### *Anthropogenic increase*

The concentrations of sulfate aerosols over the past 1000 years is recorded in several ice cores. In Figure 2.6 we show an example from Greenland. Contrary to the trend of long-lived greenhouse gases, which are nearly evenly distributed over the globe, the sulfate trend is only valid on the regional scale due to the relatively short residence time of sulfur-species in the atmosphere.

Anthropogenic $SO_3$ and $NO_2$ gas emissions have drastically increased the total number of aerosols in the atmosphere as compared to earlier Holocene times. Today, most aerosol particles, at least in the northern hemisphere, are derived from coal and oil burning in populated areas. Unlike agricultural dust, which is made of soil particles, industrial aerosol is largely composed of compounds of sulfate and nitrate. These gas-derived particles are transported by wind throughout the hemisphere into Arctic areas, far away from their source regions. Sulfate, nitrate and acidity measurements in ice cores demonstrate the rapid increase of industrial aerosols over the past two centuries, but this trend is presently confined to the northern hemisphere.

## 2.7 Conclusions, a view in the context of future changes

We are well aware that the past will not provide a precise analogue of the future, but it provides lessons to be learned from real experiments that the earth-climate system has undergone. Thus the paleo-record shows how the system reacts under different climatic conditions. Furthermore it provides the context for the dramatic change in atmospheric composition induced by the growing anthropogenic perturbation over the last 200 years. In the case of greenhouse trace gases the record impressively demonstrates that present-day atmospheric concentrations are unprecedented for over ~400,000 years.

The $CO_2$ and $\delta^{13}C$ ice-core records provide important boundary conditions and constraints for carbon models (oceanic and biospheric) used to estimate the uptake of anthropogenic carbon by terrestrial and oceanic sinks. The Holocene results point out the potential of past $CO_2$ and $\delta^{13}C$ records (when used in inverse modeling) to provide further constraints on the global carbon budget under different climatic conditions and to provide tests for climate models intended to simulate future responses to increasing concentrations in greenhouse gases. More resolved $CO_2$ and $N_2O$ isotopic records are needed to better evaluate the marine and continental contributions to the natural changes in atmospheric concentrations of these trace gases under various climatic conditions. Similarly, from new analytical developments in mass spectrometry, we are expecting in the near future to use the isotopic signatures of the $CH_4$ paleo-record to discriminate between different sources such as wetland emissions and clathrate decomposition.

The paleorecord of aerosols and trace gases (in particular $CH_4$) should provide information about the evolution of atmospheric chemistry and the oxidizing capacity of the atmosphere in the past.

However, we are still far from getting a clear picture of OH concentration changes under different climatic conditions. The aerosol record documents various processes such as: aerosol transport in the past, changes in sources due to surface environmental changes, possible role of continental dust on marine productivity (see Chapter 4, Section 4.2.2) and atmospheric chemistry (Dentener et al. 1996), and atmospheric and climatic impact of short term events (volcanic eruptions, meteorites). We can still learn much more from the past and large efforts should be made to evaluate the climatic role of aerosols. Aerosols are a major uncertainty when evaluating the anthropogenic perturbation to the radiative mass balance of the earth (cf. Chapter 5, Houghton et al. 2001).

Finally, ice core results clearly demonstrate the overall correlation between greenhouse trace-gases and climate over glacial-interglacial cycles. This highlights the potential of past records to inform our understanding of climate sensitivity to greenhouse gases under different climatic conditions. The correlation of greenhouse trace gases and climate provides tests for climate models intended to simulate future responses to increasing concentrations of greenhouse gases.

# The History of Climate Dynamics in the Late Quaternary

L. Labeyrie
Laboratoire Mixte CEA-CNRS, Laboratoire des Sciences du Climat et de l'Environnement, Domaine du CNRS, Batiment 12, av. de la Terrasse, FR-91198 Gif sur Yvette Cedex, France

J. Cole
Department of Geosciences, University of Arizona, Gould-Simpson, room 208, Tuscon, AZ 85721-0077, United States of America

K. Alverson
PAGES International Project Office, Bärenplatz 2, CH 3011, Bern, Switzerland

T. Stocker
Climate and Environmental Physics, University of Bern Sidlerstrasse 5, 3012 Bern, Switzerland

Contributors:
J. Allen, E. Balbon, T. Blunier, E. Cook, E. Cortijo, R. D'Arrigo, Z. Gedalov, K. Lambeck, D. Paillard, J.L. Turon, C. Waelbroeck, U. Yokohama

## 3.1 Introduction

Climate variability, defined as changes in integral properties of the atmosphere, is only one small realization of the workings of the much larger earth system. Parts of the other components (ice, ocean, continents) have much slower response times (decadal to millennia). True understanding of climate dynamics and prediction of future changes will come only with an understanding of the workings of the earth system as a whole, and over both the past and present time scales. Such understanding requires, as a first step, identification of the patterns of climate change on those time scales, and their relationships to known forcing. As a second step, models must be developed to simulate the evolution of the climate system on these same time scales. Within the last few decades, a significant number of long time series have become available that describe paleoclimate variability with resolution better than about 1000 years. Global general circulation models lag however, in that they have yet to be successfully integrated for more than a few hundred years. Because of these data and model limitations, the study of past climate change in the geologically recent past (the late Quaternary) has, for the most part, been limited to two basic strategies: (1) detailed description of the mean climate during specific climatic extremes (the "time slice" strategy), and (2) process modeling studies that seek to explain available time series in relation to specific external forcing and internal system dynamics. An excellent example of this latter strategy is the evaluation of the role of the insolation forcing as a driver of glacial/interglacial cycles by the SPECMAP group (Imbrie et al. 1992, Imbrie et al. 1993). Until recently there has been relatively little knowledge of past climatic variability on decadal to millennial time scales, a major problem considering that these are time scales on which climate is greatly affected by energy and material exchanges between the atmosphere, ocean, cryosphere, and biosphere.

The field has evolved rapidly in recent years. Some of the main areas of progress include:

• A much improved knowledge of the decadal variability of the surface ocean, ice and continental system over the last millennium.

• The development of an interhemispheric ice-core stratigraphy covering more than a full glacial cycle with a temporal resolution of about 100 to 1000 years, based on both ice parameters ($\delta D$, $\delta^{18}O$, dust) and concentrations and isotopic compositions of trapped atmospheric gases (see Chapter 2).

• The acquisition of oceanic time series with sufficient temporal resolution to allow the study of century-scale variability in surface and deep water properties over the last few glacial cycles. Of special importance in this regard have been the discovery and study of oceanic records that capture the millennial scale oscillations known from ice-core records as Dansgaard-Oeschger cycles (Dansgaard et al. 1984, Broecker et al. 1992, Dansgaard et al. 1993), and the ice-collapse episodes known as Heinrich events (after Heinrich 1988, Bond et al. 1992, Bond et al. 1993). Progress is being made on a common stratigraphy for these events, but it is not yet global.

• The improved calibration of calendar time scales using [14]C and U/Th radiochronological methods, and comparison with incremental dating approaches such as annual layer counting. This enables the establishment of precise relationships between external forcing and climatic response, and direct comparisons between ice, ocean and continental paleoclimate records (Stuiver and Braziunas 1993, Sarnthein et al. 1994, Wang et al. 2001).

• The identification of linked chronostratigraphic markers in marine, terrestrial and ice records. The classic approach was based on the detection of specific volcanic ash layers. The newly developed high resolution reconstruction of the earth dipolar magnetic moment, NAPIS 75 (Laj et al. 2000) and the associated evolution of cosmogenic nucleides recorded in the ice (Baumgartner et al. 1998) offer the first possibility of truly global correlation at millennial scale resolution.

• The development of models of intermediate complexity. These can be integrated for thousands of years and facilitate numerical experiments and data-model comparisons that can help to identify key processes involved in past climatic changes (Berger et al. 1994, Rahmstorf 1995, Stocker 2000, Ganopolski and Rahmstorf 2001).

The first section of this chapter discusses climate dynamics on orbital time scales: sea level and glacial/interglacial cycles, monsoon variability, interhemispheric connections, low versus high latitude insolation forcing and the Last Glacial Maximum. The second covers millennial scale variability and associated climatic processes. The final section focuses on interannual to decadal variability. We make no attempt here to review the progress made on proxy development and quantification of local climatic and environmental responses to climate changes. Much of that work may be found in the special issue of Quaternary Science Reviews based on the first PAGES Open Science conference (Alverson and Oldfield 2000).

## 3.2 Climate change under orbital forcing

The seasonal and latitudinal distribution of energy received from the Sun is modulated by oscillations of the earth's orbital parameters. The major changes derive from precession of the equinoxes (at 19 and 23 ka/cycle) and changes in the eccentricity of the earth's orbit (main periodicities around 400 and 101 ka/cycle). High latitude summer insolation and mean annual insolation are also particularly sensitive to the changes in the Sun's elevation above the horizon (obliquity) which varies with a periodicity of 41 ka/cycle (Berger 1977, Laskar 1990). The record of past climate change thus provides key information about the sensitivity of the earth system to energy balance changes.

### 3.2.1 Developing a chronology of past climatic change

The study of the sensitivity of the earth's climate to insolation forcing requires a reliable chronology. The first timescale for Pleistocene glacial cycles was established by joint application of magnetostratigraphy and changes in the $\delta^{18}O$ in fossil *Foraminifera* in ocean sediments, a proxy for ice volume (Broecker and Donk 1970, Shackleton and Opdyke 1973). This early chronology linked reversals of the earth's magnetic field recorded by ocean sediments to those recorded in dated volcanic rocks, and showed that the main periods of orbital oscillation are apparent in $\delta^{18}O$ records. Direct links between high northern-latitude summer insolation and $\delta^{18}O$ based records of high sea stands of the last interglacial were dated by Broeker et al. (1968) using $^{234,238}U/^{230}Th$ analysis of Barbados corals. That study, among others, led Imbrie et al. (1984) to propose a revised chronology of the last 800 ka, obtained by tuning paleorecords to orbital frequencies (the so-called SPECMAP method of Imbrie et al. 1989 and Martinson et al. 1987). The success of this method has been demonstrated by the reevaluation of the K/Ar dating of the last several reversals of the earth's dipolar magnetic field (Shackleton et al. 1990) and further refinements of the chronostratigraphy of the Pleistocene (Bassinot et al. 1994). Such orbital tuning of isotopic stratigraphy has a significant drawback for climatic studies. It presupposes that the interactions between the main components of the climatic system that react with response times similar to the orbital periods (thousand of years or longer) operate with constant phase lags or leads with respect to insolation (Imbrie et al. 1992). It is probable that this supposition will have to be relaxed in order to obtain a better understanding of the interactions that may occur between processes operating on different time scales, such as the influence of ice sheet extent on greenhouse gases and thermohaline circulation.

Uncertainties in interpretation, lack of precision in the reference series, and the presence of higher frequency variability generate intrinsic errors in orbitally tuned chronologies of about 1/4 of the precession period, or ±5 ka. Despite this relatively low resolution, the method has generated considerable progress in our understanding of long term climatic processes. Spectral and cross spectral analysis of orbitally tuned paleorecords have helped to improve the astronomical theory of climate and to elucidate the main interactions between slowly reacting climatic components (Imbrie et al. 1992, 1993). Indeed, the orbitally tuned oceanic time scale and its associated marine isotopic stages (MIS) remains the best chronostratigraphic time scale available for the last several million years. It is the reference for all long marine paleorecords and, by extension, for late Quaternary ice and continental records.

Numerous recent and ongoing studies have been developed to improve the absolute chronology of sea level changes, using $^{238,234}U/^{230}Th$ dating of coral reefs (Figure 3.1). Both benthic *Foraminifera* $\delta^{18}O$ records and analyses of other proxies from the same sediment cores may be linked to these chronological markers, because the growth and decay of continental ice sheets change the $\delta^{18}O$ of seawater and of the *Foraminifera* which grow in it. However, local changes in water temperature and salinity account for part of the foraminiferal $\delta^{18}O$ changes, and these effects are not currently independently estimated with sufficient precision (Adkins and Schrag 2001, Duplessy et al. 2001). Another cause of uncertainty derives from the U-series dating. Uranium can diffuse in and out of coral aragonite, a mineral that is also prone to dissolution and recrystallization if exposed to fresh water, which is usually the case given that most sampling is conducted above sea level. The geochemical community has yet to reach a consensus on the correct interpretation of U/Th ages when the $^{234}U/^{238}U$ differs significantly from the modern mean sea water value of 1.149. Such a discrepancy is evident for most coral samples older than 130,000 years. A final major uncertainty derives from local tectonic activity and isostatic readjustment to sea level changes. There are independent estimates for these motions for the last 10 to 20 ka (e.g. Bard et al. 1996, Lambeck and Chappell 2001) but they are much more difficult to constrain for older sea level records derived from raised coral deposits. The problem is especially crucial for the reconstruction of sea levels from glacial periods since most records that span such time intervals are obtained from areas with rapid vertical tectonic motions, such as the Huon peninsula (Chappell et al. 1996).

Fossil remains less than about 40,000 years old may also be directly dated by accelerator mass spectrometry of $^{14}C$. Ka calibrations to calendar scales have been developed to correct for the changes with time in the initial amount of $^{14}C$. With the help of tree ring measurements, an optimum resolution of 20 years has been obtained for the last 10 ka (Stuiver and Reimer 1993). Beyond this, in

the absence of long tree-ring sequences, floating calibrations have been proposed based on radiocarbon dating of macrofossils in annually laminated lacustrine sediments (Goslar et al. 1995, Kitagawa and Plicht 1998) and marine sediments from the Cariaco trench (Hughen et al. 1998, 2000). Lower resolution marine calibration curves have been proposed for the Late Glacial period by comparing $^{14}C$ and U/Th dating of coral samples (Bard et al. 1990a, 1990b, Bard et al. 1993), or by applying model corrections derived from recorded changes in the earth's dipolar moment (Laj et al. 1996). The uncertainty is roughly 1 ka or slightly more for the period before 18 ka BP. It may exceed 2 to 3 ka around 35 ka BP, a time when the earth's dipolar magnetic moment was significantly smaller than at present (Laj et al. 2000). An additional uncertainty, which complicates precise comparison between oceanic and continental records, comes from the $^{14}C/^{12}C$ isotopic disequilibrium between the atmosphere and ocean surface water. Because of the size of the ocean carbon reservoir and its relatively slow rate of ventilation, this difference is presently a few hundred years to one thousand years. This is the so-called ventilation age, or reservoir effect of surface waters. Estimates of this age have been proposed for specific periods such as the Younger Dryas (Bard et al. 1994) and the last glacial period and deglaciation (Sikes et al. 2000, Siani et al. 2001) using as time markers volcanic ash layers which have been bracketed by $^{14}C$ AMS dates on continental organic matter and on *Foraminifera* from oceanic sediments. These comparisons indicate a much larger variability on the ventilation age in the past, between 300 and 2000 years, than has been generally considered.

Dating of Greenland ice is performed either by combining ice flow and accumulation models (Johnsen et al. 1992) or by counting of annual layers (Greenland Summit 1997). The GRIP and GISP2 timescales agree very well (within 200 years or so) back to the Bølling/Allerød transition. But this precision deteriorates rapidly for older parts of the records, due to uncertainties in the counting methods for GISP-2 and GRIP, in the accumulation rate model for GRIP, and irregularities in deposition between the two sites. The absolute precision of the Greenland ice chronostratigraphy is not better than a few ka during the period from 30 to 100 ka BP. Other uncertainties are introduced in the comparison of the ice and gas proxies, because the gas from a given level in the ice core is younger than the ice by several hundred years to a few ka, depending on accumulation rate (see Chapter 2).

The dating uncertainties discussed above are particularly important for the study of millennial scale global climate variability. Any detailed comparison of climatic significance must rely on independent stratigraphic and dating tools. Such tools do exist within a regional framework, and include detailed *Foraminifera* $\delta^{18}O$ records and AMS $^{14}C$ dating for ocean sediment records, identification of common regional climatic patterns in oceanic and continental (or ice) records, and characteristic stratigraphic markers of known ages (ash layers, magnetic anomalies). One good example of such a technique is the strategy developed for the comparison of Antarctic and Greenland ice records, using comparison of the changes in trapped air $\delta^{18}O$ (Bender et al. 1994) and $CH_4$ concentration from both cores (Blunier et al. 1998, Blunier and Brook 2001).

Nevertheless, much progress is still required in order to generate a common absolute time scale and improve on the current uncertainties of about ±100 yr for the last 10 ka, ±0.5-1 ka until 15 ka, ±1-2 ka until 30 ka and about ± 5 ka (inherent uncertainty within the SPECMAP time scale), for the more distant past.

### 3.2.2 Understanding glacial cycles

Glacial cycles are recorded in a wide range of oceanic, cryospheric and continental paleorecords (Figure 3.2). They present a high level of similarity and are, in most cases, significantly correlated with the $\delta^{18}O$ record of ice volume changes (Figure 3.1) and its large amplitude, approximately 100 ka period, over the last 400-800 ka. On this long time scale, global climate variability appears to be forced by high latitude Northern Hemisphere summer insolation with a major climatic feedback associated with the waxing and waning of the northern continental ice sheets as predicted 70 years ago by Milankovitch (1930) (Imbrie et al. 1992). Prior to 800 ka ago, 100 ka periodicity was not a dominant mode of variability. Rather 41 ka periodicity was predominant, with climate apparently responding linearly to insolation changes associated with variations in obliquity (Tiedemann et al. 1994). Climate modulation by the precession of the equinoxes (about 20 ka/cycle) is well recorded at low latitudes, in particular by proxies linked to the evolution of the monsoon and trade winds (Rossignol-Strick 1985, Prell and Kutzbach 1987, Imbrie et al. 1989, Bassinot et al. 1994, McIntyre and Molfino 1996, Beaufort et al. 1997). There is thus little doubt that the modulation of insolation by precession and obliquity plays a major role in climatic changes. The interrelations between low and high latitude processes in both hemispheres on these time scales is however still a matter of debate. One approach which has been recently developed is to study, with

the same high resolution as for the recent past, older periods which are known to have had very different insolation characteristics. The period around 400-500 ka is a particularly interesting target, since the earth's orbital eccentricity was small, and the only significant modulation of insolation was through the oscillations of obliquity. Isotope Stage 11, at the end of that period, is a major interglacial interval and is prominent in most of the paleorecords (e.g. Rossignol-Strick et al. 1998, Droxler 2000).

Understanding the predominance of a 100 ka periodicity when most of the insolation forcing occurs in the precession and obliquity bands has been an important challenge of the last 10 years. It is generally accepted now that several processes may lead to this low frequency oscillation, either together or independently. These include non-linear threshold responses to insolation forcing from the continental ice sheet and other from climate components, including atmospheric greenhouse gases (Imbrie et al. 1993, Berger et al. 1994, Beaufort et al. 1997, Paillard 1998, Shackleton 2000). The strong non linearity of ice sheet dynamics during growth and decay is probably also involved in the large and rapid amplitude climatic oscillations associated with glacial terminations. These have been described in detail for the last deglaciation, but not for those before. Several independent studies based on coral terraces (Gallup et al. 1994) and ocean sediments (Henderson and Slowey 2000) suggest that the sea level rise associated with the penultimate deglaciation actually preceded its presumed Northern Hemisphere summer insolation forcing maximum by some 15 ka. However, these records are not dated with sufficiently high resolution to permit unambiguous interpretation. The majority of published results support the idea that during glacial terminations, sea level increased rapidly just after, or synchronously with the increase in Northern Hemisphere summer insolation.

For the penultimate termination, the mid-transition is about 5-7 ka prior to the insolation maximum, at 132-135 ka (Gallup et al. 1994, Stirling et al. 1995), compared to 129 ±5 ka in the SPECMAP age scale (Imbrie et al. 1984). There is also a significant lead at that time between the low latitude Devil's Hole (Nevada) vein calcite isotopic record and the global oceanic proxies (Winograd et al. 1988, Coplen et al. 1994). However, ongoing studies suggest an explanation for this lead. Kreitz et al. (2000) show that the proximal south-west California coast became warmer, with a slowing down (or an interruption) of the California current and associated coastal upwelling, before the end of each glacial cycle, prior to the glacial maximum.

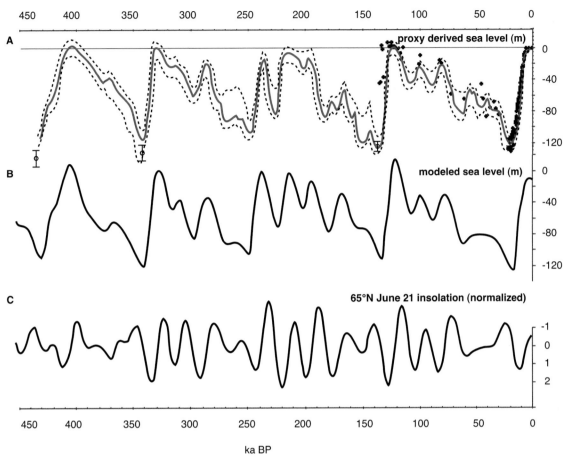

**Fig. 3.1.** Sea level change over four glacial cycles **A.** Relative Sea Level (RSL) over the last 450 kyr and U/Th dated estimates retrieved from coral terraces and [14]C-dated sediments from continental shelves (Bard et al. 1990, Stein et al. 1993, Zhu et al. 1993, Gallup et al. 1994, Stirling et al. 1995, Bard et al. 1996, Chappell et al. 1996, Stirling et al. 1998, Hanebuth et al. 2000, Yokohama et al. 2000, Yokohama et al. 2001). Open circles: RSL low stands estimated by Rohling et al. (1998). Red line and associated dashed lines: composite RSL curve of Waelbroeck et al. (2001) obtained after correction of the effect of deep water temperature changes on the benthic *Foraminifera* $\delta^{18}O$ records of marine-sediment core sites ODP 780 (North Atlantic, McManus et al. 1999) and MD94-101 (Southern Ocean, Gif data base), and a stack of ODP Site 677 and V19-30 for the Pacific Ocean (Shackleton et al. 1983, Shackleton et al. 1990) **B.** Sea level changes over the same time period, as estimated using a simple threshold function of the insolation (Paillard 1998). **C.** June 21 insolation at 65°N (Berger 1977) expressed as deviation to the mean insolation and scaled proportionally to the mean deviation.

Such changes may have resulted from a large southern expansion of the Alaskan Gyre during glacial maxima, associated with southerly winds along the California borderland, in agreement with the COHMAP simulation of the Last Glacial Maximum (COHMAP 1988). Such climatology would sufficiently affect the atmospheric hydrological cycle, at least on the regional scale, to explain the isotopic changes observed in the Devil's hole calcite prior to the termination.

Time dependent modeling of the glacial cycle has to date been undertaken primarily with relatively simplified models capable of long integration with relatively minimal computational requirements (Saltzman et al. 1984, Gallée et al. 1991, Gallée et al. 1992, Paillard 1998). One result common to many of these models is that pre-

scribed carbon dioxide variability (Tarasov and Peltier 1997) or destabilization of large continental ice sheets, either through albedo changes (Gallée et al. 1992) or other ice sheet processes, is required to model a full glacial cycle. Improvements in the development of coupled models of intermediate complexity has led to a situation where modeling the full glacial cycle with somewhat complex models, perhaps even with prognostic atmospheric $CO_2$, may become possible. General circulation modeling on these timescales has been, and continues to be, beyond the limits of computing power although asynchronous coupling schemes and other model simplifications will perhaps allow modeling of the glacial cycle with modified general circulation models in the next decade.

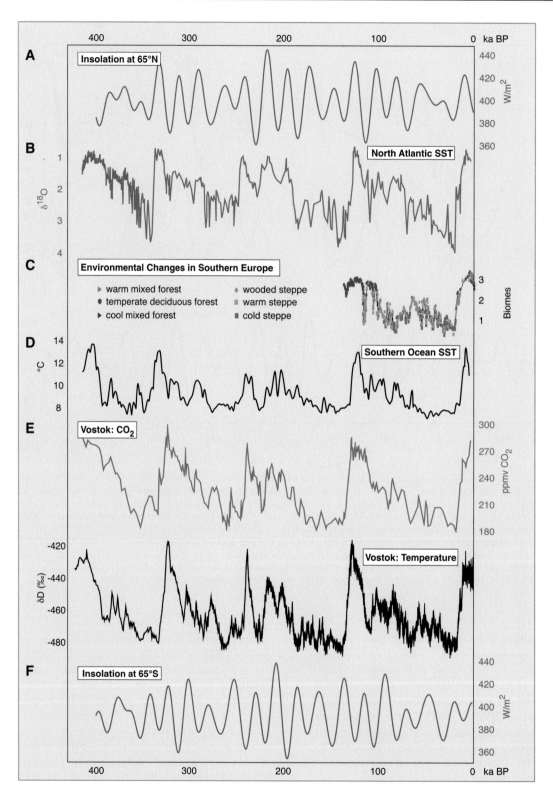

**Fig. 3.2.** Climatic proxies from both hemispheres over four glacial cycles. The last four glacial cycles as recorded in a variety of paleo-climate records. **A.** Summer insolation (21 June) at 65°N (Berger 1977). **B.** North Atlantic Sea Surface Temperature (SST) record of ODP site 980 (McManus et al. 1999). **C.** Allen et al. (1999) biomes record from lake Monticchio, southern Italy, over the last 100 ka. Biomes are statistical representations of vegetation derived from continental pollen records. **D.** South Indian ocean SST record from core MD 94-101 (Salvignac 1998). Temperatures are obtained by foraminiferal transfer functions after Imbrie and Kipp (1971). **E.** Vostok $CO_2$ and $\delta D$ (expressed as temperature) records (Petit et al. 1999). **F.** Summer insolation (21 dec.) at 65°S (Berger 1977).

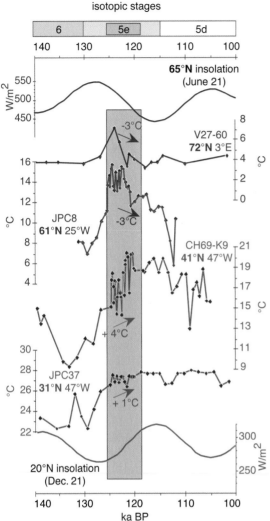

**Fig. 3.3.** The Last Interglacial (135 to 110 ka BP) in the North Atlantic and Norwegian Sea. High latitude June insolation and summer sea surface temperature versus absolute age for four ocean sediment cores located between 31°N and 72°N (Cortijo et al. 1999). Temperatures are obtained using the modern analog method (Prell 1985) with the North Atlantic data base from Pflaumann et al. (1996). Sea surface temperature and salinity began to decrease at high northern latitudes (72°N) simultaneously with decreases in high latitude summer insolation around 128 ka BP. However between 62 and 55°N, SST stayed high until about 118 ka BP, coincident with initiation of a major sea-level drop. During that period (and until the sea level rise associated with the next increase in insolation at 113 ka BP), lower latitude surface temperature (at 30-40°N) either did not change or increased slightly, suggesting a possible increase in meridional heat flux resulting from the increase in low latitude winter insolation (bottom curve).

### 3.2.3 Glacial inception

Uncertainties in absolute time scales do not preclude detailed studies of the chain of events. which follow insolation changes, as long as such studies rely on sufficiently precise relative dating based on global stratigraphies. Benthic *Foraminifera* δ¹⁸O in

ocean sediments (Martinson et al. 1987), δ¹⁸O of the atmospheric oxygen extracted from ice cores (Sowers et al. 1993, Bender et al. 1994, Shackleton 2000) and methane content in the ice cores (Blunier et al. 1997, Blunier and Brook 2001) have proven their value in this respect.

Let us first consider the inception of the last glaciation, following the decrease in summer insolation at high Northern latitudes after 129 ka BP. Ruddiman (1979) was the first to propose that the pervasive warmth of the surface North Atlantic during periods of ice sheet growth provides a strong positive feedback to glacial growth by inducing precipitation over the ice sheets. Despite some controversy (Gallup et al. 1994, Henderson and Slowey 2000), the end of the high sea level stand is relatively well dated now at 118 ± 2 ka BP (Figure 3.2 and references therein), which constrains the chrono-stratigraphy of the associated changes in ocean δ¹⁸O. The suite of North Atlantic and Norwegian sea cores studied by Cortijo et al. (1997) has recently been completed and extended to a common temporal scale by Balbon (2000) (Figure 3.3). These cores show that high-latitude (Norwegian Sea) sea surface temperature (SST) decreased in parallel with summer insolation around 120 ka BP; lower latitude SST stayed warm during that period. These results provide evidence that high-latitude summer insolation controls the development of ice sheets. However, they also highlight the importance of latent heat transport to high latitudes during periods of maximum winter insolation at Northern Hemisphere low and mid latitudes. These periods correspond to the maximum latitudinal gradients in insolation, which should enhance transport of heat and moisture to high latitudes (Young and Bradley 1984, Rind 2000).

Khodri et al. (2001) recently published a 100 year long simulation in a fully coupled ocean-atmosphere general circulation model (GCM) forced with the 115 ka insolation, which supports these hypotheses. Their model produces a build-up of perennial snow cover for ice sheet growth in response to the orbital forcing, with a strong high latitude ocean sea-ice feedback and an increase in atmospheric moisture transport from the low latitudes. De Noblet et al. (1996) have shown, using an atmospheric GCM coupled to a global biome model, that changes in vegetation at high latitudes (forest to tundra) which accompany cooling may provide an additional important feedback enhancing snow accumulation.

It is interesting to compare the last interglacial results to the mean evolution of climate during the Holocene, during which high latitude northern summer insolation decreases and low latitude winter insolation increases. The amplitude of these changes is smaller than during the last interglacial, however, because the earth's orbital eccentricity is smaller now than it was 120 ka ago. Since 6 ka, high northern latitude temperatures have decreased (Johnsen et al. 1995, Koç et al. 1996, Bauch et al. in press) but no significant change has occurred in the tropical Atlantic (Rühlemann et al. 1999).

### 3.2.4 The Last Glacial Maximum

Describing the climatic state of surface oceans, continents and ice sheets during the Last Glacial Maximum (LGM) was the first large scale coordinated objective of the paleoclimatic community, as exemplified by the international CLIMAP project (Climate: Long-Range Investigation, Mapping, and Prediction (CLIMAP 1976, CLIMAP 1981). Sea surface temperature was derived from the statistical analysis of micro-fossil species distribution in sediments (Imbrie and Kipp 1971). The project also focused on chronostratigraphy to ensure reasonable temporal resolution of the Glacial Maximum (around 18 $^{14}$C ka BP, or 21 calendar ka ). The reconstructions were published as global maps of the distribution of sea surface temperature, sea ice, continental ice sheets and albedo (August and December) (CLIMAP 1981). One of the major conclusions was that most of the cooling during LGM occurred at high latitudes, with only small changes over the tropical oceans. Although since reevaluated in detail (Mix et al. 2002), the CLIMAP data set is still the basic reference for the understanding of ice-age climate. CLIMAP results have been used for both forcing and validating the recent intercomparison study of atmospheric GCM's and data, by the Paleoclimate Model Intercomparison Project (PMIP) (PMIP 2000). Two types of simulations have been carried out in these studies, the first forced with the CLIMAP global sea surface reconstruction (CLIMAP 1976, CLIMAP 1981) and the second using a coupled slab-ocean (Broccoli 2000). Significant differences exist between the two sets of simulations (Pinot et al. 1999, PMIP 2000). The coupled simulations predict surface ocean temperature and sea-ice distributions in relatively good agreement with recent reevaluations of the CLIMAP results. CLIMAP largely overestimated high- latitude summer sea ice coverage, and underestimated surface ocean temperature in the Northern Atlantic and Norwegian Sea (Sarnthein et al. 1994, 1995, De Vernal and Hillaire-Marcel 2000), as well as in the Southern Ocean (Crosta et al. 1998). Low-latitude sea surface temperatures may also have been over-estimated in the CLIMAP reconstructions. This is particularly apparent in the equatorial

Atlantic and Pacific oceans, where recent SST hindcasts suggest LGM temperatures 2 to 5°C cooler than modern at the eastern side of the ocean basins, where equatorial currents interact with boundary currents (Hostetler and Mix 1999, Mix et al. 1999). For the subtropical gyres, however, recent results support the original CLIMAP inference of relatively unchanged SSTs compared to modern.

New proxies for reconstruction of sea surface temperature, such as ketone unsaturation ratios (Uk$^{37}$) (Brassell et al. 1986, Prahl and Wakeham 1987, Müller et al. 1998), or Mg/Ca ratios in foraminiferal shells (Elderfield and Ganssen 2000, Lea et al. 2000, Nürnberg et al. 2000) have not changed the overall picture significantly. The large amplitude (6°C) LGM cooling derived for Barbados from changes in coral Sr/Ca ratio (Guilderson et al. 1994) is not supported by these more recent studies.

Comparison of PMIP simulations with continental records for the LGM period are in progress (Kageyama et al. 1999, Joussaume and Taylor 2000, PMIP 2000, Kageyama et al. 2001). Often, however, the comparison between ocean and continental records is made difficult due to a lack of common chrono-stratigraphy. High resolution studies in ocean sediments show that the LGM was immediately preceded and followed by strong cold events (Heinrich events H-2 around 23 ka BP and H-1 around 17 ka BP). Similar variability is observed for lake levels in tropical Africa (Gasse 2000) and for the Indian (Leuschner and Sirocko 2000) and the East Asian monsoons (Wang et al. 2001). In low resolution records, an analogous series of events cannot be distinguished from a mean LGM climate. For this reason, a precise definition of the LGM period is important. It has been recently proposed by the EPILOG IMAGES Working Group that the LGM is best defined at 21±2 calendar ka BP (calendar scale) (Mix et al. 2002). This definition is in agreement with the sea level record of Yokohama et al. (2000) from Western Australia showing a –125 m relative sea level minimum between 22 and 19 ka. Because of a paucity of accurate dating or resolution, it remains difficult to integrate many records within the context of the high amplitude millennial scale climate variability surrounding the LGM period. These include observations such as the glacial lowering of snow lines (Klein et al. 1999, Porter 2001, Seltzer 2001) and ground temperature estimates derived from noble gas concentration in aquifers. The noble gas thermometry implies a mean annual air temperature decrease in northern Brazil of 5°C (Stute et al. 1995) at ~16-19 calendar ka, and 6.5°C in Oman (Weyhenmeyer et al. 2000), averaged over the period 16 to 27 ka BP. Both periods are dated with several ka uncertainty. Nevertheless, a general consensus exists that the LGM climate was not only much colder, but also, for the most part, more arid than the present day. Overall, currently available evidence suggests that intertropical areas probably cooled 1 to 3 °C in the surface ocean, and about 4 to 6°C at moderate altitude on the continents. These results are in agreement with recent fully coupled atmosphere-ocean simulations of the LGM climate. A drier continental LGM also explains the significantly lower atmospheric $CH_4$ concentrations observed in the Greenland and Antarctic ice records for that period (Chappellaz et al. 1993 and Chapter 2).

If we consider the glacial climate in more detail, using coupled GCM simulations, the system appears more complex and is harder to understand. Ganopolski et al. (1998) utilized a coupled ocean-atmosphere GCM of intermediate complexity (dynamical three-dimensional atmosphere of low spatial resolution coupled to a zonally averaged multi-basin ocean), and predicted a mean LGM cooling of about 2°C for the intertropical oceans (4.6°C for the continents). This was driven by sea-ice and a southward shift in deep-water formation, with no significant drop in the overall thermohaline transport. Weaver et al. (1998) using an oceanic GCM in equilibrium with an energy-moisture balance model for the atmosphere, also estimated a mean tropical ocean temperature LGM decrease by about 2.2°C which is consistent with a low to medium climate sensitivity to radiative perturbations. The large cooling over North America and the northern Atlantic is linked in this work to a large drop in the rate of North Atlantic deep water formation. Bush and Philander (1998) used a fully-coupled three–dimensional GCM configured for the LGM and found, in contradiction to the results of Ganopolski and Weaver, a large amplitude cooling of the tropical Pacific Ocean (6°C for the western Pacific) driven by enhanced trade winds, equatorial upwelling and equatorward flow of cold water in the thermocline. However, Bush and Philander (1998) limited their run to 15 years, thus the ocean was not in equilibrium with the atmosphere, except for low latitude surface waters.

Atmospheric circulation was strongly affected by glacial ice sheets over northern mid-latitudes. Model results suggest that an anticyclonic circulation developed over the ice sheets and that planetary waves were enhanced. Baroclinicity increased as a result of stronger meridional temperature gradients and storm tracks experienced an eastward shift especially over the North Atlantic (Kageyama et al. 1999). Such circulation changes played a key role in determining regional climate change patterns. For example, along the west coast of the Americas, an equatorward shift in the westerlies in both hemispheres during the LGM lead to relatively wetter conditions in regions that are quite dry in today's climate (Bradbury et al. 2001). The equatorial shift of the westerlies in the Southern Hemisphere is more controversial than that in the North. If it did occur, it may have been related to a northward shift of the winter sea ice belt (Crosta et al. 1998) and a coastal expansion of the Antarctic ice sheets over the continental margin, since additional forcing by large continental ice sheets must be excluded there. Models are able to reproduce the main regional trends over Eurasia except over western Europe where they suggest that conditions were relatively warm and wet. Reconstructions deduced from pollen data (Peyron et al. 1998, Kageyama et al. 2001) indicate cooler conditions suggesting that the models underestimate the meridional temperature gradient over Africa-Europe during the LGM.

In Greenland, it is generally accepted (since the work of Cuffey et al. (1995) and Dahl-Jensen et al. (1998) using bore-hole temperatures as an independent temperature proxy) that the LGM air temperature reconstruction derived from $\delta^{18}O$ on the basis of Dansgaard's (1964) spatial calibration underestimated the LGM cooling by about 10-15°C. The reasons for this are multiple, but for the most part are linked to changes in the sources, transport and seasonality of snow precipitation over the northern latitudes. These effects are probably smaller in Antarctica (Jouzel et al. 2000).

Over the last several decades, significant strides have been made in understanding the changes in ocean circulation during the last glacial period, using proxies that reflect the timescale of deep ocean ventilation and nutrient content. It has been known for more than 10 years (after the work by Boyle and Keigwin 1982, Curry and Lohmann 1983, Boyle and Keigwin 1985, Boyle and Keigwin 1987, Oppo and Fairbanks 1987, Curry et al. 1988, Duplessy et al. 1988) that the thermohaline circulation was significantly altered during the LGM. The tropical thermocline is thought to have been shallower during the Last Glacial Maximum, at least in the Western Atlantic (Slowey and Curry 1987), which helps to explain the large SST cooling observed in low latitude upwelling areas by Mix et al. (1999). Glacial North Atlantic Deep Water was located above, and not below, a deep-water equivalent of Modern Antarctic Intermediate Waters. A similar inversion of water masses with ventilated nutrient-poor waters above 2000 m, and generally poorly ventilated waters below was also present in both the Indian (Kallel et al. 1988) and Pacific oceans (Duplessy et al. 1988). Deep water temperatures, estimated by comparison of benthic *Foraminifera* $\delta^{18}O$ records from the Norwegian sea, North Atlantic and Pacific Oceans, were about 2 to 4°C colder (Labeyrie et al. 1987, Labeyrie et al. 1992). A similar cooling has been estimated, using the Mg/Ca ratio in ostracodes (Cronin et al. 2000). Benthic $\delta^{18}O$ records also provide constraints on the Glacial-Holocene changes in deep water $\delta^{18}O$ following the melting of continental ice and sea level increase (0.9‰ for the Atlantic and 1.1‰ for the Pacific oceans, Waelbroeck et al. 2001). These values are not significantly different from those derived by Schrag et al. (1996) and Adkins and Schrag (2001) from pore water $\delta^{18}O$ and chlorinity. However, LGM deep water salinity may be more difficult to extract from changes in water $\delta^{18}O$ than previously thought, as sea-ice formation (which does not fractionate $H_2^{18}O$ vs. $H_2^{16}O$) was probably a large source of salt for deep water production (Dokken and Jansen 1999, Adkins and Schrag 2001).

Several outstanding questions exist regarding the dynamics of ocean circulation at the LGM. For example some data imply a rate of meridional overturning at LGM similar to that of the present day (Yu et al. 1996). In addition, although proxy data inform us about the location of water masses at the LGM, it is difficult to quantify them in terms of the volume of meridional overturning or the quantum of meridional heat transfer (LeGrand and Wunsch 1995). In the modern ocean the wind driven circulation carries enormous amounts of heat into the subpolar latitudes, and in a generally more windy glacial period with a larger equator to pole temperature gradient, this heat transfer would have been enhanced irrespective of what was happening to the thermohaline circulation. The proxy that would best constrain rates of deep water ventilation is clearly $^{14}C$, because of the inherent "clock" in its radioactive decay. Initial efforts to estimate overturning rates at the LGM from the $^{14}C$ difference between planktonic and benthic *Foraminifera* (Broecker et al. 1988) were hampered by relatively large uncertainties. $^{14}C$ measurements in deep sea corals (Adkins et al. 1998) are a relatively new proxy development and one which may provide the ability to resolve millennial scale changes in deep ocean ventilation rates (Boyle 2000), but few such data are yet available for the LGM period. Despite great progress in mapping the location of water masses at the LGM, there remains much uncertainty in quantifying the rate of deep water ventilation and meridional heat flux in the oceans during this period.

## 3.2.5 Glacial termination

Rapid terminations of glacial periods have attracted much attention: melting $54 \times 10^6$ $km^3$ of continental ice (Yokohama et al. 2000) in about 10,000 years is not a small thing. Ice-core records, which include information on both atmospheric composition and temperature at high latitudes (Chapter 2) provide strong constraints on the possible mechanisms responsible for deglaciation. The increases in atmospheric greenhouse gases, $CO_2$ and $CH_4$, and of ice $\delta D$ (a proxy for air temperature over Antarctica), occur quasi synchronously (within the available resolution) during each of the last four terminations in the Vostok ice record (Blunier et al. 1997, Petit et al. 1999, Pépin 2000, Pépin et al. in press). Long-term trends are difficult to separate from millennial variability, even across the last termination where in the Byrd record the timescale is more precise (Blunier et al. 1997, 1998). Thus, we will first present the general picture of the deglaciation, before discussing in more detail millenial variability and associated climate dynamics.

Warming started both in Greenland and Antarctica at about 23 ka BP (Blunier et al. 1997, Alley and Clark 1999, Alley 2000, Blunier and Brook 2001), in phase with the increase in Northern insolation (Figure 3.4), but prior to the LGM. The warming accentuated after 19 ka BP when the Byrd $\delta^{18}O$ signal drifted out of its glacial range of variability. The rise in sea level due to the melting of northern ice sheets started at 19 ka BP (Yokohama et al. 2000), but moved out of its glacial range only at 15 to 14 ka BP, as is the case for the Greenland ice $\delta^{18}O$ signal. Severinghaus and Brook (1999) precisely dated the corresponding warming at 14.7 ka in the GISP ice core, using as proxies the changes in nitrogen and argon isotopic ratios (see Chapter 2, Section 2.1). They were able to show that the $CH_4$ increase in the same air samples in fact lagged temperature by about 50 years. If $CH_4$ is interpreted as a proxy for warmer or wetter tropics, such regions could not have been the trigger for initial warming. Within $^{14}C$ ventilation age uncertainties, deglacial warming in the surface waters of the low-latitude Cariaco Basin off northern Venezuela has been shown to be synchronous with the 14.7 ka rise in temperature in Greenland (Hughen et al. 1996), implying a close coupling between climate change in the tropics and high latitudes of the Atlantic.

The period of most rapid sea level rise, meltwater pulse 1A, follows by a few hundred years the initiation of the rapid warming over Greenland (Fairbanks 1989, Bard et al. 1996). This meltwater pulse, which raised sea level by about 20 m between 14.2 and 13.8 ka BP, corresponds in fact to a

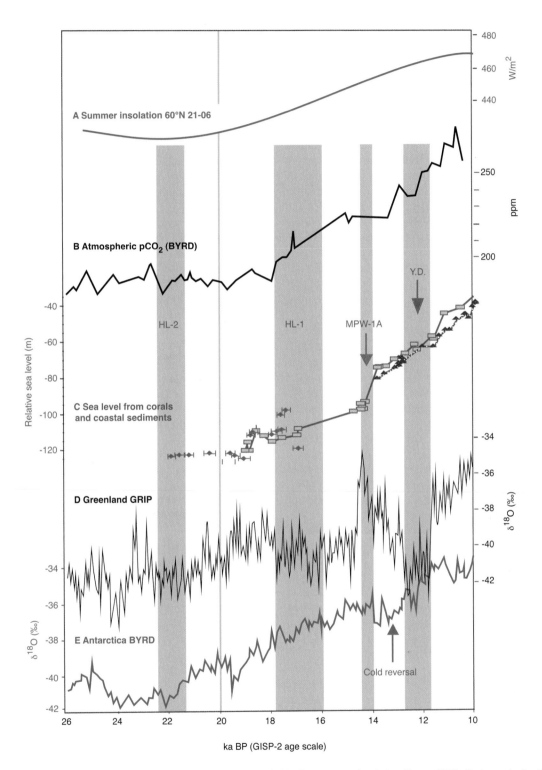

**Fig. 3.4.** Last deglaciation (25-10 ka BP) through various proxies. **A.** Northern summer insolation (Berger 1977). **B.** Atmospheric pCO$_2$ from the BYRD ice core (Blunier et al. 1998). **C.** Relative sea level from Barbados coral (boxes) (Fairbanks 1989), Tahiti coral (triangles) (Bard et al. 1996), and sedimentary facies (diamonds) from North-West Australian shelf (Yokohama et al. 2001) and Sunda shelf (Hanebuth et al. 2000). **D.** GRIP ice δ$^{18}$O (Johnsen et al. 1992). **E.** Byrd ice δ$^{18}$O (Blunier et al. 1997). The GRIP and BYRD records are shown here on the GISP-2 time scale based on annual layer counting (Grootes et al. 1993, Meese et al. 1994, Stuiver et al. 1995, Grootes and Stuiver 1997). Correlation of the GISP-2 and GRIP δ$^{18}$O signals was done with Analyseries (Paillard et al. 1996), and transferred to BYRD using the GRIP-BYRD CH$_4$ stratigraphy of Blunier et al. (1997, 1998).

short cooling phase in Greenland (known as the older Dryas in Europe, Mangerud et al. 1974).

Temperatures from the Byrd record do not show significant inflexions during meltwater pulse 1A, but decrease later (the Antarctic Cold Reversal), during a slightly warmer phase at GRIP. The cold reversal terminated during the Younger Dryas, at about 12 ka BP. These events illustrate the typical asynchronous (or anti-correlated) North/South warm/cold modulations that appear to have occurred on the millennial scale over a large part of the of the last glacial period and deglaciation (Blunier and Brook 2001). Warming was for the most part complete by about 11.5 ka in Byrd. In the Northern Hemisphere, warming continued into the Holocene (10 ka or later). The northern Arctic seas, in particular the northern Norwegian (Koç et al. 1996) and Barents seas (Duplessy et al. 2001), became ice-free in summer, with progressive warming, only after 11-10 ka.

In North Atlantic surface temperature records, the drastic cooling that culminated in Heinrich event 1 (about 18 to 15.5 ka BP) was a dramatic interruption of the overall warming trend. The Younger Dryas (13 to 11 ka BP) had a similar effect. Available data and models suggest that these cooling phases were caused by catastrophic input of iceberg-derived meltwater over the northern Atlantic. This meltwater flux (about 0.5 to 1 Sv) induced a large decrease in thermohaline heat transport (Broecker et al. 1992, Paillard and Labeyrie 1994, Sarnthein et al. 1995) or a change in the zones of deep convection (Rahmstorf 1995). Active "modern" North Atlantic deep water production appears to have been initiated as late as about 15 ka BP (Bølling/Allerød) and persisted for about 1 ka before diminishing again during the Younger Dryas only to recommence at about 10 ka BP (Sarnthein et al. 1994, Marchitto et al. 1998). Thus, North Atlantic thermohaline circulation may have acted as a direct positive climate feedback only during the main Northern Hemisphere warming phases at ~15 and 10 ka BP.

The European continent reacted directly to these changes (e.g. Björk et al. 1998, von Grafenstein et al. 1998). In contrast, the loess series of China suggests that conditions actually became more humid there during the Younger Dryas interval, though temperatures remained low (Zhou et al. 1996). This probably resulted dynamically from an increased summer monsoon associated with the summer insolation maximum, but a more active polar vortex, with frequent cold northerly winds and active loess formation and transport (An and Porter 1997).

In the Atlantic and Indian sectors of the Southern ocean, surface temperature increases lead the ben-

thic $\delta^{18}O$ record of sea level by 2 to 4 ka during the whole of the deglaciation (Labracherie et al. 1989, Labeyrie et al. 1996, Lemoine 1998). Poor constraints on the ventilation age introduce uncertainties of 0.5 to 1 ka in this estimate. Early warming is also indicated by continental southern hemisphere records. Pendall et al. (2001) recently published a well-dated peat-bog record ($\delta D$ and pollen) from Patagonia (55°S) that shows climate changes similar to those seen in the Taylor Dome and Byrd records, including a well-defined early warming at about 17 ka. Southern Ocean deep water ventilation, as inferred from benthic foraminiferal $\delta^{13}C$ started very early during the deglaciation process (less than 1 kyr after the initial warming), thus providing evidence for an early renewal of convection and reestablishment of the active exchange between the deep Southern Ocean and the global ocean (Charles and Fairbanks 1992, Lemoine 1998).

Atmospheric $pCO_2$ presents characteristics similar to the $\delta^{18}O$ signal at Byrd Station, with a progressive increase after 19 ka, and a plateau between 15 and 13 ka BP, during the cold reversal. Such correlations support the idea that the Southern Ocean played a key role in atmospheric $pCO_2$ changes during the deglaciation (Blunier and Brook 2001, see Chapter 4).

## 3.3 Interaction among climate system components on millennial time scales

### 3.3.1 Millennial scale variability in proxy data: high latitude signals

When the rapid and large amplitude temperature oscillations recorded during the glacial period in the Camp Century Greenland ice core were first published (Dansgaard et al. 1984), they did not attract much interest in the climatological community. The delayed reaction (Broecker et al. 1992) occurred a few years after Heinrich's (1988) publication of his interpretation of the succession of sandy layers observed in a sediment core from the northeast Atlantic Ocean. Broecker and his colleagues interpreted these so-called Heinrich events as layers rich in ice-rafted detritus (IRD) that resulted from the catastrophic collapse of ice sheets into the North Atlantic via fast ice streams. These surges would have perturbed the hydrology of the North Atlantic, stopping the thermohaline conveyor belt and cooling regional climate. Simultaneously, Dansgaard et al. (1993) and Grootes et al. (1993) confirmed the presence of about 21 large-amplitude changes in air temperature, now called Dansgaard-Oeschger events (D-O), in the new GRIP and GISP2 Greenland ice cores. Bond et al. (1993) observed that each of the six Heinrich events identified from 60 to 15 ka BP occur at the end of a several kyr long cooling phase and appear to be simultaneous with the coldest of the Greenland stadial events. The (logical) hypothesis of synchronous cooling and warming over northern Atlantic and Greenland has been strengthened by the discovery of two peaks of cosmogenic $^{36}Cl$ within the GRIP ice core, one located between interstadials 10 and 8 and the other prior to interstadial stadial 6 (Figure 3.6). These correspond, in agreement with the proposed correlation, to the two periods of low-level paleomagnetic field intensity (the Laschamp and Mono lake events) recorded in North Atlantic sediments cores on each side of Heinrich event H-4 (Laj et al. 2000, Wagner et al. 2000). The presence of detrital carbonate (Bond et al. 1992) and very old (over 1 Gyr) detrital silicates (Huon and Jantschik 1993) within the IRD point to the Laurentide ice sheet as the major contributor of these sedimentary deposits. Geochemical studies confirm these results (Gwiazda et al. 1996, Revel et al. 1996) for all the Heinrich events of the last glacial period, save H-3 (around 30 ka BP). Grousset et al. (1993) and Gwiazda et al. (1996) have shown that this particular event probably derived from European or Greenland sources.

Labeyrie and his colleagues (Labeyrie et al. 1995, Cortijo et al. 1997, 2000) mapped the surface sea water $\delta^{18}O$ anomaly that resulted from melting icebergs during Heinrich events. From this they derived a rough figure for the ice volume change, which correspond to a meltwater flux of about 0.5 Sv, probably discontinuous over several hundred years, and an integrated sea level change of about 5 m. Available data from the Huon Peninsula (Lambeck and Chappell 2001) indicate that the sea level change may have been even larger, as much as 10-15 m during some of the Heinrich events. Such shifts correspond to both catastrophic collapses of the Laurentide ice sheet and additional input from grounded ice shelves destabilized by the initial sea level increase. North Atlantic deep-water ventilation significantly decreased during these events (Vidal et al. 1997). Ventilation of intermediate water may have increased in parallel (Marchitto et al. 1998).

Several authors have described the existence of a similar millenial variability at low latitudes in both the Northern (Little et al. 1997, Hendy and Kennett 1999, Sachs and Lehman 1999, Vidal et al. 1999, Peterson et al. 2000) and the Southern hemispheres (Charles et al. 1996, Kanfoush et al. 2000). Southern Hemisphere cooling episodes, possibly triggered by increased trade wind intensity (Little et al. 1997), are marked by an increased flux of ice rafted detritus between 41 and 53°S (Labeyrie et al. 1986,

Kanfoush et al. 2000). They appear to have been approximately in phase with periods of warmth and active NADW formation in the Northern Hemisphere. In addition, Southern Hemisphere surface temperatures may have been warmer at the time the Northern Hemisphere was cold prior to and during Heinrich events. This asynchronous climatic behavior as recorded in ocean sediments is similar to that described in the Greenland and Antarctic ice cores (Blunier et al. 1998, Blunier and Brook 2001). However, unlike the ice records which are narrowly tied by their $CH_4$ signals, most of the millenial-scale oceanic and continental records currently available are not sufficiently well correlated from region to region nor with ice records to define precise inter-relationships.

The cause of the ice sheet collapse associated with Heinrich events is still a matter of active debate between the proponents of internal ice-sheet dynamics (the binge-purge hypothesis of MacAyeal 1993), and those who favor an external origin (see discussion in Clarke et al. 1999). Since the interval between events decreases from about 10 kyr (between H5 and H4), to 5 kyr between H2 and H1, they clearly do not occur in direct response to insolation forcing, as originally hypothesized by Heinrich (1988) and McIntyre and Molfino (1996). However, taking into account the massive disruptions in the atmospheric and oceanic circulation linked to these events, it is evident that multiple positive and negative feedbacks must have been operating on a range of different time scales.

The Younger Dryas (YD), as discussed in the previous section, corresponds to the period at about 12 ka BP when Northern Atlantic temperatures returned to glacial levels for more than 1 kyr, despite the fact that Northern summer insolation was at its maximum. Broecker et al. (1989) suggested that the YD event signaled a major rerouting of the Laurentide meltwater from the Mississippi Delta, through which it was flowing until about 13 ka BP (Kennett and Shackleton 1975), to the St Lawrence estuary. The addition of meltwater near the sources of deep water formation would have directly affected the thermohaline circulation. However, De Vernal and her colleagues (1996) presented evidence that during the Younger Dryas, the St-Lawrence estuary was sea-ice covered most of the time, with very limited output of fresh water, and no indication of a significant meltwater spike. Broecker's hypothesis is therefore not clearly supported. Available data indicate, in fact, that continental ice melting decreased significantly during the YD (Fairbanks 1989). Andrews et al. (1995) attributed the YD cooling to a Heinrich-like event (H0). Interestingly, the characteristic sedimentary signature of Heinrich events (with detrital carbonate) is limited to the Labrador Sea at this time, and is not associated with a significant meltwater anomaly. It is therefore not at all proven that the YD corresponds to an ice sheet instability. However, as shown by Fairbanks (1989) and confirmed by Bard et al. (1996), the major meltwater period of the deglaciation occurred at 14 ka BP (Figure 3.4), at the beginning of the cooling that culminated in the YD proper, 1500 yr later. We may therefore imagine that a decoupling occurred, with an initial high latitude cooling linked to a decrease in thermohaline heat transfer, accumulation of snow, ice sheet regrowth, and subsequent collapse. The low latitude western Atlantic Ocean was warmer between 13 and 11 ka BP (which straddles the YD) than between 15 and 13 ka BP (the Bølling-Allerød) (Rühlemann et al. 1999). This accumulation of heat at low latitudes (also observed by Rühlemann and his colleagues between 16 and 15 ka BP) would be a consequence of lower thermohaline heat transport to the Northern Atlantic. But it would also be expected to promote enhanced evaporation and atmospheric water transport to the Laurentide and Fenno-Scandian ice sheets (Labeyrie 2000).

The origin of the relatively rapid D-O cycles is even less well understood than Heinrich events. Patterns of temporal variability with similar frequencies and durations (a few hundred to a few thousand years) have been recorded by numerous paleo-proxies over the Northern Hemisphere and low latitudes from both hemispheres (Grimm et al. 1993, Guiot et al. 1993, Chen et al. 1997, Hatté et al. 1998, Hatté et al. 2001, Wang et al. 2001) as well as in the oceans (Rasmussen et al. 1996a, 1996b, Curry and Oppo 1997, Moros et al. 1997, Kissel et al. 1998, Schulz et al. 1998, Wang and Oba 1998, Cacho et al. 1999, Hendy and Kennett 1999, Sachs and Lehman 1999, Tada et al. 1999, Peterson et al. 2000, Shackleton 2000, van Krevelt et al. 2000, among others). The signal is also clearly apparent in methane records from ice cores (Chapter 2). The characteristic signature of the Greenland ice $\delta^{18}O$ records with 21 large amplitude oscillations over the Last Glacial, each composed of a rapid shift to warm temperature (in few decades), and a slower cooling (several centuries) may help to identify the climatic processes that are directly linked to the D-O episodes. The rapid warming phases are especially significant, because any climatic mechanism, which operates with a rate constant longer than few decades, such as the dynamics of intermediate or deep ocean and ice sheets would have smoothed this signal. All the events and their characteristic temporal evolution are found in proxies which record some aspects of the Northern

Hemisphere atmospheric circulation (Mayewski et al. 1994) and wind-driven surface ocean circulation (Peterson et al. 2000, Shackleton et al. 2000, van Krevelt et al. 2000). The new high resolution records of atmospheric $CH_4$ content (Blunier and Brook 2001) have the same signature, indicating that $CH_4$ is modulated by processes occurring on the millenial timescale on the Northern Hemisphere continents and possibly along continental margins (Kennett et al. 2000).

Thus, available data would point to large North-South oscillations of the North Atlantic Polar Front, of the westerly wind belt and maybe of the ITCZ and associated trade winds as direct modulators of the temperature over Greenland and the northern Atlantic (Peterson et al. 2000). The Fennoscandian and maybe other Arctic ice sheets were also affected. In sediment cores from the northern North Atlantic, Southern Norwegian and Greenland seas, the amount of ice rafted detritus (IRD) from Scandinavian sources increased (Blamart et al. 1999) and the foraminiferal $\delta^{18}O$ decreased (acting as a tracer of continental ice meltwater (Labeyrie et al. 1995) in apparent phase with each cold stadial (Bond and Lotti 1995, Rasmussen et al. 1996, Rasmussen et al. 1996, Elliot et al. 1998, Vidal et al. 1998, Dokken and Jansen 1999, Grousset et al. 2000, van Krevelt et al. 2000).

We suggest from these observations a simple cause and effect relationship whereby D-O oscillations result from a direct coupling between atmospheric circulation, coastal ice sheets and ice shelves. During interstadials, when relatively warm waters (10-12°C summer sea surface temperature) reached as far as 60°N (Manthé 1998, van Krevelt et al. 2000), heavy snow would have fallen over Iceland, Scandinavia and Southern Greenland, causing fast growth of the ice sheet periphery, development of ice shelves, and expansion of the polar vortex. This would correspond to the progressive cooling phase towards a stadial, and to the southern shift of polar waters to about 45°N (Shackleton et al. 2000b). In parallel, snowfall over the ice sheets would have decreased, while active ice streams would have eroded coastal ice sheets, in turn leading to the disappearance of the ice shelves, the decay of the polar vortex, and the start of a new cycle with the rapid warming of the high latitude ocean.

Recent results may help to establish possible links between Heinrich events and D-O oscillations. IRD and isotope records from the Norwegian Sea (Rasmussen et al. 1996a, 1996b, Dokken and Jansen 1999) and northern North Atlantic (Elliot et al. 1998) show a typical succession of meltwater events with peaks in IRD from the Arctic ice sheets (Blamart et al. 1999) that follow the rhythm of the

D-O events. Magnetic susceptibility measurements on these cores present a temporal succession which is very similar to the Greenland $\delta^{18}O$ record. This signal is interpreted as resulting from the formation and transport of nepheloid layers formed by resuspended magnetic microparticles along the Norwegian Sea volcanic ridges during periods of active deep water convection and thermohaline circulation (Kissel et al. 1999). There is no direct indication of the influence of Heinrich events in these records. By contrast, in the mid-latitude North Atlantic Ocean, both Heinrich and D-O oscillations appear superposed, each with their own typical rhythm and morphology (Bond et al. 1992, Grousset et al. 1993, Weeks et al. 1995, Grousset et al. 2000), as if the longer scale oscillations of the Laurentide and the shorter scale oscillations of the Fennoscandia and other Arctic ice sheets were operating, at least in part, independently. Yet, there is evidence for interrelationships. A major connection may be found by comparing the millennial variability signals, recorded in the Greenland and Antarctic ice cores. A correlation between the GISP-2 and Byrd ice records is now available (Blunier and Brook 2001), using higher resolution atmospheric $CH_4$ records than those published in Blunier et al. (1998). During each of the major Northern Hemisphere D-O coolings (those corresponding to the Heinrich events), air temperature over Antarctica gradually warmed, peaking exactly at the end of the Greenland cold stadials. The following cooling reached a minimum approximately in phase with the beginning of the next large stadial in Greenland, after which Antarctic air warmed again as part of the next cycle. Such oppposite temperature trends in the Northern and Southern Hemisphere records is also seen for D-O 20 and 21, at the transition between MIS 5.1 and 4 (at about 70-80 ka BP) (Blunier and Brook 2001), as well as for the YD. These contrasting trends may result from a shift between a thermohaline circulation regime similar to the modern one, with active NADW formation and transfer of heat from the Southern to the Northern Hemisphere, and a regime with enhanced deep water formation in the high southern latitudes. This is the so-called bi-polar seesaw hypothesis of Broecker (1998) and Stocker (1998). Meltwater excess at high northern latitudes (in particular during Heinrich events) would slow down NADW formation and activate deep water transport from the south (Figure 3.5). At the end of the Heinrich event, thermohaline circulation would start again in the north, bringing warm waters near ice-sheets, thus facilitating snow accumulation and rapid ice sheet growth until the next Heinrich event. This hypothesis is also supported by the results of Marchitto et al. (1998) who observed in a set of

North Atlantic cores from different water depths that during deglaciation and the YD, the nutrient contents of intermediate and deep waters evolved in opposition. The YD was associated with a shallow-nutrient-poor "NADW-like" intermediate water, and nutrient rich "AABW-like" deep waters. The reverse is true for the warm Bølling/Allerød period.

Interestingly, Shackleton and colleagues (2000) recently published a study of an IMAGES high resolution core from about 2000 m depth on the Iberian margin (MD95-2042), which shows in its planktic *Foraminifera* $\delta^{18}O$ record the typical signature of the D-O events, but in its benthic *Foraminifera* $\delta^{18}O$ record precisely the same signal that is recorded in the Antarctic ice (Figure 3.6). The planktic vs benthic foraminiferal oxygen isotope profiles from that sediment core present the same phase relationship as the Greenland vs Antarctic ice records discussed by Blunier et al. (1998). The millenial variability of the Vostok $\delta^{18}O$ is thus apparently a signal of global significance. This provides a strong argument in support of the hypothesis that the millenial variability in $\delta^{18}O$ in Vostok (or Byrd) ice is linked to oscillations in interhemispheric thermohaline heat transport.

### 3.3.2 Millenial variability of climate at low latitudes

The fundamental direct response of the Asian and African monsoons to changing seasonality of insolation forcing has been well known for nearly two decades. When perihelion coincides with summer, as it did in the Northern Hemisphere 11,000 years ago, the seasonal insolation contrast increases, and monsoonal circulation intensifies. Across North Africa and monsoon Asia, reconstructed lake levels, vegetation, and lake chemistry support the inference of a much-intensified monsoon in the early Holocene (e.g. Street and Grove 1979, Gasse and Van Campo 1994, Lamb et al. 1995, Bradbury in press); see summaries in Winkler and Wang (1993), Overpeck et al. (1996) and Gasse (2000). Paleoceanographic records from the northern Indian Ocean and eastern Mediterranean complement this picture with evidence of increased monsoon-related upwelling and enhanced riverine deposition of terrigenous material and diagnostic pollen types during this period (Rossignol-Strick 1983, Prell 1984, Prell and Campo 1986, Overpeck et al. 1996). The driving force behind these large changes (increased seasonality of solar radiation), has been confirmed with a series of GCM studies (Kutzbach 1981, Kutzbach and Otto-Bliesner 1982, Kutzbach and Street-Perrott 1985). Although the response of the monsoon to orbital forcing was first characterized

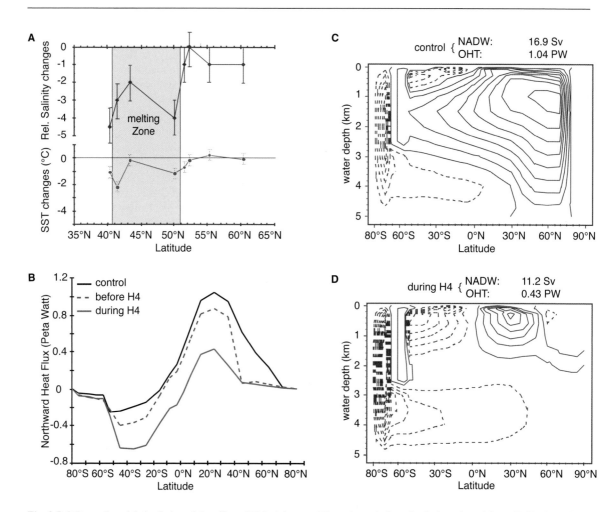

**Fig. 3.5.** 2-D zonal model simulation of the effect of Heinrich event H4 on thermohaline circulation adapted from (Paillard and Cortijo 1999). **A.** Zonal mean of the anomaly (H4 – before H4) for Sea Surface Salinity and Sea Surface Temperature, versus latitude. These data are used as input for the model. **B.** Changes in modeled northward oceanic heat transport for the control run, before H4 and during H4. **C.** Vertical distribution of the ocean heat fluxes for the control run, versus latitude. Flux of North Atlantic deep water (NADW) at its maximum is 16.9 Sv and oceanic heat transport (OHT) 1.04 PW. **D.** During H4: The model predicts a lower northern heat flux (by more that 50%) with a smaller NADW flux (11.2 Sv) and much shallower thermohaline circulation. Low latitude heat is preferentially transported to the Southern Hemisphere.

as linear (Prell and Kutzbach 1987), records from terrestrial and marine systems that have greater resolution, better chronological control, and improved spatial coverage demonstrate that the monsoon does not respond gradually to gradual insolation forcing. The warm and wet monsoon maximum in N. Africa and West Asia (about 13 to 6 ka BP) is interrupted at least twice by dry and/or cool spells (Younger Dryas 12-10 ka BP and 8.5-8 ka BP). The Mediterranean Sea hydrology, also influenced by the evolution of the African monsoon, similarly presents a large deficit in evaporation (and/or excess in precipitation) peaking at 9 ± 0.5 ka BP and 7.5 ±0.5 ka BP, and a drier interval between 8.5 to 8 k yr BP (Mercone et al. 2000). The precipitation excess may have disrupted deep-water ventilation

of the Mediterranean Sea, and contributed to the development of anoxic bottom waters (Sapropel S1) (Rossignol-Strick 1985, Fontugne et al. 1994). At about the same time, an abrupt cooling affected a large part of the Northern Hemisphere, the so-called "8.2 ka event". This is the only notable event in the Greenland ice isotopic record for that period (Alley et al. 1997), and is recorded as a 1.5‰ negative shift in ice $\delta^{18}O$ (equivalent to a 4-8°C drop in air temperature) which lasted for about 200 yr. A cold event of similar timing is recorded in the Lake Ammersee (Germany) record of von Grafenstein et al. (1998), and as a color shift (interpreted as drier and cold climate with stronger trade winds) in the Cariaco Basin record of Hughen et al. (1996). An increase in IRD deposits occurred at that time

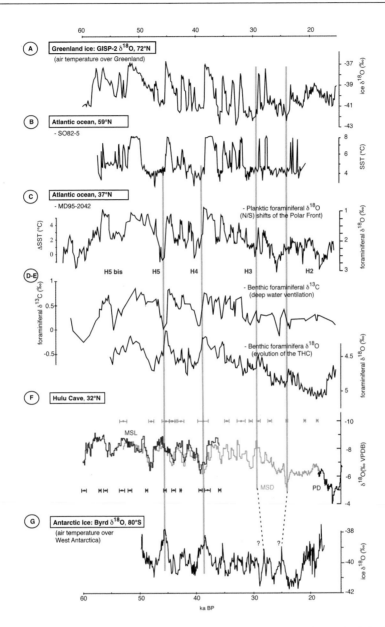

**Fig. 3.6.** Millenial variability during the Last Glacial period (15 to 60 ka BP). Ocean sediment time scales (calendar ka) are derived from AMS [14]C dating and correlation with the GISP-2 δ[18]O record. Timing of the large meltwater Heinrich events is indicated by the labels H2 to H5, while the vertical grey lines mark the initiation of each subsequent warming in the North Atlantic system. **A.** GISP ice δ[18]O record, a proxy for air temperature over Greenland (Grootes and Stuiver 1997). **B.** North Atlantic SST record from SO82-5 (59°31'W) (van Krevelt et al. 2000), derived from planktonic foraminiferal counts by the SIMMAX modern analog method (Pflaumann et al. 1996). **C.** δ[18]O of planktic *Foraminifera* from IMAGES core MD95-2042 (37°48'N 10°10'W 3100 m depth ) along the Iberian margin (Shackleton et al. 2000). The corresponding surface water temperature changes are plotted relative to the left axis. Signals **B** and **C** record the North-South oscillations of the surface hydrological system (in particular the polar front) **D.** δ[13]C (7 pt Gaussian smoothing) of the benthic *Foraminifera* (*C. wuellerstorfi*) from the same core, a proxy for deep water ventilation (Shackleton et al. 2000). This signal reproduces, with more noise, the δ[18]O of the planktonic *Foraminifera*, thus linking deepwater ventilation and surface hydrology in the North Atlantic. **E.** δ[18]O of the benthic *Foraminifera* in the same core (Shackleton et al. 2000), which tracks the temperature and salinity changes of deep water at the core location. **F.** Speleothem δ[18]O record from Hulu Cave (East China, 32°N), a proxy for the relative proportion of precipitation derived from winter vs. summer monsoon (Wang et al. 2001). The record is dated by [234]U/[230]Th disequilibrium directly in calendar years. **G.** Byrd ice core δ[18]O, a proxy for air temperature over West Antarctica (Blunier et al. 1998, Blunier and Brook 2001),placed in the GISP-2 timescale by correlation of the CH4 records. The good analogy between the Chinese monsoon record F and the North Atlantic records **A** to **C** supports the idea of a general northern hemisphere reaction to the events occurring around North Atlantic. The so-called North/South <anti-phasing> of the climatic response during the main Heinrich events, which is apparent between the Greenland and Antarctic ice records **A** and **G** is also well apparent within Atlantic core MD95-2042 between the planktic **C** and benthic **E** δ[18]O signals. The strong analogy between records **E** and **G** suggests a common causal relationship for the deep ocean and Antarctic records, linked to oscillations of the THC (see sections 2.1 and 2.2) and initiated, about 1 ka before, by the changes in North Atlantic surface hydrology (records **A** to **C**).

in the Northern Atlantic Ocean (South of Iceland to the Newfoundland margin) (Bond et al. 1997, Labeyrie et al. 1999), which probably derived from increased transport of icebergs from northern Greenland and other remnant ice sheets from the Arctic periphery. The atmosphere-surface ocean system was thus affected over much of the North Atlantic Ocean and surrounding continents in a similar way (although with a lower amplitude), than during the Younger Dryas period. There is no direct indication for an associated meltwater spike from the ocean sediment records of the Northern Atlantic. Yet, Barber et al. (1999) have shown evidence supporting a catastrophic drainage of glacial lakes Agassiz and Ojibway at about 8.47 ka BP, just prior to the cold event. That result supports the hypothesis that, once more, a freshwater anomaly could have caused a major breakdown of the North Atlantic thermohaline circulation and a major cold spell that had global effects. The 8.2 ka event appears in the available northern Atlantic records as one of several periods during the Holocene when low-salinity polar water and the iceberg melting zone penetrated southward (Duplessy et al. 1993, Bond et al. 1997, Labeyrie et al. 1999).

These results support the idea that low latitude phenomena such as monsoons are sensitive to the interaction between gradually changing insolation and high-latitude changes (Overpeck et al. 1996, Sirocko et al. 1996, Gasse 2000) and surface feedbacks such as vegetation (De Menocal et al. 2000) and SST (Gasse 2000). A recent synthesis of well-dated, high-resolution monsoon records (Morrill et al. submitted) identifies statistically significant monsoon transitions at 1300, 4500-5000, 11,500, and 13,500 years before present. The earliest two shifts likely relate to North Atlantic changes during deglaciation. The mid-Holocene shift would relate more directly to the non-linear interaction between surface ocean, the hydrological cycle and continental albedo (with its vegetation feedback) to changes in insolation (De Menocal et al. 2000). The more recent shift remains unexplained. Climate model studies of varying complexity support the idea that surface ocean and vegetation feedbacks can add nonlinearity to the response of the monsoon to insolation forcing (De Menocal and Rind 1993, Overpeck et al. 1996, Kutzbach and Liu 1997, Claussen et al. 1999). The abrupt shifts in monsoon intensity that result from these nonlinearities can occur within centuries or even decades. In some cases, paleoclimate records of the terrestrial hydrological cycle may themselves reflect threshold effects. For example, dust records in deep-sea cores may reflect the migration of source areas or the changing location of a particularly effective delivery vector (river

plume or wind belt), rather than regional climate patterns. Lake sediment records may likewise show threshold effects, as a lake nears desiccation or reaches a depth at which its chemistry, biota, or sedimentation regime can change abruptly. Despite potential uncertainties in individual records, regional changes in moisture balance can be clearly seen when many individual records are pooled. There are clear examples of regional changes in moisture balance that do not correspond to any obvious forcings, yet are clearly expressed. For example, widespread dry events occur in Asia/Africa around 7ky and 4.5ky BP (summarized in Gasse 2000).

Despite a plethora of new results (only partially presented here), we still lack the temporal resolution and chronological constraints necessary for a full understanding of the processes involved in these oscillations and abrupt climatic changes. Our understanding will increase dramatically with the availability of better continental and marine records, as illustrated by the exceptional isotopic record compiled from speleothems from Hulu cave (near Shanghai, China) recently published by Wang et al. (2001). The record comprises long annually-banded series and spans the last 10-70 ka period with century-scale resolution, with time control provided by more than 50 U/Th dates (Wang et al. 2001), The overall morphology of this record mimics perfectly the GISP, GRIP or the North Atlantic MD95-2042 records of Shackleton et al. (2000). The relative intensity of the summer and winter East Asian monsoons appear to have oscillated over the whole period in parallel with the changes in North Atlantic climate, with each cold North Atlantic event being associated with a proportional reinforcement of the winter Asian monsoon. But what are the climatic links that explain such direct connections? It is evident that the repeated succession of large amplitude, rapid climatic fluctuations in comparative studies of this type can provide a wealth of information relevant to understanding and quantifying the inter-relationships between climatic and environmental changes. This field is in its infancy, and will no doubt offer many surprises in the coming years.

### 3.3.3 Modeling millennial scale climate variability

The rapidly expanding suite of high-resolution paleoclimatic records that exhibit millennial-scale variability discussed above has provided an enormous stimulus to numerical modeling efforts over the last decade. While model simulations exist that contain variability on this time scale, it should be noted that no model has yet been shown capable of generating millennial variability with the characteristic temporal structure seen in records such as the isotopic temperature from the Greenland ice cores or high-resolution marine sediments. One reason for this is that the time scale typical for the recurrence of D-O events (about 1-2 thousand years) or Heinrich events (5-7 thousand years) is longer than the characteristic time scales of the ocean-atmosphere system. Although the renewal time associated with the deep circulation in the global ocean, about 1500 years, is often called upon as a mechanism of millennial scale variability, none of the models of the thermohaline circulation (2D, 3D, coupled A/O GCM) exhibit natural cycles on these time scales. Rather, the models suggest that oscillations in which the thermohaline circulation is involved have time scales of 3-5 hundred years at most (Mikolajewicz and Maier-Reimer 1990, Pierce et al. 1995, Aeberhardt et al. 2000). This time scale is the typical renewal time of the modern Atlantic basin only and is consistent with the fact that the other ocean basins have so far not been identified as centers of action for this variability.

The apparent absence of a natural time scale in the atmosphere/ocean system strongly suggests that the cryosphere plays a central role in pacing or exciting this variability. Indeed, as discussed above, paleoclimatic records do show, for each of the D-O events, traces of ice rafted debris which have been interpreted as a sign of rapid discharge from circum-Atlantic ice sheets (Bond and Lotti 1995). A minimal model of millennial-scale variability should therefore include the ocean, atmosphere and ice sheets. Entire sequences of millennial-scale changes involving the thermohaline circulation (THC) have been simulated only in simplified models which employ ad-hoc parameterizations of ice sheet discharge (Paillard and Labeyrie 1994, Stocker and Wright 1998, Ganopolski and Rahmstorf 2001). In these models, changes occur on a very regular time scale and are characterized by self-sustained oscillations. In contrast, the paleoclimatic record exhibits millennial variability in a band of time scales.

Recent dynamical modeling of the Laurentide ice sheet (Clarke et al. 1999) confirms the plausibility of earlier suggestions (MacAyeal 1993) of successive destabilizations of ice streams. While the timing of these events in the model (about 1 in 4500 years) is roughly consistent with the paleorecord, the amount of freshwater associated with the discharge of for example, the Hudson Bay ice stream, is rather small (order 0.01 Sv). Based on most ocean model results, such a freshwater perturbation is too small to induce significant changes of the Atlantic thermohaline circulation.

As noted earlier, these ice-sheet modeling fresh water fluxes do not agree with values derived from high-resolution dating of the Huon peninsula sea level record (figure 3.1). The reconstructions suggest sea level changes of the order of 10 m, which corresponds roughly to about 1 Sv sustained for 100 years or more. This implies a globally integrated freshwater flux several orders of magnitude larger than the regional values simulated by an ice sheet model (Clarke et al. 1999). Hence, there must be additional mechanisms in the climate system that multiply the effect of individual ice stream discharges. One possibility would be sea level rise itself, which would act as a synchronizer for discharge from various sites around the North Atlantic and Antarctica. Grounded marine ice sheets are very sensitive to sea level rise and can be effectively destabilized in this manner.

Perturbations of the thermohaline circulation with freshwater discharge have been tested with numerous coupled climate models of different complexity. The degree of collapse depends on the amount and location of freshwater discharge, and is sensitive to the model parameters. The model experiments indicate that perturbations in the order of 0.1 Sv or more are generally needed to induce significant changes of the THC (Manabe and Stouffer 1993, Wright and Stocker 1993, Mikolajewicz and Maier-Reimer 1994, 1997). Indeed, the above estimate of 1 Sv for 100 years, were it to occur in the northern North Atlantic, would be largely sufficient to collapse the Atlantic thermohaline circulation and induce strong cooling in the North Atlantic region (with amplitudes in near surface air temperature of up to 15 °C). In compensation, a warming is seen in the south, as proposed by Crowley (1992). Model simulations suggest that in that case, a bipolar seesaw is in action, with northern cooling associated with southern warming (Broecker 1998, Stocker 1998). For partial collapse, changes are more limited to the North Atlantic with little influence on the Southern Hemisphere. However, model simulations suggest that changes in the convection patterns of the Southern Ocean can strongly influence regional temperature response. As noted above (Section 3.1), although the concept of the bipolar seesaw is straightforward, its verification in paleoclimatic records remains controversial.

Coupled climate models therefore appear to contain sufficiently different circulation modes in the ocean to explain hemispheric to global scale reorganisation suggested by the paleoclimatic records. It is evident that rapid warming, as well as rapid cooling, can in principle, be realized by these models. The timescale of decades for these changes (as suggested by some paleorecords) are not inconsistent with simulations. Models make predictions regarding the spatial expression of such changes which are not always the same. Nonetheless, they do generally agree that climate changes are strongest in the North Atlantic region and that they are transmitted to other locations of the Northern Hemisphere via the atmosphere and through oceanic teleconnections (Schiller et al. 1997).

Of course, the goal would be to simulate entire climate cycles such as a series of Dansgaard-Oeschger events. If the climate system, for some reason, happens to be marginally stable, any small perturbation could trigger a mode change. In principle it is possible to construct a climate model that is very close to a bifurcation point at which very small freshwater perturbations can induce mode switches of the thermohaline circulation. The Climber-2 intermediate complexity model (Petoukhov et al. 2000, Ganopolski and Rahmstorf 2001), for example, presents a large hysteresis loop for the response of NADW formation to changes in freshwater flux over Northern Atlantic high latitudes. The "width" of that loop (along the freshwater flux axis) is directly proportional to the meridional oceanic transport of heat. The buoyancy gained by release of heat to the atmosphere has to exceed the buoyancy loss by freshwater input to sustain convection (Ganopolski and Rahmstorf 2001). In this model, the modern $10^{15}$ W meridional heat flux at 20°N corresponds to a width of the hysteresis loop of 0.22 Sv of fresh water, and all convection occurs north of Iceland. This defines the stability of the modern "warm mode" thermohaline conveyor to freshwater perturbations. During the last glacial period, the main site of convection occurs south of Iceland, and may shift progressively south during periods of ice-sheet extension. There is no clear bifurcation. However, during the intermediate climatic stage of MIS 3, the system appears particularly unstable, with deep convection moving north or south of Iceland under minimal changes of the fresh water flux (0.01 to 0.02 Sv.) (Paillard 2001). It is currently impossible to determine how close the system is to possible bifurcation points during the glacial, because the conditions at the atmosphere-ocean interface and in the ocean interior are not well known. At present, there is no easy way out of this dilemma.

Another question that has been treated only in a cursory fashion is the possible role of oceanic tidal cycles. Tides provide more than half of the total power for vertical mixing in the oceans (Munk and Wunsch 1998). Indeed, Keeling and Whorf (2000) propose that the 1-2 kyr oscillation observed on Last Glacial and Holocene northern Atlantic Ocean paleorecords by Bond et al. (1997, 1999) derives from the 1800 year cycle of the gradually shifting lunar declination from one episode of maximum tidal forcing on the centennial time scale to the next.

A multitude of parameters is recorded in ice, marine sediments and terrestrial records. Among these are greenhouse gases and their isotopic composition, isotopes and trace elements in marine organisms, assemblages and higher organic compounds. Although they are often hard to tie directly to physically relevant parameters, they contain invaluable information about how the climate system components reacted to millennial climate change. The goal of modeling is therefore not only to faithfully simulate certain phases of past records, but also to provide quantitative and dynamically consistent interpretations of these records.

## 3.4 Climate modes on interannual to centennial scales

Paleoclimate studies of the past few millennia, and in particular the past few centuries, have seen a tremendous expansion in the past decade and have made substantial new contributions to our understanding of natural climate variability. One approach to studying the climate of the past millennium, the synthesis of temperature-sensitive data to produce regional-global temperature histories, is discussed in detail in Chapter 6. Here we describe how paleoclimate reconstructions have contributed to an expanded view of climate dynamics on interannual to century time scales, with a particular focus on how modern patterns or modes of variability have changed through time.

Observational and modeling studies confirm that a substantial portion of modern climate variability can be described in terms of modes with distinctive temporal scales and spatial patterns. In the broadest sense, these modes result from the interactions of the ocean and atmosphere over a spatially heterogeneous surface boundary. The spatial scales are set by fundamental features of the earth's surface (such as ocean/continent geometry and gradients in insolation receipt), and they derive their time scales from the scales of the forcings and responses within the system and the degree to which the slower components of climate (ice, deep ocean, and vegetation) become entrained along with the more rapid variations of the surface ocean and atmosphere. They appear to represent fundamental physical aspects of the modern circulation. To the extent that climate varies by perturbations of these modes, they offer a framework for interpreting the past and possibly anticipating the future.

It is important to distinguish between the fundamental aspects of these modes, which are relatively well described, and their impacts, which occur outside the region of well-described physics. The El Niño/Southern Oscillation (ENSO), for example, is a well-described physical mode in the equatorial Pacific, with characteristic impacts outside this narrow region that are less consistent (Kumar and Hoerling 1997, Trenberth 1998). Changes both within and beyond the tropical Pacific may significantly alter how ENSO influences remote regions; changed teleconnections do not necessarily mean changed ENSO physics in the equatorial Pacific. Both paleoclimatic and model-based studies show that the teleconnections of ENSO are not always stable as background climate changes (Meehl and Branstator 1992, Cole and Cook 1998, Gershunov and Barnett 1998, Otto-Bliesner 1999, Moore et al. 2001). Attempting to reconstruct ENSO based on teleconnected patterns may lead to erroneous conclusions; this caution also holds true for other modes (and probably other timescales).

Both paleoclimate and instrumentally based climate studies have focused on physical modes of climate variability and their impacts as targets for sampling, modeling, and greater understanding. The following sections focus on recent results that bear on the variability and sensitivity of specific modes.

### 3.4.1 The tropical Pacific: El Niño/Southern Oscillation

*ENSO in recent centuries*

The well-known interannual variations of the ENSO system arise due to coupled interactions among the atmosphere and the surface and thermocline waters of the equatorial Pacific. Bjerknes (1969) first described the positive feedback mechanisms at the heart of ENSO, and these have subsequently been refined, described and summarized by others, including Philander (1990), Battisti and Sarachik (1995), and Wallace et al. (1998). In the case of El Niño warm phases, weaker trade winds lead to a drop in the normal sea level pressure gradient across the Pacific and a consequent deepening of the thermocline on the eastern side. This leads to weaker upwelling, warm SSTs, a further reduction in the zonal SST gradient, and further weakening of the trades. The opposite set of feedbacks acts to maintain cold La Niña events: stronger trades enhance and shallow the eastern equatorial Pacific thermocline, driving colder and more intense upwelling, strengthening the zonal SST gradient and consequently the trades. Warm SST anomalies in the east are initiated by the Kelvin wave response of the thermocline to seasonal wind anomalies in the

western Pacific, and are terminated by slow adjustment processes that act between the ocean surface and the thermocline. These carry the oscillation into a cool phase. The positive feedbacks that maintain anomaly states may not be restricted to the interannual time scale, but could prolong anomalies for longer periods and even operate as a mode of response to external forcing (Clement and Cane 1999).

Interannual variations between extreme states of the ENSO system in the tropical Pacific orchestrate year-to-year climate variability in many parts of the world (Kiladis and Diaz 1989, Trenberth et al. 1998). Teleconnections throughout the tropical oceans, in the western Americas, and in the Indian and African monsoon regions are among the many well-documented impacts of the modern ENSO system. ENSO also influences high-latitude processes, including Antarctic sea ice extent and ice chemistry (Simmonds and Jacka 1995, White et al. 1999), as well as aspects of Atlantic hydrography relevant to thermohaline circulation, e.g. North Atlantic SST and the subtropical freshwater balance (Schmittner et al. 2000, Latif 2001). Interannual changes in the rates of atmospheric $pCO_2$ increase are attributable to ENSO variability, through changes in surface ocean temperature and upwelling patterns (Chavez et al. 1999, Feely et al. 1999) together with the response of terrestrial productivity to changes in water balance and nutrient feedbacks (Keeling et al. 1995, Braswell et al. 1997, Tian et al. 1998, Rayner et al. 1999, Asner et al. 2000). ENSO-related rainfall changes strongly influence fire frequency in many regions, both naturally (Swetnam and Betancourt 1998) and by abetting anthropogenic burning (Nepstad et al. 1999). ENSO variability is thus imprinted in an extensive suite of parameters relating to modern and past physical climate, biogeochemical cycles, and ecosystem dynamics; changes in ENSO have the potential for diverse and substantial global impacts.

Although modern instrumental studies of ENSO characterize its dominant frequency as interannual, a recent shift in 1976 to warmer and wetter conditions in the tropical Pacific has drawn attention to decadal variability in ENSO (e.g. Zhang et al. 1997). Extratropical decadal variability is discussed in section 3.3. A growing body of paleoclimatic evidence indicates that in the 19th century, variance in the tropical Pacific was weaker on interannual and stronger on decadal timescales, relative to the 20th century. This pattern is seen both in the impact of ENSO inferred from records largely outside the tropical Pacific (Mann et al. 1998, Stahle et al. 1998) and in the characteristic frequency of anomalies within the central equatorial Pacific (Dunbar et

al. 1994, Urban et al. 2000, Figure 3.7). This and
other changes in the dominant frequency of ENSO
appear to be related to the background climate of
the tropical Pacific; in the central Pacific, decadal
ENSO variability is strongest when the central
Pacific is relatively cool/dry (Urban et al. 2000).
This decadal variance was felt beyond the equatorial Pacific regions. At least one of the prolonged
La Niñas of the 19th century (1855-62) likely played
a role in an extended drought in the central US,
along the lines of what would be expected from
modern climate relationships (Cole et al. submitted).

Another "style" of decadal ENSO variability consists of modulating the frequency of extreme events;
some decades have stronger interannual variability
than others (Trenberth and Shea 1987). Studies
based on geochemical records from long-lived Pacific corals clearly show decadal modulation of the
frequency of interannual ENSO extremes (Cole et
al. 1993, Dunbar et al. 1994, Tudhope et al. 1995,
Urban et al. 2000). The teleconnections of ENSO
experience substantial decadal modulation as well.
The well-known link between ENSO and the Indian
monsoon has virtually disappeared since 1976
(Kumar et al. 1999b, 1999a).

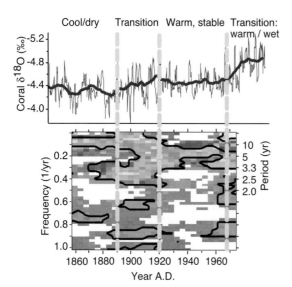

Fig. 3.7. Coral δ[18]O data from Maiana Atoll (central Pacific) and
evolutionary spectrum of the data plotted on the same horizontal
axis (Urban et al. 2000). The top panel shows bimonthly values
with a superimposed 21-yr running mean. The bottom panel
maps the changing concentrations of variance revealed by evolutionary spectral analysis, in which 40-year segments were analyzed, offset by 4 years. Colored regions are significant above
the median (50%) level, and the dark line encloses variance
significantly different from a red noise background spectrum at
90%. Changes in the mean of the time series correspond to
changes in the frequency domain characteristics of the record,
particularly in the correspondence of strong decadal variance and
weak interannual variance to cooler/drier background conditions

in the 19th century. Vertical lines separate intervals where apparently different background conditions prevailed locally; these
coincide with transitions in the spectrum.

Patterns of North American drought that correlate
with ENSO variability changed during the twentieth
century, a consequence of both changing ENSO
strength and interactions with midlatitude systems
(Cole and Cook 1998, Figure 3.8). Correlations
among paleoclimate records sensitive to ENSO wax
and wane (Michaelsen and Thompson 1992), suggesting nonstationarity in spatial patterns and/or
intensity of teleconnections. Ice-core based precipitation records on Mt Logan in the Pacific
Northwest change the sign of their correlation with
ENSO on decadal scales (Moore et al. 2001).

Several candidate mechanisms exist for decadal
modulation of ENSO variability (Kleeman and
Power 2000). First, the physics responsible for
interannual variability may be invoked for longer-
period variability (Clement et al. 1999); the positive
feedbacks that maintain La Niña and El Niño
anomalies could become more persistent than today,
due to changes in the background state (e.g. in the
Pacific thermocline). Second, decadal changes in
subtropical latitudes can propagate along isopycnal
surfaces into the tropical thermocline, thereby influencing the temperature of upwelled water (Gu and
Philander 1997). Subsurface temperature observations support this idea (Zhang et al. 1998). However, geochemical data from equatorial Pacific
corals indicate that the 1976 shift cannot have been
caused solely by changes in the thermocline source
waters in the northern subtropics (Guilderson and
Schrag 1998). Additionally, decadal variations in
ENSO may result from physical processes not yet
identified; they may originate from external forcing,
or they may be stochastic. Cane and Evans (2000)
argue that decadal variability in the tropical Pacific
is not necessarily created by a single mechanism
specific to that time scale, and that it may well
result from multiple processes acting over a range
of time scales.

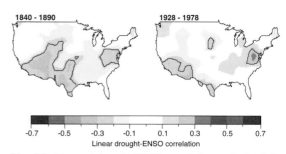

Fig. 3.8. Maps of drought-ENSO correlations calculated for
1840-90 and 1928-78 using tree-ring reconstructed drought
indices (Cook et al. 1999) and a coral record of ENSO (Urban et
al. 2000). Contours outline regions where correlations are significant at 90%.

### ENSO in the late Quaternary

To assess the sensitivity of ENSO to changes in background climate, paleoclimatic studies of ENSO have begun to focus on more distant periods of recent earth history. Perhaps the most complete look at past ENSO variability to date uses a suite of coral records from northern New Guinea, where modern corals and instrumental data show a strong influence of ENSO variability (Tudhope et al. 2001). Tudhope et al. analyzed, at near-monthly resolution, multidecadal sections of coral in well-dated time windows over the past 130,000 years (Figure 3.9), and the results show clearly how the strength of interannual variability at this site has varied. They identify two aspects of long-term climate variability that appear to influence ENSO strength, precessional forcing and glacial background climates. Both act to dampen ENSO variability in past periods; modern samples (even those that predate substantial anthropogenic greenhouse forcing) show the strongest ENSO variability of any interval sampled.

The precessional influence on ENSO has been physically described using climate models of varying complexity. In a simple model of the equatorial Pacific Ocean and atmosphere, seasonal insolation changes associated with the precession of the earth's equinoxes influence the seasonal strength of the trade winds. When perihelion falls in the boreal summer/fall, stronger trades in that season inhibit the development of warm El Niño anomalies. This response is sufficient to generate significant changes in ENSO frequency and recurrence over the late Quaternary (Clement et al. 1999). In a global coupled ocean-atmosphere model, the intensified Asian monsoon at this phase of the precession cycle further enhances Pacific summertime trade winds, cooling the equatorial Pacific and reducing interannual variability (Liu et al. 2000).

The mechanisms for glacial ENSO attenuation are not nearly as well described. Possibilities include weaker ocean-atmosphere interactions in a cooler Pacific and intensified trade winds resulting from a stronger temperature gradient across the Pacific. A lower sea level exposing shallow continental shelves in the western Pacific may also stabilize variability by anchoring the Indonesian Low convection system. Possible mechanisms for strengthening ENSO in a glacial world also exist. For example, a shallower, steeper thermocline in the eastern Pacific could allow for greater interannual variability. The NCAR coupled climate model simulates stronger ENSO variability during the last glacial maximum (Otto-Bliesner 1999). The inference of weaker glacial ENSO from the coral data does

not necessarily conflict with this simulation, however. The "glacial" intervals in the Tudhope et al. (2001) study are from less cold periods than the last glacial maximum, when precessional forcing also differs. Additional data will be needed to resolve which of the potentially competing influences ultimately determines ENSO strength, and when.

### ENSO in the mid-Holocene

It is becoming clear that ENSO operated very differently prior to the mid-Holocene. Tudhope et al. (2001) document the weakest interannual variability of any time in their record at 6500 yr BP. Interannual variability associated with ENSO along the Great Barrier Reef was absent in a coral record from 5300 yr BP (Gagan et al. 1998). Debris flow deposits in an Ecuador lake, which today occur during El Niño rains, occur with a period of approximately 15 years before about 7ky BP, and show the establishment of modern ENSO periodicities (2-8 yrs) only around 5ky BP (Rodbell et al. 1999). Sediment profiles from archaeological sites along the Peru coast indicate a lack of strong flood events (interpreted as El Niño's) between 8900-5700 years ago at Quebrada Tacahuay (Keefer et al. 1998) and between 8900-3380 years ago at nearby Quebrada de los Burros (Fontugne et al. 1999). Pollen records from both South America and New Zealand/Australia indicate that early Holocene vegetation did not include types adapted to periodic droughts that occur today, associated with interannual ENSO. Such vegetation types became established in these areas only in the late Holocene (McGlone et al. 1992, Shulmeister and Lees 1995). These and other lines of evidence suggest that the global imprint of ENSO was very different before the mid-Holocene (Markgraf and Diaz 2001). Numerical modeling studies provide insight into mechanisms of such changes. Several coupled GCM simulations bear on the question of ENSO in the mid-Holocene. Using the NCAR coupled Climate System Model, Otto-Bliesner (1999) found a cooler equatorial Pacific and no substantial change in ENSO variability. She also noted that teleconnections significant in the modern system (modeled and observed) are absent in the 6K simulation, due to the stronger influence of regional climate changes. Liu et al. (2000) used the Fast Ocean-Atmosphere Model coupled with a low-resolution atmospheric GCM and forced with early and mid-Holocene insolation (6ky BP and 11ky BP). They found a weaker ENSO system in both cases, with a tendency for slightly stronger La Niña events and even weaker El Niños, a consequence of two mechanisms. First, stronger trade winds

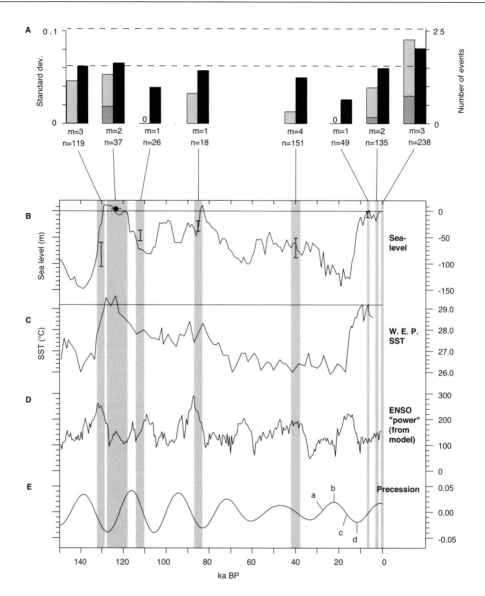

**Fig. 3.9.** Interannual ENSO variability over the past 130,000 years. **A.** Strength of ENSO variability in coral $\delta^{18}$O records for eight time periods. Solid black bars show the standard deviation (scaled on the left axis) of the 2.5 to 7 year bandpass-filtered coral $\delta^{18}$O records from each time period. Shaded bars denote high-amplitude events for each period. The darker bars indicate the percentage of the data in the ENSO bandpass-filtered data that exceeds 0.15 absolute amplitude; lighter bars indicate the percentage of data that exceeds 0.10 amplitude (both scaled relative to the left vertical axis). The number of corals for each group is given by m, and the total number of years represented by all corals in each group is given by n. The horizontal dashed lines indicate the maximum and minimum standard deviation for sliding 30-year increments of modern coral $\delta^{18}$O 2.5- to 7-year bandpass-filtered time series. **B.** Estimate of global sea level (plotted as meters below present sea level) derived from benthic foraminiferal $\delta^{18}$O (Shackleton 2000). Bars indicate paleo-sea level estimated from the elevation, age, and uplift rate of corals analyzed in this study. These bars include uncertainty in the water depth in which the corals grew. Estimates of uplift rate are based on an assumed sea level of +5 m at 123 ka (circle and bar). **C.** Sea surface temperature record for the western equatorial Pacific (ODP Hole 806B, 159°22'E, 0°19'N, 2520-m water depth) based on Mg/Ca composition of planktonic *Foraminifera* (Lea et al. 2000). The horizontal line indicates modern SST. **D.** ENSO variability estimated from application of the Zebiak-Cane coupled ocean-atmosphere model forced only by changing orbital parameters (Clement et al. 1999). Shown here is power in the 2- to 7-year (ENSO) band from multitaper spectral analysis of nonoverlapping 512-year segments of the modeled NINO3 SST index. Power is approximately equal to 100x variance. Although there is considerable variation at sub-orbital wavelengths (2s of power estimates ~±71 based on a control run with no change in orbital parameters), the main precessionally related features, including the trend of increasing ENSO amplitude and frequency through the Holocene, are statistically significant (Clement et al. 1999, 2000). **E.** The precessional component of orbital forcing (Berger 1978). For one cycle, the timing of perihelion is indicated as follows: a, boreal autumn; b, boreal winter; c, boreal spring; d, boreal summer.

(resulting from enhanced deep convection in the Asian monsoon region) force a shallower thermocline in the eastern Pacific. Second, a warm anom-aly in the equatorial thermocline weakens the verti-cal temperature gradient and hence the El Niño-La Niña contrast. Bush (1999), using the GFDL atmopheric GCM coupled with a primitive equation ocean model, found an intensified equatorial Pacific cool tongue at 6000 years ago, with greater seasonality (changes in interannual variability were not discussed).

Mid-Holocene model results and paleoclimatic observation are beginning to yield a semi-consistent picture (Clement et al. 2000, Cole 2001). A cooler (more La Niña-like) eastern and central Pacific is shown by all coupled models (Bush 1999, Otto-Bliesner 1999, Liu et al. 2000). Regional warming suggested in the westernmost Pacific (Gagan et al. 1998) supports a picture of a more La Niña-like average state. Reduced interannual variability in northern Australia (McGlone et al. 1992, Shulmeister and Lees 1995, Gagan et al. 1998) and New Guinea (Tudhope et al. 2001) implies a weaker ENSO overall and attenuated teleconnections. Weaker interannual variance around a cooler background mean would be consistent with fewer flood events in an Ecuador lake record (Rodbell et al. 1999), as warm anomalies large enough to generate intense rainfall would be rarer.

The presence of warm water molluscs of mid-Holocene age on the Peru coast (Sandweiss et al. 1996) disagrees with this picture, although local geomorphological complications (de Vries et al. 1997) or regional oceanic influences (Liu et al. 2000) may explain those observations. The inference of comparable ENSO variance at 6ky BP in one modeling study (Otto-Bliesner 1999) does not correspond to the generally observed picture of reduced variance at that time. Well-calibrated data from the regions of strongest ENSO influence are still sparse, and models have yet to simulate modern ENSO variance perfectly, so there is room for improvement on both sides. Records from outside the tropical Pacific may confuse this analysis, as both data and models increasingly reveal the nonstationarity of ENSO teleconnection patterns, particularly as background climates change (Meehl and Branstator 1992, Cole and Cook 1998, Kumar et al. 1999, Kumar et al. 1999, Otto-Bliesner 1999, Moore et al. 2001)

### 3.4.2 Decadal variability in the extratropical Pacific

Instrumental data offer clear evidence of coherent decadal variability over much of the Pacific Ocean. The spatial pattern associated with this variance is latitudinally broader than ENSO, and the time scale of variability is longer, but the impacts on climate variability around the Pacific are often similar. Many instrumental and modeling studies have described this pattern (e.g. Latif and Barnett 1996,

Mantua et al. 1997, Minobe 1997, Zhang et al. 1997, Garreaud and Battisti 1999). Pacific decadal variability clearly has a strong influence on many aspects of natural systems in the Pacific (Ebbesmeyer et al. 1991) including salmon fisheries and glacial mass balance in the northwestern US (Mantua et al. 1997, Bitz and Battisti 1999), the strength and pattern of ENSO teleconnections in North America (Cole and Cook 1998, Gershunov and Barnett 1998, Figure 3.10), and the predictability of Australian rainfall (Power et al. 1999).

Pacific decadal variability is typically described in terms of SST variations in the north Pacific. Trenberth and Hurrell (1994) use a North Pacific Index (NPI) of monthly SLP anomalies over the domain 30-65°N, 160°E-140°W; other studies refer to this index as the North Pacific Oscillation (NPO Gershunov and Barnett 1998). Mantua et al. (1997) describe the Pacific Decadal Oscillation as the time series of the first EOF of Pacific SST north of 20°N; others have redubbed this the Pacific Interdecadal Oscillation (retaining the acronym PDO). Power et al. (1999) refer to the phenomenon as the Interdecadal Pacific Oscillation (IPO) and include both Northern and Southern Hemisphere SST in its definition. The spatial pattern of a positive PDO or IPO is similar to that of El Niño, but with a greater latitudinal spread of warming centered on the equator and a stronger cooling in the western mid-latitudes. When the PDO is positive, low pressure anomalies (a deepened Aleutian Low) exist over the north central Pacific and the NPI is negative.

Time series of the NPI and the PDO calculated from various datasets show a multidecadal time scale since the 1920's, with prominent transitions around 1976, 1946, and 1924 (e.g. Mantua et al. 1997, Minobe 1997, Chao et al. 2000). Before 1925, however, instrumental records of Pacific decadal variability tend to disagree, as data become increasingly sparse (Figure 3-10a). With so few iterations of Pacific decadal variability available in the instrumental record, numerous questions remain open about its fundamental nature. Given the long time scale of this mode relative to most instrumental records, paleoclimatic efforts to reconstruct this phenomenon in the preindustrial era (e.g. from subtropical corals and tree-ring data along the Pacific rim) have important contributions to make in understanding its persistence, stability, and role in regional-hemispheric anomalies.

Tree ring-based reconstructions of Pacific decadal variability typically proceed by either averaging over multiple sites or extracting patterns of common variance from networks of sites in the western Americas. Because this phenomenon influences both hydrologic balance in the subtropical latitudes

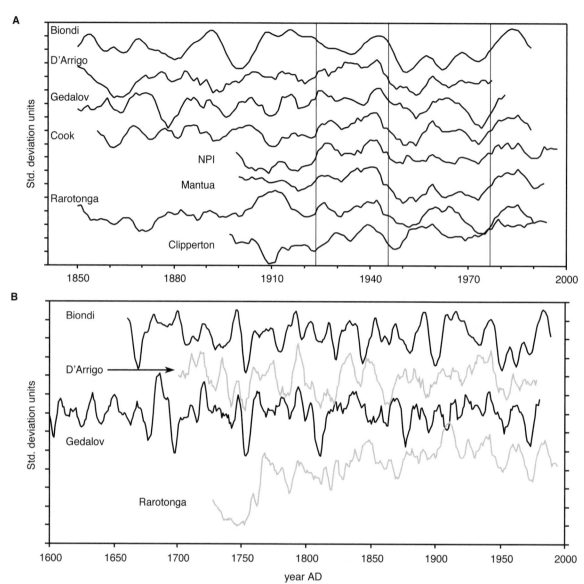

**Fig. 3.10.** Time series of Pacific decadal variability from instrumental and paleoclimatic sources. Contours outline regions where correlations are significant at 90%. The data sources are, from top: (Trenberth and Hurrell 1994, Mantua et al. 1997, Linsley et al. 2000a, 2000b, Biondi et al. 2001, Gedalov and Smith 2001, D'Arrigo et al. submitted). Records have been normalized to a standard deviation of 1 in the 1900-1979 common interval and smoothed using a 5-point running mean, to enhance the low-frequency variance. Vertical axes are marked every 1 standard deviation. The NPI (North Pacific Index of Trenberth and Hurrell (1994)) and Rarotonga (Linsley et al. 2000b) SST reconstructions are plotted inversely to facilitate comparison. The Clipperton (Linsley et al. 2000) record indicates the alternation between relatively warm/fresh (upwards) and cool/salty (downwards) conditions; isotopic data do not distinguish temperature and salinity. **A.** Instrumental and paleoclimatic records from 1850 onwards show transitions in the mid-1920's, mid-1940's, and mid-1970's. Before about 1925, however, the records agree less well – even the two instrumental SST-based PDO indices differ, due to sparse observational coverage and different methods for infilling missing values. **B.** Pre-instrumental records of Pacific decadal variability from tree-ring and Rarotonga coral data appear to move in and out of phase, suggesting a variable footprint (and perhaps variable mechanisms) for Pacific decadal variability.

and temperatures at higher latitudes, sites in one or both of these regions are commonly used. Biondi et al. (2001) used a network of precipitation-sensitive sites in southern and Baja California to develop a PDO reconstruction since AD 1661. Gedalov and Smith (2001) reconstructed north Pacific decadal variability since 1599 using six tree-ring chronologies from Oregon to Alaska. Another tree ring-based PDO reconstruction dating to 1700 (D'Arrigo

et al. submitted) used sites in coastal Alaska, the Pacific northwest, and northern Mexico to capture both the temperature and hydrologic aspects of this variability. Other studies identified a broad pattern of Pacific decadal variability in tree-ring data from regions near the Pacific coast of North America (Ware and Thomson 2000) or in reconstructions of temperature and precipitation in North and South America (Villalba et al. 2001). The latter study strongly supports the extension of this mode to the Southern Hemisphere.

Although many coral records from the Pacific exhibit a decadal time scale, few describe potential links to the PDO. A coral oxygen isotope record from Clipperton Atoll (10°N, 109°W) shows decadal variations that may reflect changes in the strength of the low-salinity equatorial countercurrent; the correspondence between this coral record and the PDO implies a stronger ECC and northward-displaced ITCZ during positive PDO phases (Linsley et al. 2000). Recent oscillations in the PDO are also present in an SST reconstruction based on coral Sr/Ca records from Rarotonga (21.5°S, 159.5°W) which extend to 1726 (Linsley et al. 2000b). This study provides additional evidence for the symmetry of decadal temperature responses across the hemispheres, at least during some intervals.

These records tend to confirm the oscillations seen in the mid-late 20th century instrumental data, but typically show much less agreement in earlier times (Figure 3.10b). The disagreement likely stems from several sources. First, if taken at face value, these records imply that the Pacific decadal variability in the 20th century is anomalously coherent across a broad latitudinal range, compared to previous centuries. The disagreement in earlier periods may suggest that Pacific decadal variability is not a single fundamental phenomenon, as it appears from the 20th century, but instead involves multiple independent influences, which are more clearly seen when records are extended. Second, the sensitivity of the sites used may not be ideal for representing the PDO. As initially defined, the PDO core region lies in the extratropical North Pacific; all terrestrial data are essentially teleconnections, which may change through time. Finally, the records are located in widely varying locations, and each represents the superposition of local variability on top of large-scale patterns. Local variability may be more of an issue for coral data, which are single-site geochemical records, than for the tree-ring data, which are combinations of records that span larger regions.

We can potentially learn much from comparing these records, despite their dissimilarities. Comparison of the Rarotonga coral record with the Northern Hemisphere data, for example yields strong agreement for the 20th century and certain intervals of the 18th century, but weak agreement in between. Tropical forcing may be most important during those times when the southern and Northern Hemispheres agree (e.g. the 20th century), and during other times, the hemispheres may behave more independently. Also notable is the tendency for the hydrology-sensitive and temperature-sensitive reconstructions to behave differently at times; the PDO has apparently not always caused these anomalies to move synchronously, which calls into question its use in prediction. Finally, there are intervals where most long records do agree (e.g. the mid-late 18th century), which points to promising intervals for further study of the PDO phenomenon. Periods when the greatest covariation is seen among multiple records may provide a useful focus for more in-depth analysis (e.g. mapping of annual anomalies). An event-based analysis may prove useful in identifying how consistent the PDO's impacts are over time; more data from the Southern Hemisphere (e.g. Villalba et al. 2001) are sorely needed for such work.

Paleoclimatic records of Pacific decadal variability also indicate diverse frequency-domain characteristics. Typical time scales include one around a decadal period (identified variously as 12, 14-15, and 10-20 yrs) and one in the multidecadal range (identified as 23, 25-50 and 30-70 yrs). The tree-ring studies address the stationarity of these components to varying extents, with no clear consensus: the 23-yr period of Biondi et al. (2001) strengthens post-1850; the 12-yr period in D'Arrigo et al. (submitted) weakens post-1850, and Gedalov and Smith (2001) found that the multidecadal (30-70) period is concentrated before 1840, while their 10-20 year period is strong throughout. Spectral analysis of instrumental indices of Pacific decadal variability also shows decadal and multidecadal periodic components whose strengths depend on the time interval analyzed (Cole, unpublished analysis). Either there is no clear frequency-domain signal of this "decadal" phenomenon, or the available records have yet to identify it unambiguously.

Although paleoclimatic studies of the PDO have not yet answered the questions raised by instrumental records, we note that this approach is still very new (the reconstructions shown here were all published in 2000 and 2001). Reconciling discrepancies among existing reconstructions, and developing better ones is a focus of substantial ongoing effort in the tree-ring and coral paleoclimate communities. Basic questions on the spatial and temporal pattern and coherency of Pacific decadal variability should soon be answerable with much greater confidence.

### 3.4.3 North Atlantic Oscillation

Interannual and decadal variations in North Atlantic climate are closely tied to the state of the North Atlantic Oscillation (NAO), broadly defined as the meridional pressure gradient between the high pressure system that lies around the Azores and the low pressure system that covers a broad swath of the Atlantic Arctic centered around Iceland (Hurrell 1995). The state of the NAO governs the strength of the westerly flow across Europe and thus the regional moisture balance and the tendency for severe winter storms.. The NAO is typically monitored by indices of the sea level pressure difference between stations in Iceland and stations near the Mediterranean (e.g. Lisbon, the Azores, or Gibraltar). However, NAO physics have not been well described, and the oscillation itself may reflect the combination of semi-independent processes that influence the strength and extent of the Icelandic Low and the Azores High. The NAO also influences the formation of North Atlantic Deep Water (Dickson et al. 1996, Curry et al. 1998), giving it a potential role in broader patterns and longer time scales of climate change.

The late 20th century experienced exceptional variability in the NAO, including an unprecedented series of high-NAO-index winters between 1988-1995, which led to unusual warmth and storminess over Europe. Paleoclimatic reconstructions have attempted to address whether the recent variability is unusual in the context of longer records. Several reconstructions have been proposed, using various combinations of tree-ring, ice core, and long instrumental data (D'Arrigo et al. 1993, D'Arrigo and Jacoby 1993, Appenzeller et al. 1998, Cook et al. 1998, Luterbacher et al. 1999, Cullen et al. 2001, Glueck and Stockton in press, Mann in press). Recent comparisons of NAO reconstructions (Schmutz et al. 2000, Cullen et al. 2001) suggest that the various reconstructions have different sensitivities and tend to correlate only moderately among themselves. The best-calibrated reconstructions include long instrumental data from European sites, and extend the NAO record back to 1700-1750; the late 20th century appears anomalous in that long-term context (Cullen et al. 2001). The best calibrated reconstructions may not necessarily be the best reflection of the NAO, however. Instrumental SLP records against which paleo data are compared may not be ideally situated to capture variance in the NAO variance (Deser 2000), while the paleoclimatic records have the potential to include a spatially broader NAO signal (Cullen et al. 2001).

Analyses of high-latitude climate variability have identified a mode that is potentially of broader significance (Thompson et al. 1998). The Arctic Oscillation (AO) describes an oscillation that spans the troposphere and lower stratosphere, and raises pressures alternately at the polar cap and along a zonal ring at about 55°N. In its positive phase, the pressure along 55°N is high, which strengthens westerly

winds there and steers oceanic storms along a more northerly path. The NAO may be a regional manifestation of the AO; they correlate well, and the AO explains a greater fraction of variance in European climate than the NAO. Alternatively, Deser (2000) has noted that the NAO is strongly linked to Arctic variability, but Pacific-Arctic links are weaker and the AO is dominated by the NAO. Debate about the AO centers in part on whether climate patterns along the full circumference of the southern extent of the AO are actually well correlated (Kerr 1997), a question that paleodata could address, but has not yet, at least directly. Arctic data appear relatively coherent (Jacoby and D'Arrigo 1989, D'Arrigo et al. 1993, Overpeck 1997), but a direct comparison of Pacific and Atlantic regions along the southern extent of the AO is needed.

### 3.4.4 Tropical Atlantic: the dipole and extratropical links

The main mode of decadal SST variability in the tropical Atlantic has been described as a dipole across the equator (Servain 1991). The northern and southern tropical SSTs are generally uncorrelated (Houghton and Tourre 1992, Rajagopalan et al. 1997, among many others); however the cross-equatorial SST gradient has an important influence on rainfall on adjacent continents (Hastenrath and Greischar 1993). Analyses of instrumental data show strong coherence between this gradient and the North Atlantic Oscillation (Rajagopalan et al. 1998). Rajagopalan et al. speculate that the North Atlantic may respond to tropical heating anomalies and that the cross-equatorial gradient is a reasonable indicator of extratropical linkages. The mechanisms for a dipole in tropical Atlantic SST have been described in a modeling study (Chang et al. 1997) that identified a characteristic dipole time scale of 13 years.

Paleoclimatic records support these inferences over longer periods. An 800-year varved sediment record of trade wind-driven upwelling from the Cariaco Basin, Venezuela, exhibits significant variance at a 13-year period (Black et al. 1999), although the period is nonstationary. Longer-period variance in this record suggests an upwelling response to solar forcing (see Chapter 6, Section 7 for discussion). The strongest correlation with Cariaco Basin upwelling is found with North Atlantic SST. Stronger trade winds and upwelling are associated with colder North Atlantic conditions. This relationship holds true over longer time scales as well, with multidecadal North Atlantic anomalies clearly visible in laminated sections of the Cariaco record during deglaciation (Hughen et al. 1996).

The influence of the dipole on Sahel rainfall may hold a clue to explaining a mystery of past climate - the "Green Sahara" of the early-mid Holocene. Climate models forced with appropriate insolation values fail to explain the magnitude of increased moisture in north Africa reconstructed from lake level and other paleoclimatic data (Joussaume et al. 1999). Even when albedo and moisture recycling feedbacks associated with expanded wetlands and vegetation cover are included, climate models fail to simulate the degree of monsoon intensification observed (Doherty et al. 2000). But when ocean temperatures are calculated to be in equilibrium with the radiative forcings and incorporated into the model, the African monsoon response is significantly enhanced (Kutzbach and Liu 1997).

### 3.4.5 Global teleconnectivity

As research proceeds on the dynamics of modern climate systems – ENSO, monsoons, the high and low latitude Atlantic – couplings among them, often spanning time scales, are becoming clearer. For example, ENSO influences the freshwater balance in the tropical Atlantic; by changing the salinity of the source waters that eventually feed convection in the North Atlantic, long-term changes in the frequency of ENSO extremes can feed back on thermohaline circulation (Schmittner et al. 2000, Latif 2001). A lagged correlation between ENSO and the NAO supports this connection and implies predictability for the NAO based on earlier ENSO behavior (Latif 2001). Alternatively, recent unusual behavior in the NAO has been attributed to Pacific changes as a direct (immediate) atmospheric response to Indo-Pacific warming (Hoerling et al. 2001). North Atlantic variability influences the hydrologic balance over the Asian landmass, with consequences for the monsoons of the subsequent year. As noted earlier, the well-known linkage between ENSO and the Asian monsoon appears to have broken down as the Eurasian landmass has warmed in the 20th century (Kumar et al. 1999). These and other studies of global teleconnectivity offer targets for paleoclimatic investigation, yet caution us against assuming that such linkages should be permanent in the face of changing background climate.

# The Late Quaternary History of Biogeochemical Cycling of Carbon

T.F. Pedersen
School of Earth and Ocean Sciences, 6270 University of Victoria, PO Box 3055 STN CSC, Victoria, B.C.
Canada V8W 3P6

R. François
Woods Hole Oceanographic Institution, Dept. of Marine, Chemistry & Geochemistry, Woods Hole, MA
2543, U. S. A

L. François
CICT, 118, route de Narbonne, FR-31062 Toulouse CEDEX 4, France

K. Alverson
PAGES International Project Office, Bärenplatz 2, CH-3011 Bern, Switzerland

J. McManus
Woods Hole Oceanographic Institution, 121 Clark Laboratory, MS#23, Woods Hole, MA 2543, U. S. A.

## 4.1. Introduction

The cycling of carbon between the atmosphere, ocean and continents controls the concentration of atmospheric $CO_2$ ($pCO_2$), which in turn impacts the earth's radiation balance. Correlations between the concentration of atmospheric $pCO_2$ measured in air bubbles in ice cores with climatic indices such as inferred global temperature (Chapter 3, Figure 4.3) strongly suggest that some link exists between $pCO_2$, "greenhouse" warming and global climate (Shackleton 2000, and others). Late Quaternary climatic variability occurs predominantly at the 100 ky frequency associated with orbital eccentricity, $\varepsilon$. Since changes in $\varepsilon$ result in relatively small changes in incoming solar radiation, non-linear internal feedback mechanisms must be invoked to explain climate variations at this frequency. In this context, the rapid collapse but slow growth of northern hemisphere continental ice sheets has often been suggested as an important influence (Imbrie and Imbrie 1980). The spectral analysis of Shackleton (2000) and the direct multiple-parameter comparisons of ice-core data recently published by Alley et al. (2002) reveal however that most indices of global climate changes (deep water temperature; inferred Vostok air temperature) precede ice volume by up to several thousand years. Although ice sheet dynamics are clearly an important component of the climate system on these timescales, such observations rule them out as the trigger of deglaciations.

On the other hand, $CO_2$ changes are in phase with orbital eccentricity, which reinforces the potential importance of $CO_2$ in influencing global climate changes, and highlights the importance of under-

standing the mechanisms that control the concentration of this gas in the atmosphere. Such mechanisms are the focus of this chapter.

The global carbon cycle can be viewed as a series of nested loops, all of which include atmospheric $CO_2$ (Figure 4.1). The amplitude of changes in $pCO_2$ is controlled by feedbacks that effect the magnitudes of fluxes between reservoirs. Paleodata, on the other hand, tend to constrain the size of carbon reservoirs rather than fluxes. This disparity between processes that are primarily distinguished by their influence on *fluxes* and data that primarily tell us about reservoirs is one of the fundamental reasons why developing an understanding of the long term operation of the carbon cycle remains a daunting task (LeGrand and Alverson 2001). The rapidity of the changes is dictated by the intrinsic response time of the processes involved and the rates of change of the external forcings that drive them (e.g. insolation intensity and distribution, or tectonics).

We can distinguish variability on five temporal scales, listed from short to long:

1) Changes on seasonal timescales, which mainly involve the insolation-driven cycle of respiration/photosynthesis of land biomass.

2) Changes on annual to decadal timescales, which involve climatically-driven exchanges between atmosphere, land biomass, litter, and the upper ocean.

3) Changes on decadal to centennial timescales. These involve transfers between the atmosphere, the upper and intermediate ocean, and

soil carbon, as well as climatically forced changes in land biomass.

4) Changes on millennial timescales. These comprise exchanges between the atmosphere, the entire ocean, the upper sediment layer, and the "reactive" crust, and include climatically-forced changes in land biomass and soil carbon.

5) Changes on million year timescales controlled by tectonically-driven (orogeny, volcanism, subduction) exchange of carbon with the lithosphere.

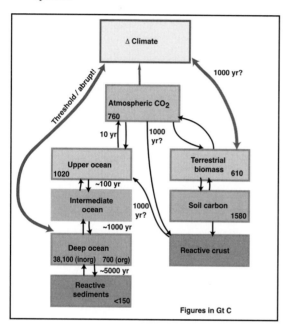

**Fig. 4.1.** Schematic showing the major reservoirs that exchange carbon on millenial and sub-millenial timescales, thereby participating in climate change (represented by the uppermost box). Black arrows indicate the directions of the exchanges with mean time constants. Thick double-headed arrows are meant to convey the potential for abrupt impacts on $pCO_2$ and thus climate from oceanic processes (left side) and more gradual impacts from continental processes. Numbers inside the boxes indicate estimated reservoir sizes in Gt C.

We discuss here only the terrestrial and marine reservoirs and processes that affect the cycling of carbon on century to multimillennial timescales. Within this temporal context, the steady-state level of atmospheric $CO_2$ is forced by exchanges between the atmosphere and the largest reservoirs (i.e. the deep ocean, reactive sediments and the reactive crust). In contrast, the smaller reservoirs (terrestrial biomass, soil carbon and the upper ocean) have only limited influence on the long-term steady-state level of atmospheric $CO_2$, but they can potentially respond much faster and produce transient excursions in atmospheric $CO_2$ levels on decadal to millennial timescales.

## 4.2. Continental processes and their impact on atmospheric $CO_2$

### 4.2.1. Biospheric carbon

During the LGM, the climate was colder and drier. These conditions induced a dramatic reorganisation of the vegetation at the Earth's surface (Chapter 5). Ice sheets prevented the growth of plants in the northern hemisphere at high latitudes. At lower latitudes, vegetation colonized the exposed shelf (following the ~125-129 m sea level drop; Peltier 2002; see also other papers in Mix and Clark 2002, and Chapter 3, Section 4, Figures 3.2 and 3.5) and deserts were much more widespread (eg. Sarnthein 1978, Petit-Maire 2000). Under the prevailing lower atmospheric $CO_2$ level, C4 plants were favoured. C3 plants did not thrive during times of low $pCO_2$ as they lacked the strong fertilization effect that $CO_2$ exerts on them, and they suffered from a decline in water use efficiency due to enhanced stomatal openings at low partial pressures of $CO_2$ (Polley et al. 1993). Grasslands, steppes and savannas expanded at the expense of forests (Adams et al. 1990). This is the general view of the LGM world that can be drawn from palynological and other continental data (Figure 4.2), although these are still too sparse to provide a precise picture.

Many attempts to reconstruct the carbon budget associated with the terrestrial biosphere at the LGM have been performed during the last decade. These studies agree that the land biosphere contained less carbon at the LGM than today, indicating that the terrestrial biospheric carbon reservoir must have been a source, rather than a sink for atmospheric $CO_2$. The magnitude of the deglacial increase is still intensively debated, however. Studies based on palynological and sedimentological data provide the highest estimates (740-1500 Gt C) of the increase in biospheric storage between the LGM and the present (Adams et al. 1990, Van Campo et al. 1993, Crowley 1995, Peng et al. 1995, Adams and Faure 1998). However, these estimates have been obtained by extrapolating a very limited set of data. Moreover, they usually rest on the assumption that the average carbon storage density (kg C m$^{-2}$) of a given biome was the same at the LGM as it is today. Most of these estimates include uptake of C by peatlands, which formed during the mid and late Holocene (Franzén 1994) and today cover roughly 2.3-5.0 x $10^6$ km$^2$ of the Earth's surface. These deposits are estimated to store 110-455 Gt C (Botch et al. 1995), with a best guess of roughly 200 Gt C. If the 0.3-0.4 ‰ rise in the global value of oceanic $\delta^{13}$C recorded in foraminifera from the LGM to the present (Curry et al. 1988, Duplessy et al. 1988,

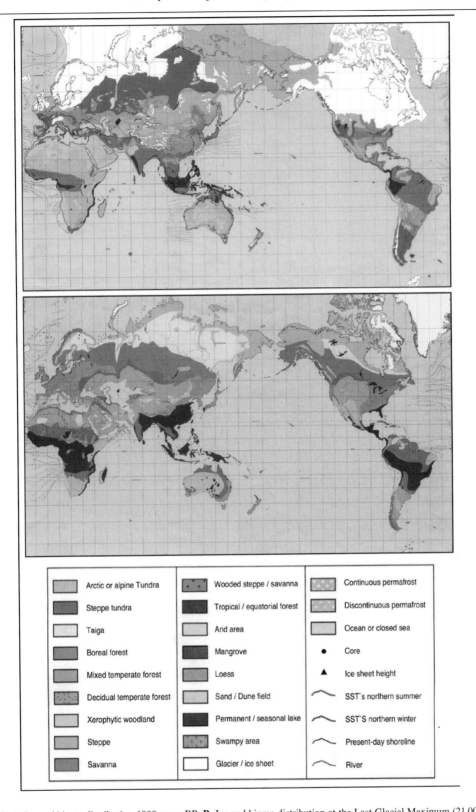

**Fig. 4.2**. **A.** Ice and biome distribution 6000 years BP. **B.** Ice and biome distribution at the Last Glacial Maximum (21,000 years ago). From Petit-Maire et al. (1999).

Crowley 1995) is fully attributed to carbon reservoir changes, it suggests a somewhat lower range for the increase in biospheric carbon stock of 270-720 Gt C (Bird et al. 1994). This range incorporates the possibility that isotope discrimination in C3 plants was reduced at the LGM (Van de Water et al. 1994). Bird et al. argue that the true change was probably at the lower end of this range, since C4 plants were more widespread during glacial times (Cole and Monger 1994). However, laboratory experiments on two species of foraminifera by Spero et al. (1997) showed that the $\delta^{13}C$ of carbonate shells decreases linearly with increasing concentration of carbonate ion in the water in which the organisms grow. This observation suggests that a significant portion of the observed 0.3-0.4 ‰ rise in the $\delta^{13}C$ of benthic foraminifera might reflect a decrease of seawater carbonate ion concentration ($[CO_3^=]_{sw}$) from the LGM to the present, rather than land-sea reservoir shifts. Using the estimate of Spero et al. of 40 $\mu$mol kg$^{-1}$ for this drop in $[CO_3^=]_{sw}$, the "true" range of seawater $\delta^{13}C$ rise from the LGM to the present becomes 0.06-0.16 ‰. Based on the methodology of Bird et al. (1994), we estimate that the range of biospheric carbon change between the last glacial and the present would then span from −160 Gt C (i.e. lower carbon storage at the LGM) to +240 Gt C. The implications of the work by Spero et al. are thus that the biospheric carbon stock at LGM may have been quite close to, or even somewhat higher than, its present value. However, these isotope-based estimates rest on the hypothesis that the land biosphere was the only reservoir that exchanged significant amounts of isotopically-light carbon with the ocean-atmosphere system over the last deglaciation. Exchanges with isotopically light soil carbonate could also have played a role, as discussed below.

Other attempts to estimate the post-glacial uptake of carbon by the biosphere have used GCM climate reconstructions for the LGM as inputs to land biosphere models of varying complexity. The range of estimates yielded by these studies is 0-700 Gt C yr$^{-1}$ (Prentice and Fung 1990, Friedlingstein et al. 1992, 1995, Prentice et al. 1993, Esser and Lautenschlager 1994, François et al. 1998, 1999). The uncertainties associated with this model-based methodology are well illustrated through the results of the recent PMIP intercomparison (*PMIP-Carbon*; François et al. 2000). In this work, three land biosphere models (ALBIOC, Roelandt 1998, BIOME4, Kaplan 2000, CARAIB, Warnant et al. 1994) were forced with a subset of the LGM climatic fields reconstructed within PMIP. The advantage of this intercomparison was that all participating biosphere models were forced with an identical set of climate reconstruc-

tions, so as to determine if discrepancies originated from differences in the adopted climate forcings or in the biosphere models themselves. The first experiment tested the model sensitivity to the atmospheric $CO_2$ increase that occurred between the LGM (200 ppmv) and the Pre-industrial (280 ppmv). $CO_2$ fertilization results in higher rates of photosynthesis, biomass growth and hence higher amounts of carbon in vegetation, litter and soil reservoirs during the Holocene. This trend was reproduced by all models (Figure 4.3), but the amplitude of the increase in biosphere carbon storage varied considerably among them (from 230 Gt C in ALBIOC to 890 Gt C in BIOME4). Such a wide range in sensitivities to $CO_2$ probably reflects our poor quantitative understanding of the mechanisms involved in $CO_2$ fertilization and the lack of observational data to constrain this process. This gap must be filled to produce reliable reconstructions of LGM carbon storage. This problem is not inherent only to model-based reconstructions, but also to those based on palynological and sedimentological data, since the biome carbon densities used to translate biome areas into carbon stocks are also dependent on $CO_2$ fertilisation.

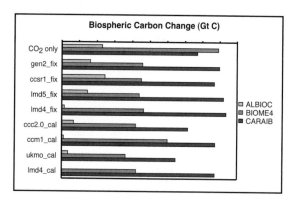

**Fig. 4.3.** Estimates of the increase in biosphere carbon storage since the Last Glacial Maximum for three land-biospheric models used in the Paleoclimate Model Intercomparison Project (PMIP). Boundary conditions in each case are described in the text. Data from François et al. (2000).

Other experiments tested the model response to the combined $CO_2$, land area (sea level) and climate changes that occurred from the LGM to the Pre-industrial. Again, the responses to these changes were strongly model-dependent. For example, land area and climate shifts strongly mitigate the effects of the $CO_2$ fertilisation in BIOME4, while an opposite but smaller trend is usually observed with CARAIB. Furthermore, the variability of carbon storage estimates associated with the use of different LGM climate reconstructions was much smaller than the intrinsic variability among the biosphere models. Indeed, the overall range of carbon storage changes predicted in the intercomparison (an increase of 9-935 Gt C from the LGM to pre-industrial times) was larger than the range exhibited by all previous model reconstructions. Presumably, one reason for such large contrasts among present models is that the calculated LGM biosphere carbon stock is the net result of several factors with opposite effects. For instance, the LGM cool climate results in an overall decrease of carbon storage in living vegetation, but the same cooling reduces organic matter decomposition rates by bacteria and, hence, tends to increase carbon storage in the soil.

In summary, Figure 4.4 shows a synthesis of all existing estimates of the increase in carbon storage by the biosphere between the LGM ~21,000 years ago and the present. These estimates range from –160 to 1900 Gt C. The lower end of this range corresponds to the isotopic estimate with inclusion of $[CO_3^=]_{sw}$ effects or to model-based estimates without $CO_2$ fertilization, while the upper end relates to the largest increase derived from continental data. Inclusion of $[CO_3^=]_{sw}$ effects on carbonate shell $\delta^{13}C$ reinforces the discrepancy between isotope-based and continental data-based estimates of the LGM-to-present shift in biospheric carbon storage. It is essential to find the reasons for this discrepancy in order to obtain more precise constraints on long term temporal variability in carbon cycling. A value of 600 Gt C represents the intersection of three different types of estimate (paleodata, models, and carbon isotopic budget excluding $[CO_3^=]_{sw}$ effects) and can be considered to be the 'best current guess'. The assimilation of this net amount of C by the terrestrial biosphere during the last deglaciation could have provided significant compensation for the $CO_2$ released during the deglacial reorganization of the ocean (Section 4 below).

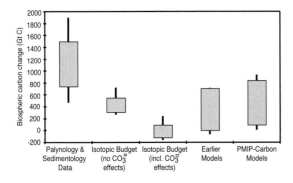

**Fig. 4.4.** Compilations of existing estimates of the increase in biospheric carbon storage between the Last Glacial Maximum and the present. Note that the range exceeds three orders of magnitude. Data sources are listed in François et al. (2000).

A few estimates of the change in biospheric carbon storage between the mid-Holocene (6000 yr B.P.) and the present have also been derived. The magnitude is much smaller than the shift that occurred between the LGM and the present. Using continental data, Adams and Faure (1998) estimated that biosphere carbon storage decreased by 27 Gt C from the mid-Holocene to the present. This is similar to the 36 Gt C decrease calculated by Foley (1994) from a biosphere model forced with GCM climatic fields. These estimates can be compared to the 195 Gt C decrease in biosphere carbon stock between 7000 and 1000 years BP obtained by Indermühle et al. (1999) by inversion of the $pCO_2$ and $\delta^{13}C_{air}$ signals archived by ice cores (see Chapter 2, Section 5). It must be noted, however, that the estimate by Indermühle et al. implicitly contains a contribution from the development of agriculture, while the present reference state in the other studies is defined by potential natural vegetation. Moreover, Adams and Faure (1998) and Foley (1994) both neglected $CO_2$ fertilisation in response to the 20 ppmv increase in $pCO_2$ from the mid-Holocene to the immediate pre-industrial period. This may have induced an increase in biospheric carbon storage of ~120 Gt C (François et al. 1999), which if added to these continental estimates would bring them up to the range suggested by Indermühle et al. (1999).

### 4.2.2 Soil carbonate

A major implication of the larger desert areas during the LGM is that the inventory of soil carbonate carbon (calcrete) in arid soils should have been higher at that time. Calcrete often occurs as layers and nodules, and represents an often-overlooked reservoir in the global budget. The current carbon content of calcrete and other soil carbonates is estimated to be some 700 Gt C (Batjes 1996). A first-order estimate of the change in magnitude of this reservoir since the LGM has been produced by Adams and Post (1999) using paleovegetation maps from the Quaternary Environment Network Atlas (QEN 1995, Adams and Faure 1997). They estimate that there was a decline in the storage of inorganic C in soils of ~500 Gt C from the LGM to the mid-Holocene, and an increase in storage of ~100 Gt C from the mid-Holocene to the present, yielding a net change of ~-400 Gt C in this reservoir over the last 21,000 years. This transfer of carbon to the ocean-atmosphere system over the last 21,000 years presumably had very limited impact on atmospheric $CO_2$ evolution, since carbon is transferred together with alkalinity. Moreover, a substantial fraction of the mobilized C probably re-precipitated as calcium carbonate on the ocean floor. However, calcrete and soil carbonates may have played a non negligible role on the isotopic budget of the ocean-atmosphere system. Precipitation of soil carbonates uses soil $CO_2$ with a negative carbon isotopic signature that is derived at least partly from vegetation. Indeed, the $\delta^{13}C$ of soil carbonates ranges from $-2$ to $-10$ ‰ (Cerling 1991). Dissolution of soil carbonates over the last deglaciation and their addition to the sea would then have lowered the $\delta^{13}C$ of the ocean, thus reducing the rise in oceanic $\delta^{13}C$ induced by the expansion of the biosphere by roughly 0.02 to 0.1 ‰. This effect is in the right direction to help reconcile continental data- and isotope-based estimates of the LGM biospheric carbon stock (Section 2.1), although it is far too small to explain the full difference between both estimates.

### 4.2.3. Weathering and river transport

Ludwig et al. (1999) have estimated using models that the delivery of riverine particulate and dissolved organic carbon to the ocean by rivers during the LGM was reduced by at least 10%. About two-thirds of the modern flux of riverine DOC and POC is respired to $CO_2$ in the coastal region and returned to the atmosphere (Smith and Hollibaugh 1993); thus, the lower riverine input of carbon during the LGM coupled with reduced glacial-age temperatures (and associated respiration rates) presumably represented a slight net decline in the $CO_2$ flux to the atmosphere from this source during glacial times. However, the smaller shelf area during the LGM limited the burial of particulate organic carbon in relatively rapidly accumulating nearshore or shelf deposits; this would have countered the reduced $CO_2$ flux. Ludwig and Probst estimate that the shelf-area effect would have dominated. Thus, reduced burial of the riverine terrestrial particulate C flux in marine sediments would have had a net effect of marginally increasing the $CO_2$ flux to the atmosphere.

The residence time of $CO_2$ in the atmosphere, if weathering were the only process operating to remove it, would be ~7000 to 8000 years (Kasting and Walker 1992). Thus, changes in continental weathering rates could have influenced glacial-interglacial $pCO_2$ directly, through either more or less consumption of carbon dioxide. An indirect effect would have been associated with any change in the alkalinity flux to the ocean (see Section 3.3). However isotopic tracers suggest that there was little change in weathering rates between the LGM and today. The high-precision $^{87}Sr/^{86}Sr$ measurements of Henderson et al. (1994) constrain glacial-interglacial changes in the riverine flux of Sr to $\leq 30\%$, while the $^{187}Os/^{186}Os$ data of Oxburgh (1998) show no evidence of enhanced chemical weathering during periods of intense continental glaciation. These results imply collectively that changes in chemical weathering intensity were unlikely to have exerted a significant impact on atmospheric $pCO_2$ on the timescales of interest here.

## 4.3 Marine processes that affect atmospheric $CO_2$

The ocean can be subdivided into four carbon reservoirs (Figure 4.1) each of which influences atmospheric $CO_2$ on a different timescale:

(1) *The upper ocean:* This is a comparatively small reservoir that exchanges $CO_2$ with the atmosphere on a subannual to decadal timescale. When integrated over space and time, localized areas of supersaturation, as in upwelling regions, are compensated by regions of undersaturation, often associated with regions of seasonally high productivity.

(2) *The intermediate ocean:* On longer timescales, the $PCO_2$ of surface water and the $pCO_2$ of the atmosphere are primarily controlled by the exchange of total dissolved inorganic carbon ($\Sigma CO_2$) and carbonate alkalinity (CA) between surface and deeper waters. We can define an "intermediate ocean" reservoir that supplies $CO_2$, CA and nutrients to the upper ocean. The rate of supply is controlled by vertical mixing and upwelling. Atmospheric $CO_2$ exchanges with this reservoir on century to millennial timescales.

(3) *The deep ocean:* The deep ocean is the largest reservoir of inorganic carbon relevant to the Quaternary (the much larger sedimentary carbon reservoirs exchange with atmospheric $CO_2$ on much longer timescales). It encompasses most of the ocean, and controls the steady-state level of atmospheric $CO_2$ on a timescale equivalent to the turnover period of the entire ocean (ca. 1000 y). The latter is a function of the rate of thermohaline circulation.

(4) *The reactive surficial carbonate sediment:* The balance between carbonate burial and dissolution in surface sediment also affects carbonate alkalinity and $\Sigma CO_2$ of deep water. Carbonate sediment in contact with bottom water is thus also a comparatively large reservoir of inorganic carbon that can affect atmospheric $CO_2$ with a multi-millenial response time (Catubig et al. 1998).

### 4.3.1 Air-sea flux

Transfer of carbon between the atmosphere and the ocean requires that there be a difference between the partial pressures of carbon dioxide in the lower troposphere ($pCO_2$) and the surface water of the ocean ($PCO_2$). The $PCO_2$ of seawater is determined primarily by its carbonate alkalinity (CA, $= [HCO_3^-] + 2[CO_3^{2-}]$) and total $CO_2$ ($\Sigma CO_2$) concentrations, via:

$$PCO_2 = \frac{(2\Sigma CO_2\text{-}CA)^2}{K'(CA - \Sigma CO_2)}$$

where K' is related to the first and second dissociation constants for carbonic acid, and can be summarized as $K' = ([HCO_3^-]^2)/(pCO_2[CO_3^{2-}])$. At S = 35 and 1 atm. pressure, K' decreases with increasing T, from ~11.3 x $10^4$ mmol $L^{-1}atm^{-1}$ at 0° C to ~3 x $10^4$ mmol $L^{-1}atm^{-1}$ at 30° C (Broecker and Peng 1982, Table 3.7); thus, atmospheric $pCO_2$ increases significantly as water temperature rises, by about 4 % per °C. Salinity has a more restricted effect on $CO_2$ solubility in the surface ocean – a 1 psu increase in S (from say 34 to 35) would decrease $pCO_2$ by only ~10 $\mu$atmospheres. Furthermore, $pCO_2$ is also influenced by the carbonate buffering system in surface seawater, via the summary reaction $CO_2 + H_2O + CO_3^{2-} = 2HCO_3^-$. Simply stated, for a given $\delta CO_2$, an increase in the carbonate ion content in surface water will decrease $pCO_2$.

In addition to depending on the difference ($\Delta pCO_2$), the air-sea exchange of carbon has a strong, nonlinear and poorly understood dependence on wind speed. Calibrations in the modern context are summarized by Wanninkhof (1992), who suggests a quadratic wind speed dependence. The bulk formulae used in numerical models to parameterize the air-sea flux associated with a given $\Delta pCO_2$ and wind speed are accurate to within about 25%. On a global scale, the generally accepted uncertainty in modern air-sea $CO_2$ flux climatologies, which includes uncertainties in both wind speed effects and the air-sea $CO_2$ partial pressure difference, is of order 100% (Doney et al. 2000). For the LGM, when wind speeds were likely to have been significantly higher, the situation is even worse. For example, if, over a given region, modern average wind speed is 5 m $s^{-1}$, and the LGM average was 7 m $s^{-1}$, the air-sea flux at LGM in this region, as a result of the necessity to square these values, would be approximately double its modern value, based on wind speed change alone. Of course regional wind speeds may have been less during the LGM, further complicating things. As a result, even if glacial $\Delta pCO_2$ fields were perfectly known, the resulting pattern of air-sea carbon fluxes could not be estimated to within a factor of two, because of uncertainties in the wind fields. Sea ice can also influence the exchange of $CO_2$ between ocean and atmosphere by almost entirely blocking gas exchange across the interface, although greatly enhanced fluxes in leads (large gaps that open periodically in sea ice cover due to its dynamic nature) counter this effect somewhat. Although sea ice is restricted to high latitude areas it may nonetheless be important for the global carbon cycle. Stephens and Keeling (2000), for example, attribute the bulk of the increase in $pCO_2$ between glacial and preindustrial times to a decrease in the extent of sea ice in the Southern Ocean, thereby allowing more efficient outgassing of $CO_2$ from this high-latitude region.

In summary, the exchange of $CO_2$ between atmosphere and ocean can be driven by changes in sea surface temperature (SST) and salinity (SSS), changes in the supply and removal of total $CO_2$ and alkalinity to and from surface water, changes in the surface winds and variations in sea ice cover. These various influences and their operation on a spectrum of timescales will be assessed in the following sections.

### 4.3.2 SST and SSS control (the solubility pump)

The solubility pump is classically defined as the physical transfer of $CO_2$ into the deep ocean driven by solubility and meridional overturning. Because $CO_2$ solubility in seawater decreases with increasing temperature and salinity, as noted in the previous section, the effects on $pCO_2$ induced by an average decrease in sea-surface temperature and increase in sea-surface salinity, as would occur during a glacial period, would be opposite. The meridional overturning circulation also influences solubility-driven $CO_2$ drawdown. $CO_2$ fluxes into the ocean interior are enhanced when by warm low-latitude surface water advects poleward. As this water cools it absorbs $CO_2$ from the atmosphere and, as it sinks, this $CO_2$ is sequestered in the deep ocean. Variability in the deep and intermediate-water circulation through time could then have impacted $pCO_2$, as discussed in Section 4.3 below.

### 4.3.3 Removal of $\Sigma CO_2$ from surface waters by sinking $C_{org}$ (the biological pump)

The biologically-mediated export of particulate organic matter from surface waters into the ocean interior promotes invasion of atmospheric $CO_2$ into the sea, while the resupply of regenerated $CO_2$ to the surface (and thus the atmosphere) by vertical mixing and upwelling counters this flux (e.g. Karl et al. 1997). The efficiency of the biological pump thus depends on the balance between the rate at which organic carbon is exported from surface waters and the rate at which regenerated $CO_2$ is resupplied from deep water. While the latter is controlled on a global scale by the rate of ventilation of the deep ocean, export production is controlled on regional scales and is often restricted by the rate of supply of a limiting nutrient to the euphotic zone.

Although light limitation and grazing pressure have also been invoked as alternative controls on primary production, only in restricted coastal upwelling regions such as along the Peru and Namibian margins are surface waters likely to be replete in all nutrients.

Four limiting nutrients, nitrate, phosphate, silicate and iron, have so far been clearly identified and their effect recognized in different areas of the ocean. They are all supplied to the ocean by river runoff, but iron and nitrate have important additional sources. Whereas Fe transported by rivers is readily removed in estuaries (Boyle et al. 1977), restricting its impact to coastal waters, eolian (dust) inputs add iron to the surface ocean, particularly in the open sea. Nitrate is added to seawater by nitrogen fixation, which predominantly occurs in the warm, stratified nutrient-poor (oligotrophic) regions of the ocean (Karl et al. 1997, Gruber and Sarmiento 1997). In these areas Cyanobacteria such as *Trichodesmium* spp. fix dissolved dinitrogen ($N_2$) into organic N, which is subsequently released to seawater as dissolved organic N species or ammonium and oxidized to $NO_3^-$.

The factors controlling the rates of supply of limiting nutrients to the euphotic zone depend on the limiting nutrient and vary from region to region. For nitrate, phosphate and silicate, rates of supply are mainly controlled by their concentration in intermediate waters and rates of vertical mixing or upwelling. In turn, their concentrations in subsurface waters are controlled by their inventories in the ocean, their regeneration rates, and deep water circulation. In contrast, while some Fe can also be supplied by vertical mixing (e.g. Watson et al. 2000), its main source in the open ocean is often dust deposition onto the surface. Likewise, in oligotrophic regions, nitrate supply from below is typically supplemented by direct fixation of $N_2$ (Karl et al. 1997; Gruber and Sarmiento 1997). These two modes of nutrient addition are distinguished by an important difference in their relative effectiveness in "pumping" $CO_2$ to the deep sea. Limiting nutrients supplied by vertical mixing are accompanied by stochiometrically equivalent amounts of $\Sigma CO_2$, so that increasing their supply rate simply by increasing vertical mixing or upwelling rates will not remove more $CO_2$ to the deep sea. Enhanced removal can only be achieved by increasing the oceanic inventory of these nutrients, thereby increasing the nutrient/$\Sigma CO_2$ ratio in deep and intermediate waters. On the other hand, aeolian iron supply and $N_2$ fixation occur independently of $\Sigma CO_2$ supply by vertical mixing and their utilization and subsequent removal from surface water with sinking organic matter amounts to a net export of $\Sigma CO_2$.

In the subtropical oligotrophic gyres, thermal stratification inhibits the upward diffusion of dissolved nutrients, thereby greatly constraining export production. Both nitrate and phosphate are added to surface waters by slow diffusion or intermittent vertical mixing associated with mesoscale eddies (McGillicuddy et al. 1998), but because phosphate is more efficiently recycled within the euphotic zone than fixed nitrogen ($NH_4^+$ and $NO_3^-$) there is often a slight excess of dissolved P in the upper water column (e.g. Gruber and Sarmiento 1997) relative to fixed nitrogen concentrations which often drop below detection limits. This makes fixed nitrogen the limiting nutrient, but only on short ("proximal", Tyrrell 1999) time scales. Over longer periods, the surface excess in phosphate sustains N fixation, so that phosphate is the ultimate limiting nutrient (Tyrrell 1999, Ganeshram et al. 2002) that dictates the export flux of organic matter supported by vertical supply of nitrate and $N_2$ fixation. This ecological feedback is thought to control the atomic ratio of these elements in average phytoplankton and seawater (the so-called Redfield ratio).

$N_2$ fixation also demands a large supply of iron, and it has been suggested that, instead of phosphate, Fe may limit N fixation and thus export production in oligotrophic regions (Falkowski 1997). While there is strong circumstantial evidence for Fe limitation of nitrogen fixation in areas of the Pacific Ocean (Wu et al. 2000, Deutsch et al. 2001), a recent study has shown that in the Atlantic, which receives a high iron input, $N_2$ fixers are in fact phosphate limited (Sanudo-Wilhelmy et al. 2001). By analogy to N vs P limitation (Tyrrell 1999), it could be argued that while iron may be the "proximal" limiting nutrient for $N_2$ fixers in oceanic regions with low iron input, P is their "ultimate" limiting nutrient, i.e. an abundant iron supply rapidly translates into P limitation.

Because of its longer regeneration scale length, the rate of supply of silicate to the surface waters of central oligotrophic gyres is even lower than those for N and P. This severely limits diatom production in such regions. Nonetheless, large diatoms can grow at the nutricline or after sporadic mixing events that supply nitrate and silicate into the euphotic zone (Siegel et al. 1999). Although these events can contribute a significant fraction of the organic material exported to the ocean interior, the mean annual rates of export in oligotrophic regions are still very low (Longhurst et al. 1995).

In stark contrast to the oligotrophic gyres, the surface waters of the equatorial and high latitude northern Pacific and the Southern Ocean are characterized by perennially large excesses of nitrate, phosphate and sometimes silicate. There is increasing evidence for Fe limitation in many of these regions (Martin et al. 1990, Kolber et al. 1994, Behrenfeld et al. 1996, Boyd et al. 1996, Coale et al. 1996, Van Leeuwe et al. 1997, Watson et al. 2000). In the subAntarctic and some areas of the equatorial Pacific, surface waters are also depleted in silicate raising the possibility that this is the limiting nutrient in these regions (Dudgale and Wilkerson 1998). The recent finding that Si/N uptake of diatoms increases with Fe limitation implies that silicate limitation could actually be induced by low Fe supply (Takeda 1998, Hutchins and Bruland 1998).

### 4.3.4 Supply of carbonate ions to surface waters (the alkalinity pump)

The $CO_3^{2-}$ content of the sea is governed by the balance between the input of calcium and carbonate ions from the chemical weathering of rocks on land and the burial of $CaCO_3$ on the sea floor. The upper open ocean is everywhere supersaturated with $CaCO_3$ but because solubility increases with pressure and decreases with temperature, and because the weathering supply of calcium and carbonate ions is less than the demand from organisms that manufacture carbonate shells, the deep ocean is undersaturated deep waters compensate for the imbalance between continental supply and demand in shallow waters. The water depth of the saturation horizon is poised so that the rate of $CaCO_3$ burial on the seafloor balances the rate of addition from weathering, and any changes in the latter is opposed by an adjustment in the level of the saturation horizon and the net area of the sea floor hostile to carbonate-shell burial. Since the residence time of Ca in seawater is ~ $10^6$ y, the dissolved Ca concentration can be considered constant on $10^3$ to $10^4$ year timescales, and the level of the saturation horizon must be controlled by carbonate ion concentration. Thus, deepening the saturation horizon to respond to higher weathering input requires a higher carbonate ion concentration in seawater, which engenders a lower $pCO_2$.

The alkalinity pump can also play an important role even in the absence of significant changes in continental weathering rates. For instance, the growth of carbonate formations in shallow waters, such as coral reefs, can be viewed as a means of preventing weathering products from reaching the deep sea and are thus equivalent to reducing weathering input, decreasing carbonate ion concentration and promoting higher $pCO_2$ (e.g. Opdyke and Walker 1992). Likewise, if the rate of carbonate production in surface waters of the open ocean increases, the resulting increase in carbonate burial

rate will temporarily exceed weathering supply, prompting a gradual decrease in seawater carbonate ion concentration, a shoaling of the saturation horizon to a new level that will again compensate for weathering input, and a rise in $pCO_2$.

The alkalinity pump is also closely linked to the biological pump through the so-called "carbonate compensation" effect (Broecker and Peng 1987, Boyle 1988). Lower export of organic matter (or higher rates of deep water ventilation) for example, decreases $CO_2$ concentration in deep water, thereby increasing $[CO_3^{2-}]_{deep}$ without immediately changing the oceanic carbonate alkalinity inventory. This is because the increase in carbonate concentration is compensated by a decrease in bicarbonate concentration. The higher $[CO_3^{2-}]_{deep}$ deepens the carbonate saturation horizon, such that carbonate burial exceeds weathering input. This perturbation results in a gradual decrease in the oceanic inventory of carbonate alkalinity until the $[CO_3^{2-}]_{deep}$ and the level of the saturation horizon regain their initial values. The gradual lowering of seawater carbonate alkalinity required to regain this initial state thus augments any increase in $pCO_2$ initiated by decreasing the biological pump.

Another link between the biological and alkalinity pumps is provided by the generation of metabolic $CO_2$ from the decomposition of organic matter in sediments, which dissolves $CaCO_3$ and decreases the rate of $CaCO_3$ burial. From initial conditions at steady state, lowering $C_{org}$ input into sediments increases carbonate burial in the deposits above that driven by weathering input, without initially changing the depth of the carbonate saturation horizon in the water column or the mean $[CO_3^{2-}]$ of seawater. In response, carbonate alkalinity gradually decreases, engendering higher $pCO_2$, lower $[CO_3^{2-}]$, and a shallower saturation horizon (Archer and Maier-Raimer 1994, Sigman et al. 1998). Thus, in contrast to the previous mechanism (accumulation of $CO_2$ in deep waters), which only produces a transient shoaling of the saturation horizon, lowering the rate of metabolically-driven dissolution of carbonate in sediment would result in a shallower steady-state saturation horizon. This result would be the same if carbonate production rates were increased or continental weathering rates were lowered.

Given known ocean mixing rates and considering how fast carbonate sediments on the sea floor can be chemically eroded, the temporal response for "carbonate compensation" has been estimated, using a coupled ocean/sediment carbon cycle model, at 5-10 ky (Catubig et al. 1998).

### 4.3.5 The export ratio (biological versus alkalinity pumps)

Oceanic productivity has opposing effects on the "biological" and "alkalinity" $CO_2$ pumps. While export of organic matter decreases $pCO_2$ and $\Sigma CO_2$ in surface water, the ensuing export of calcium carbonate from the surface ocean decreases carbonate alkalinity, thus increasing surface water and atmospheric $pCO_2$. Atmospheric $CO_2$ is thus also in part controlled by ecological factors that regulate the relative export of organic carbon and calcium carbonate, primarily through the relative contribution of carbonate producing (mainly coccolithophoridae) and non-carbonate producing (mainly diatoms) phytoplankton (e.g. Dymond and Lyle 1985).

## 4.4 Impact of marine processes on atmospheric $CO_2$

Based on the above discussion, it is apparent that a wide range of factors must have come into play to arrive at the net decrease in $pCO_2$ recorded in the glacial sections of Antarctic ice cores. Data from the Vostok ice core record indicate that increases in $pCO_2$ associated with the last several deglaciations occurred gradually over periods of 7 to 14 ka (Petit et al. 1999). Within such a time frame, mechanisms that contributed to the observed increases could include a wide combination of processes, some with rapid but relatively small impacts on $CO_2$ that could have occurred at any time during the transition, some with potentially rapid responses but forced externally at a slow rate, and some with intrinsically slow response times.

### 4.4.1 Contribution from the solubility pump

Cooler SST in areas of deep water formation during LGM (e.g. Shackleton 2000) would have enhanced the transfer of $CO_2$ to the deep sea. Using the CYCLOPS box model, Sigman and Boyle (2000) estimated that the competing effects of the global average glacial salinity increase of about 1 psu and a decrease in temperature of roughly 2.5 degrees in high-latitude surface waters and 5 degrees in the tropics would have lead to a net change in $pCO_2$ of about 20-25 $\mu$atm. This result is similar to previous estimates: Broecker and Peng (1998) for example, suggested that the combined effect of $\Delta S$ and $\Delta T$ between the LGM and today was some 10 $\mu$atmospheres, "small potatoes" in their terms. Legrand and Alverson (2001) note however that estimates of this type are heavily dependent on model assumptions and under-represent the range of possi-

ble effects of salinity and temperature variations on $pCO_2$. Their work suggests that the impact of S and T shifts between the LGM and the present cannot yet be prescribed with confidence.

### 4.4.2 Global export of $\Sigma CO_2$ from surface waters

It is now clear that over large areas of the world ocean biological productivity is limited by the supply of one or another limiting nutrient. Thus, the only way to enhance the efficiency of the biological pump in such regions to lower $pCO_2$ is to change the rate at which the limiting nutrient is supplied to the euphotic zone. On the other hand, in regions where all nutrients are available in surface waters, environmental changes that promote their utilization until one of the nutrients becomes limiting (e.g. water column stratification) would lower atmospheric $CO_2$. Both effects (i.e. increased supply of a limiting nutrient and environmental changes that promote its utilization) can also occur concurrently.

*Changes in N and P supply in oligotrophic regions*
Changes in the ocean N and P inventories may have occurred on glacial/interglacial timescales. The oceanic inventory of P is controlled by the balance between weathering input and sediment burial, but its response time is relatively long (e.g. 20 to 30 ka; Delaney 1998). On the other hand, the inventory of fixed nitrogen (mainly nitrate) is controlled by the balance between $N_2$ fixation and denitrification, and its response time to perturbations is significantly shorter (~3 ka; Codispoti 1995).

Increasing evidence suggests that the extent of water column (Ganeshram et al. 1995, Altabet et al. 1995, Altabet et al. 2002) and sediment (Delaney 1998, Ganeshram et al. 2000) denitrification was lower during LGM, which would have resulted in a higher nitrate inventory in the glacial ocean. However, if the "Redfield ratios" C/N and C/P remain fixed, an increase in nitrate inventory alone, without an equivalent increase in phosphate inventory, could not have translated into higher export production in oligotrophic regions and a lowering in atmospheric $CO_2$. Excess nitrate would simply have remained unutilized in surface waters once phosphate was depleted, and the amount of surface-water $\Sigma CO_2$ removed to the deep sea would have remained equivalent to the unchanged supply of phosphate. On the other hand, if nitrogen fixers are ultimately limited by P supply (Sanudo-Wilhelmy et al. 2001), increasing P availability will promote $N_2$ fixation and the more dynamic ocean nitrate inventory will be anchored to the P inventory (Tyrrell 1999). Thus, in an ocean with fixed "Redfield" ratios, the key to lowering pCO$_2$ by increasing export of C$_{org}$ in oligotrophic regions is to increase the phosphate inventory. The increase in the stock of both nutrients that would result would lead to a more extensive removal of $\Sigma CO_2$ from surface oligotrophic waters, contributing to the lowering of atmospheric CO$_2$. However, while the oceanic P inventory might have been somewhat higher during glacial times, as a result of reduced burial of authigenic carbonate fluorapatite in upwelling regions (Ganeshram et al. 2002), both the low amplitude of change and the slow response time of the P inventory should have contributed to minimizing its impact on glacial-interglacial changes in atmospheric CO$_2$.

It has also been proposed that if N$_2$ fixation in the modern ocean is limited by Fe instead of P, abundant dust supply during glacial periods could have enhanced export production, leading to reduced $\delta CO_2$ in surface oligotrophic waters and lower pCO$_2$ (Falkowski 1997, Broecker and Henderson 1998). But even if N$_2$ fixers in the modern ocean are Fe limited, relaxation of that limitation during glacial periods should have rapidly brought about P limitation (Sanudo-Wilhelmy et al. 2001); thus, significantly increased N$_2$-fixation during glacial periods seems unlikely (Ganeshram et al. 2002).

### Changes in Fe supply in HNLC regions

Large excesses of nitrate and phosphate remain unutilized in the surface waters of large areas of the world ocean (the so-called High Nutrients Low Chlorophyll, or HNLC, regions). These regions are currently confined to areas where vertical mixing (North Pacific and Southern Ocean) or upwelling (Eastern Equatorial Pacific) supply abundant nitrate and phosphate to the euphotic zone. Evidence is mounting that many of these regions are Fe-limited (Martin et al. 1994, Coale et al. 1996, Boyd et al. 2000, Watson et al. 2000), which prevents the complete utilization of nitrate and phosphate and further lowering of surface $\Sigma CO_2$.

Of the three main HNLC regions, the Southern Ocean has elicited the most attention to explain the low glacial atmospheric CO$_2$, because that is the only area whose surface waters are directly connected to the large reservoir of $\Sigma CO_2$ stored in the deep sea. As such, the region can be viewed as a direct CO$_2$ "leak" from the deep sea to the atmosphere that can be variably "plugged" by a more or less efficient utilization of the upwelled macronutrients (and stochiometrically equivalent $\Sigma CO_2$) by phytoplankton (e.g. Sarmiento and Toggweiler 1984, Sigman and Boyle 2000, Watson et al. 2001, Legrand and Alverson 2001).

South of the modern position of the Polar Front,

evidence based on the $\delta^{15}N$ of bulk sediment and diatom frustules supports higher surface nitrate utilization and a contribution from this region to lowering atmospheric CO$_2$ during LGM (François et al. 1997, Sigman et al. 2000, Figure 4.5). If the modern Southern Ocean is Fe-limited, higher utilization of surface nitrate and $\Sigma CO_2$ during the LGM could have been attained by increasing the supply of Fe relative to nitrate (and phosphate). This, in turn, could have been achieved by increasing the eolian input of Fe (Kumar et al. 1995), or by reducing the vertical supply rate of macronutrients by intensifying water column stratification (François et al. 1997, Stephens and Keeling 2000, Sigman and Boyle 2000). The dust record from Antarctic ice cores (Petit et al. 1999) indicates that the dust supply to the Southern Ocean was indeed higher during glacial periods. If the implied relaxation of Fe limitation occurred without a decrease in the rate of supply of macronutrients by vertical mixing, there should have been dramatically higher rates of export production in the region during the LGM. This is not unequivocally corroborated by the sedimentary record, primarily because of the difficulty in developing reliable proxies for paleoproductivity.

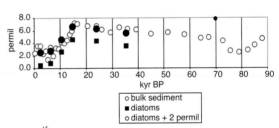

**Fig. 4.5.** $\delta^{15}N$ in bulk sediment and diatom frustules (Sigman et al. 1999) from core MD 84-552 taken south of the polar front in the Indian sector of the Southern Ocean (55°55'S; 73°50'E; 1780m)

Accumulation of sedimentary organic carbon, opal, biogenic Ba, and authigenic U, as well as $^{231}Pa/^{230}Th$ and $^{10}Be/^{230}Th$ ratio profiles have been used in attempts to define paleoproductivity history in the Southern Ocean (e.g. Kumar et al. 1995, François et al. 1997, Frank et al. 1999). However, the interpretation of each of these proxies is plagued by uncertainties that still need to be resolved. Productivity estimates from the accumulation of organic carbon in sediment are complicated by the extent of organic matter preservation in sediment and the factors that control it (e.g. Hedges and Keil 1995, Ganeshram et al. 2000). Likewise, opal preservation and changes in the C$_{org}$/opal rain rate from ecological changes (e.g. Phaeocystis vs diatoms) or Fe availability (Takeda 1998, Hutchins and Bruland 1998) obscure the interpretation of sedimentary opal flux (Anderson et al. 1998). Accumulation of biogenic Ba, which appears linked to export production (e.g. François et al. 1995), may also be affected by ecological factors (Dehairs et al. 2000) and its preservation in sediment is redox-sensitive and poorly constrained (McManus et al. 1999). Accumulation of authigenic U depends not only on the labile organic carbon flux reaching the seafloor, but also on bottom water oxygen concentration. $^{231}Pa/^{230}Th$ ratios of settling particles and sediments increase with particle flux over most of the ocean, as a result of differential scavenging between the two natural radionuclides (Bacon 1988, Chase et al. 2001, Yu et al. 2001). In regions dominated by opal, however, as in the Southern Ocean, differential scavenging is minimized and $^{231}Pa/^{230}Th$ is insensitive to changes in particle flux (Walter et al. 1997). Much remains to be done to refine our understanding of these proxies to obtain eventually reliable estimates of paleoproductivity in the Southern Ocean.

Notwithstanding the persisting uncertainty concerning the process involved (i.e. increased stratification vs increased export production), the $\delta^{15}N$ record indicates that the Southern Ocean $CO_2$ "leak" was partially "plugged" during the LGM south of the modern position of the Polar Front, and this contributed to lower $pCO_2$. Universal acceptance of this interpretation, however, still requires that it be reconciled with several seemingly divergent observations. Elderfield and Rickaby (2000) have interpreted changes in the Cd/Ca of planktonic foraminifera as an indication of higher surface phosphate concentration during glacial periods south of the modern Polar Front. This would be consistent with increased relative nitrate (and presumably phosphate) utilization only if the rate of vertical nutrient supply (and export production) were dramatically increased. The interpretation preferred by Elderfield and Rickaby is that the high glacial $\delta^{15}N$ signal is due to the predominance of ice algae, which have been shown to have high $\delta^{15}N$. This observation, however, is not consistent with recent analysis of sediment trap material collected in a region of the Southern Ocean affected by sea ice formation and melting (Altabet and François 2001). Alternatively, the planktonic Cd record could be reconciled with increased relative nutrient utilization and stratification in surface waters if the chemical composition of foraminifera shells were affected by secondary gametogenic calcification (Lohmann 1995, Kohfeld et al. 1996) within the nutrient-enriched pycnocline of the stratified glacial water column. In support of the work by Elderfield and Rickaby (2000), a preliminary Si isotope study suggests lower surface silicate utilization during the glacial periods (De la Rocha et al. 1998). But this

approach too must be qualified in light of the recent finding that the Si/N uptake-rate ratio of diatoms decreases significantly when the cells grow in Fe-replete conditions. It is therefore conceivable that relaxation of Fe limitation in the glacial Southern Ocean could have produced an increase in relative nitrate utilization concurrent with a decrease in the proportion of silicate utilized. Alternatively, such dual but opposite impacts on relative nutrient inventories could have been caused by an ecological shift in which Phaeocystis sp. dominated diatoms, although the latter appear to be favored by highly stratified conditions (e.g. D'Arrigo et al. 1999).

In contrast, north of the Polar Front, opal accumulation rates increased during the LGM (Kumar et al. 1995) providing strong evidence for the increased export of $C_{org}$, possibly as a result of Fe fertilization. However, in this region the $\delta^{15}N$ of bulk sediment was actually somewhat lower than its Holocene value (François et al. 1997, Sigman and Boyle 2000), suggesting that increased Fe supply must have been matched by increased vertical supply of macronutrients and $\Sigma CO_2$. Notwithstanding higher export production, the $\delta^{15}N$ data suggest that this region should not have contributed to lower glacial $pCO_2$, although the interpretation of the $\delta^{15}N$ record in this region is more ambiguous (Sigman et al. 1999), and should still be considered tentative.

Increases in $pCO_2$ during deglaciations occur over time intervals longer than 5 ky (Petit et al. 1999), either because of the intrinsic response time of the mechanisms involved or that of their forcing. While changes in stratification or export production can have a very rapid effect on atmospheric $CO_2$, factors that force the changes (e.g. retreating sea-ice cover) can be geographically gradual. Change in Fe input also appears to have been gradual at high southern latitudes, but it may be dangerous to extrapolate the dust input function at Vostok to the entire Southern Ocean. An accurate evaluation of the impact and timing of Southern Ocean productivity, stratification and Fe fertilization on the deglacial increase in atmospheric $CO_2$ will require synoptic mapping of the timing of changes in $\delta^{15}N$, paleoproductivity and eolian dust over the entire region. Given the existing uncertainties in such proxies and what they mean, it remains a considerable challenge to estimate quantitatively the collective impact on $pCO_2$ of the processes that have been inferred to have operated in the glacial Southern Ocean. Model-based approaches face similar uncertainties. For example, while box model simulations suggest that surface nutrient depletion in high latitude oceans could produce a large drop in $pCO_2$ (Sarmiento and Toggweiler 1984, Sigman et al. 1999, Sigman and Boyle 2000, Watson et al. 2001, Legrand and Alverson 2001) simulations using ocean general circulation models are unable to replicate these results (Heinze and Maier-Reimer 1991, Archer et al. 2000). One reason for the lack of sensitivity of General Circulation Models is the relatively large and unrealistic diapycnal mixing often employed. Until this variable is better constrained in the modern and past ocean, and better controlled in the models that are used to estimate the response of $pCO_2$ to oceanic changes, quantification of the contribution of the biological pump in the Southern Ocean to the glacial lowering of atmospheric $CO_2$ will remain prone to large uncertainties.

Like the Southern Ocean, the perennial pool of "excess" nitrate, phosphate and silicic acid concentrations that characterizes the North Pacific represents a potential oceanic $CO_2$ sink. Although mean annual primary production in the region today is similar to that of equatorial Pacific waters (some 140 g⟨ ⟩ 1999) export production is ⟨lower than in⟩ ⟨ ⟩ and comprises ⟨ ⟩ ⟨total production⟩ ion (Varela and H⟨ ⟩ ⟨regenerated produc⟩ ction dominates; ⟨ ⟩ ⟨remineralization ar⟩ e the prominent r⟨ ⟩ ⟨primary produ⟩ cers, rather than ⟨ ⟩ ⟨nitrate. The lo⟩ w new production ⟨ ⟩ ⟨mentally attrib⟩ uted to two facto⟨ ⟩ ⟨shallow halo⟩ cline of the relati⟨ ⟩ ⟨fic, which in⟩ hibits physical m⟨ ⟩ ⟨water column⟩ ; and b) iron limit⟨ ⟩ ⟨the growth ⟩ of small algal cells⟨ ⟩ ⟨ristically lo⟩ w Fe requirement⟨ ⟩ ⟨rrison 1996⟩ , Sunda and Hu⟨n ⟩ ⟨zooplankton⟩ apply essentiall⟨ ⟩ ⟨essure on the⟩ e small phytoplan⟨ ⟩ ⟨actor that pr⟩ events the occur⟨ ⟩ ⟨ast Pacific ⟩ of the spring blo⟨ ⟩ ⟨ther high-la⟩ titude oceanic reg⟨ ⟩ ⟨eres. Large ⟩ phytoplankton are ⟨present, but only as⟩ a small component of the tot⟨al assemblage, and⟩ ⟨their growth is⟩ now known to be ⟨controlled by availability⟩ of Fe - recent grazer-replete iron-enrichment ⟨experiments⟩ conducted with Gulf of Alaska surface water in bottles at sea (Boyd et al. 1999) have shown that the addition of small amounts of iron stimulate rapid increases in diatom growth rate, particularly large-celled species (e.g. *Nitzchia*).

Could utilization of the presently unused pool of nutrients in the North Pacific have contributed to the $pCO_2$ drawdown of the LGM? The answer appears to be no, at least for the eastern sector of the region. The net annual $CO_2$ exchange across the air-sea interface in the Northeast Pacific today is slightly negative; that is, the region is a minor net sink for $CO_2$ (Wong and Chan 1991, Zeng et al.

2002). Based on data published by McDonald et al. (1999), the direction and magnitude of this exchange appears to have changed little over the last 30,000 years. McDonald et al. (1999) used the sedimentary record at ODP Site 887 under the HNLC waters of the Gulf of Alaska to show that there was no significant enhancement of carbon burial during the LGM in the area. This observation is supported by lack of biobarium enrichment and isotopically light (-24 to –23 ‰) $\delta^{13}C$ values in bulk organic matter, probably indicating little relative deficiency of $CO_2$ (aq) in euphotic zone waters (McDonald et al. 1999), similar to the case today. $\delta^{15}N$ measurements made on bulk sediments across the same time interval (Figure 5.6) show light values during the LGM, suggesting that relative nutrient utilization in the area was, if anything, reduced during that time. The available data collectively imply that there was no enhancement of export production during the LGM.; indeed, it may have been reduced. This inference would appear to rule out a strengthened biological pump in the Northeast Pacific as a contributor to the glacial drawdown in $pCO_2$.

Unlike the high northern latitudes of the Pacific, the Eastern Equatorial Pacific region has long been recognized as a principal source region for $CO_2$ to the atmosphere. Supersaturation and outgassing from the surface waters in the area results from warming of newly-upwelled $CO_2$- and nutrient-rich waters in the equatorial divergence (Chavez and Barber 1987, Feely et al. 1999). Although rates of primary production along the divergence are high, the influence of upwelling and warming dominates the impact of export production of C. The modern net air-sea exchange of $CO_2$ is therefore strongly positive (i.e. out of the ocean).

But did the sign of the flux change in the past? A number of studies have implied that export production was higher in the far eastern equatorial Pacific (EEP) during glacial periods (Pedersen and Bertrand 2000, and references therein). But very recent work suggests that this pattern may not be applicable to the entire equatorial belt east of about 150° W; Loubere (1999) for example used a novel benthic-foram transfer-function approach to suggest that export production in the LGM was not significantly higher than that today, at least along the equator west of the Galapagos Islands. The critical issue here, however, is not whether export production was or was not enhanced during the LGM, but whether or not the EEP ceased to be a source region for $CO_2$ during that time. All evidence currently available, which comprises a range of isotopic, faunal and sedimentary geochemical proxies, suggests that it did not (Pedersen and Bertrand 2000), although it remains possible that the magnitude of the efflux decreased. When coupled with the results from the Northeast Pacific, this conclusion reinforces the perception that the Southern Ocean was the critical oceanic sink for $CO_2$ during the LGM.

### Changes in Si supply

Although characteristically replete in $NO_3^-$ and $HPO_4^{2-}$, some HNLC regions are low in surface silicate concentration. This is the case for the subantarctic zone of the Southern Ocean and regions of the equatorial Pacific, both of which are areas that could be silicate-limited (Dugdale and Wilkerson 1998). In such regions, increasing the silicate inventory could potentially enhance transfer of atmospheric $CO_2$ to the deep sea. Glacial/interglacial changes in the supply rate of silicate to the ocean by weathering, riverine transport and aeolian dust input have not yet been adequately constrained in order to estimate the amplitude of this effect (Froelich et al. 1992, Archer et al. 2000) although the strontium and osmium isotopic evidence published to date suggests that an increased silicate flux to the LGM ocean via weathering at least was unlikely (Section 2.1.3). Alternatively, relieving Fe-limitation would decrease the $Si/C_{org}$ export ratio, which would be equivalent to increasing the silicate supply to Si-limited regions. In addition, the thinner shells produced by Fe-replete diatoms would be more susceptible to dissolution, regenerating silicate at shallower depth and increasing the silicate/nitrate supply rate ratio by vertical mixing. These possibilities have not yet been systematically investigated and should be the subject of future studies.

### 4.4.3 Global rate of supply of $\Sigma CO_2$ to surface waters

The rate at which metabolic $CO_2$ regenerated in deep waters is brought back towards the surface depends on the strength of the meridional overturning and the rate of diapycnal mixing.

Comparing modeling results obtained with box models, z-coordinate and isopycnic general circulation models have highlighted the sensitivity of diapycnal mixing in establishing the response of atmospheric $CO_2$ to ocean perturbations (Archer et al. 2000). However, this mixing parameter is poorly constrained even in the modern ocean. Quantifying and understanding the mechanisms that control diapycnal mixing (e.g. Ledwell et al. 1993, Polzin 2000) are thus prerequisite to a quantitative evaluation of the impact of ocean processes on $pCO_2$.

The main contribution of thermohaline circulation to global climate is in transporting heat from low to high latitude as a result of North Atlantic Deep Water (NADW) formation (e.g. Stommel 1980, Gordon 1986). It may also have an indirect climatic role through the C cycle, however, as it partly controls the sequestration of $CO_2$ to the deep-sea. The global rate of thermohaline circulation in the modern ocean can be fairly well constrained from direct observations in which current meter measurements, tracer data and geostrophic flow calculations are combined (Ganachaud and Wunsch 2000). But its evolution in the past is more difficult to determine and must rely on reconstructions based on the distribution of chemical tracers that can be measured in sediments. Further, in order to establish the effect of deep circulation on the C cycle, we must reconstruct not only the rate of overturn of the global ocean but also water-mass geometry. The main tracers that have been used to estimate both rate and geometry of thermohaline circulation during the LGM are nutrient proxies, including $\delta^{13}C$, Cd/Ca and Ba/Ca in benthic foraminifera (e.g. Boyle 1992, Lea and Spero 1994, Duplessy et al. 1988). While these tracers disagree at some key locations, notably in the Southern Ocean, they clearly document an increase in nutrient concentration in the deep Atlantic during the LGM and a decrease in intermediate depths shallower than 2,000 m (Boyle 1992). This has been taken as evidence for a glacial shoaling of NADW, increased northward penetration of the Antarctic Bottom Water (AABW) and a general decrease in the rate of thermohaline circulation. Nutrient tracers, however, are not uniquely dependent on circulation, but also on biological uptake and remineralization. As a result, they are difficult to translate in terms of absolute rates of deep water formation (Legrand and Wunsch 1995). Evaluating the impact of possible change in the rate of thermohaline circulation thus still awaits the further developments of tracers that are more quantifiably linked to rates of ocean overturn, such as $^{14}C$ in foraminifera (Broecker et al. 1988, Hughen et al. 1997), $^{231}Pa/^{230}Th$ in sediments (Yu et al. 1996, Marchal et al. 2000) or "paleo" geostrophic calculations (Lynch-Stieglitz et al. 1999).

### 4.4.4 Contributions from the alkalinity pump

Changes in seawater alkalinity and attendant shifts in $pCO_2$ will ineluctably follow variations in continental weathering rate and oceanic production rates of organic carbon and carbonate. While the alkalinity of surface waters will respond rapidly to a changing rate of carbonate export, shifts in the actual rate are likely to be climatically controlled and gradual. Similarly, changes in whole-ocean alkalinity, resulting from variation in weathering rates or deep water acidity (or $\Sigma CO_2$ concentration) are inherently slow with a 5 to 10 ky response time (Boyle 1988, Catubig et al. 1998, Archer et al. 2000). Thus, changes in the operation of the oceanic alkalinity pump have the potential to drive multi-millennial variations in $pCO_2$, but not shifts on shorter time scales.

Temporal variations in sedimentary $CaCO_3$ concentration and burial rates can provide clues about past changes in the position of the carbonate saturation horizon (e.g. Catubig et al. 1998) and thereby shed light on the history of alkalinity distribution in the sea. However, there is not a unique relationship between changes in saturation depth and ocean alkalinity (Sigman et al. 1998, Sigman and Boyle 2000). Augmenting alkalinity by enhancing continental weathering rates or decreasing coral reef formation results in a gradual deepening of the saturation horizon and a concomitant drop in $pCO_2$ (by about 25 $\mu$atm for each km of deepening according to Sigman and Boyle 2000). Decreasing pelagic carbonate production relative to total primary production forces a change in the same direction, in part by extracting less alkalinity from the surface ocean (thereby increasing $CO_2$ solubility and drawing down atmospheric $CO_2$), and in part by promoting increased carbonate dissolution in the deep sea (thereby increasing the whole-ocean alkalinity inventory).

Did such changes occur? There is as yet no clear answer. Glacial/interglacial variations in the saturation depth are complex and remain poorly constrained by existing data. In very broad terms, the saturation horizon appears to have shoaled in the Atlantic and deepened in the Pacific during the LGM (e.g. Crowley 1985), while global carbonate accumulation appears to have been similar during the two periods, albeit with increases in carbonate production in the open equatorial and North Atlantic (Catubig et al. 1998). High-resolution reconstruction of carbonate burial rates in the equatorial Atlantic document a deglacial dissolution maximum, followed by a mid-Holocene preservation maximum (François et al. 1990). This progression is opposite to that observed in sediments of similar age elsewhere (e.g. the Indian Ocean: Peterson and Prell 1985). Such variations highlight the complexity associated with postulated changes in the strength of the alkalinity pump, for they could be the result of multiple influences which have variable geographic expression. The dissolution-preservation doublet in the Atlantic described by François et al. (1990) could have been caused for example by changes in the relative penetration of deep water from northern and southern sources. Such variations have not been predicted by any of the models or processes suggested in previous work. Alternatively, the doublet could reflect induction of lower alkalinity in deep waters by increasing $\Sigma CO_2$ through vertical transport of metabolizable carbon, followed by a relaxation to an initial position. A third hypothesis is that regional changes in the rain rate ratio of $CaCO_3/C_{org}$ could induce more or less dissolution of carbonate in the uppermost sediments. A lower ratio would cause increased dissolution as "excess" organic carbon was oxidized and lowered the carbonate ion concentration in shallow pore waters (Archer et al. 2000). In addition, there remain uncertainties in estimating continental weathering rates; these may have played a role in the alkalinity balance globally, although the limited evidence available suggests not (e.g. Henderson et al. 1994, Oxburgh 1998). Finally, deglacial and post-deglacial increases in coral reef and shallow carbonate accumulation did occur (Opdyke and Walker 1992), and these should have induced general shoaling of the saturation-horizon depth in the sea. Further progress in assigning weight to these various possible explanations and their geographic impact awaits detailed regional studies; it is only via this approach that the history and causation of relative movement of the saturation boundary will be defined.

### 4.4.5 Contributors to transient excursions in atmospheric $CO_2$

Significant changes in $pCO_2$ on the centennial-millennial time scale have been documented in ice cores (Chapter 2). Although much smaller in magnitude than the observed glacial-interglacial variability, such variations are nevertheless indicative of important global-scale processes. Progress in understanding controls on these higher-frequency changes has been slowed by the combination of a relatively small signal, complications arising from chronological uncertainties and the potential for naturally occurring artefacts within the glacial ice, particularly in Greenland ice cores. Nevertheless, recent studies of the less problematic Antarctic ice recovered from Byrd Station (Stauffer et al. 1998) and Taylor Dome (Indermühle et al. 2000) have confirmed that robust changes of 20±1 $\mu$atm tracked millennial-scale climate cycles during the last ice age, as noted in Chapter 2.

The relatively ephemeral persistence of these events rules out carbonate-compensation or alkalinity-pump explanations for their origin. Instead, the millennial-scale character coupled with their small magnitude points to changes in thermohaline circulation (THC) as the likely cause. Model studies indicate variability in THC can account for rapid changes in atmospheric composition, and that THC shutdowns could result in $CO_2$ increases of the observed magnitude (Marchal et al. 1998, Stocker and Marchal 2000). Modeled changes in $pCO_2$ result from Southern Ocean warming (Marchal et al. 1998) which occurs as part of the contrasting bipolar surface temperature response (Broecker 1998, Stocker 1998) when North Atlantic Deep Water production is curtailed. In addition, associated changes in alkalinity and dissolved inorganic carbon in Atlantic Ocean surface waters would amplify the $CO_2$ increase relatively rapidly (Indermühle et al. 2000).

Not all high-frequency climate changes are associated with $pCO_2$ variations, however. Some individual Dansgaard-Oeschger events (see Chapter 3) show no association with $pCO_2$ shifts (Chapter 2). This may reflect limited freshwater forcing due to smaller iceberg discharges (Indermühle et al. 2000) and lack of a thermohaline response sufficient to perturb $pCO_2$. Alternatively, the response time for $pCO_2$ to reach equilibrium with a new ocean circulation regime may be large relative to the quasi-1500 year duration of many D-O cycles (Stauffer et al. 1998), and this may have muted $\Sigma pCO_2$ shifts in the ice-core archives.

While these observations draw links between high-frequency, low-magnitude $pCO_2$ variability and physical oceanographic changes, it is also true that the warmest and longest interstadials of the D-O cycles occurred at the times of the highest concentrations of $pCO_2$ during the last ice age. This association implies a climatic response to the increased $CO_2$ and feedbacks potentially related to it, such as atmospheric water vapor content.

## 4.5 Summary and critical areas for future research

Atmospheric $CO_2$ changes on glacial/interglacial timescales are primarily controlled by oceanic processes. Although variations in continental carbon storage could have mitigated the amplitude of $pCO_2$ shifts during transition periods, such effects are likely to have been small. They are also very difficult to quantify. Estimates from isotopic mass balances, palynological data, and land biosphere models are widely divergent and need to be reconciled. Limited evidence suggests that there was little change in continental weathering rate variability on the global scale between glacial and interglacial times. If true, this rules out weathering-related effects as major controlling influences in the late Quaternary $pCO_2$ evolution.

While the importance of oceanic processes is well established, there is no consensus, as yet, on the precise mechanism or series of events responsible for the timing and amplitude of $pCO_2$ changes. The potential importance of high latitude oceans in regulating atmospheric $pCO_2$ has often been highlighted and has lead paleoceanographic carbon research to concentrate on the Southern Ocean. It is now apparent, however, that this perception is primarily model-driven (Broecker et al. 1999). The sensitivity of $pCO_2$ to high latitude surface oceanic processes, which is clearly recognized in box and zonally-averaged ocean circulation models, is not found in three-dimensional depth-coordinate ocean general circulation models, while isopycnic GCM's display an intermediate behavior (Archer in press). This discrepancy needs to be resolved and its consequences fully understood. If the 3-D models are shown to yield a closer representation of ocean circulation, a shift of focus toward lower latitude regions may be indicated.

Even though the Southern Ocean has been an area of focus, the paleoceanographic reconstruction of its carbon cycle remains uncertain. While a recent study of N isotopes in the modern Southern Ocean corroborates the interpretation of the $\delta^{15}N$ record in sediments as reflecting an increase in surface nitrate utilization during glacial periods (Altabet and François 2001), the mechanism whereby this higher depletion arises is still in doubt due to our inability to constrain properly past changes in export production. In addition, there remains a need to reconcile the sedimentary $\delta^{15}N$ record with those of the Cd/Ca ratio in planktonic foraminifera (Elderfield and Rickaby 2000) and silicon isotopes (De la Rocha et al. 1998). The history of vertical shifts in the depth of the lysocline and its relation to the thermodynamic saturation horizon in the water column and the compensation depth in sediments must also be better documented (e.g. Sigman et al. 1998, Catubig et al. 1998). Regional, high-resolution reconstruction of lysocline depth can be obtained by measuring $^{230}Th$-normalized carbonate fluxes on the flanks of topographic features (e.g. François et al. 1990). This could help distinguish between the various factors that affect ocean alkalinity which include weathering rates, shallow carbonate accumulation, pelagic carbonate production, the length scale of organic matter remineralization, and deep water circulation.

Developing tools to estimate past changes in bottom water oxygen concentration would also bring important constraints to verify the validity of proposed circulation- and productivity-change scenarios. Such proxies do not exist in a well-verified sense, but the accumulation of redox sensitive metals with different redox potentials or geochemical behaviors holds promise (Crusius et al. 1996, Nameroff et al. 2002).

Accurately documenting eolian inputs to the deep-sea has now gained new urgency with the realization that many HNLC regions are Fe limited and might not have been so during glacial periods. Measurements of the terrigenous fraction of deep-sea sediments do not always provide an accurate reflection of eolian input, as nepheloid transport of resuspended shelf and slope sediments can also be significant, particularly during glacial periods (François and Bacon 1991). This is particularly well illustrated in the southern ocean, where $^{230}$Th-normalized fluxes of terrigenous material in Holocene sediments far exceeds estimates of eolian input in this region (Kumar et al. 1995; François et al. 1997). Input of volcanic ash can also obscure the eolian fraction, or make difficult its measurement. Thus, developing a geochemical means of distinguishing eolian dust from other terrigenous sediment constituents would be very useful in reconstructing regional history of dust input and Fe fertilization.

A promising technique in this regard is the measurement of $^4$He in the non-biogenic fraction of the bulk sediments, or in isolates of specific components. Geologically old rocks and soils such as those in the deserts of north-central Asia contain a relatively high $^4$He content. The $^4$He is derived from the decay of U and Th isotopes, and a U-rich mineral phase such as zircon is particularly enriched in $^4$He because alpha particles cannot readily escape from the dense atomic structure of that mineral phase, at least when the zircon grain is larger than the typical recoil length of 10-30 $\mu$m (Farley et al. 1995). Analysis of the $^4$He content of zircons extracted from a sediment core could potentially yield information on the provenance of the grains. In a region such as the northeast Pacific, where "old" $^4$He-rich Asian dust is delivered to the sea surface (Boyd et al. 1998) along with "young" Alaskan dust, $^4$He analyses could allow specific determination of the Asian dust input through time. Such an approach has been used successfully elsewhere. Patterson et al. (1999) used $^4$He analyses to map Asian dust input through time to sediments on the Ontong-Java Plateau in the western equatorial Pacific (Patterson et al. 1999). Developing this fingerprinting technique further would be beneficial in mapping the time history of aeolian contributions to the open ocean. Having such information could prove very valuable in determining the influence of iron fertilization on the strength of the open-ocean biological pump during the Quaternary.

Finally, we suggest that the ca. 80 $\mu$atm difference in atmospheric $CO_2$ recorded in ice cores should not be used as a reference against which to appraise the significance of every proposed change in the C cycle. Many potential mechanisms may have acted in parallel or in sequence during the course of the late Quaternary glacial-interglacial seesaw, resulting in a vast reorganization of the earth's carbon cycle. Although we cannot rule out a simple one- or two-process control for the decline in pCO$_2$ seen in glacial maxima, such a spare explanation seems unlikely. The 80 $\mu$atm glacial decrease recorded in ice cores should therefore be viewed as a net effect and not a specific target that any one single process has to explain. Future progress will come from quantification of specific contributions from multiple sources and sinks, from reducing the uncertainties inherent in such estimates, and by working to gain an integrated understanding of the causalities and feedbacks among the processes involved.

# Terrestrial Biosphere Dynamics in the Climate System: Past and Future

J. Overpeck
Department of Geosciences and Institute for the Study of Planet Earth, University of Arizona, Tucson, AZ 85721, USA

C. Whitlock
Department of Geography, University of Oregon, Eugene, OR 97403, USA

B. Huntley
Environmental Research Centre, University of Durham, Durham DH1 3LE, United Kingdom

Contributors: P.J. Bartlein, Y.C. Collingham, E.C. Grimm, T. Webb III, J.W. Williams and S.G. Willis

## 5.1 Introduction

The terrestrial biosphere is one of the most critical and complex components of the climate system, regulating fluxes of energy, water and aerosols between the earth surface and atmosphere. The terrestrial biosphere also is central to the biogeochemistry of our planet, particularly with regard to the global carbon and nitrogen cycles. Prediction or assessment of future earth system change will always be limited, to some degree, by our ability to define how the terrestrial biosphere will respond to altered climatic forcing, and how this response will express itself as biophysical and biogeochemical feedbacks. The terrestrial biosphere plays more central roles in the health of the planet via biodiversity and ecosystem services (i.e. benefits to society). The exact dimensions of these roles are difficult to define, but there is no doubt that a premium must be placed on maintaining biodiversity, as well as the health of ecosystems critical for human sustainability.

Given the large magnitude of projected future (e.g. 21$^{st}$ century) climate change, it is simply not possible to make realistic assessments of future regional climate or biogeochemical ecosystem and biodiversity responses, without turning to the record of past climate and ecosystem change. Just as study of the contemporary biosphere is essential for understanding short-term biosphere dynamics, paleoenvironmental research provides the only way to observe and understand how the biosphere responds to climate change on decadal and longer timescales. Moreover, the paleoenvironmental record contains the only data relevant to understanding biosphere dynamics in the face of large climatic change and in the absence of significant anthropogenic influences. Because many aspects of future change are likely to

be without 20$^{th}$ century or even geological analogs (Crowley 1990, Webb 1992), assessments of future conditions will, by necessity, be increasingly based on numerical process models. As with other areas of environmental science (e.g. climate prediction), this means that we must also turn to the paleoenvironmental record to evaluate the realism of predictive models. It is not possible to assess future change without a complete understanding of past change, and how well we can simulate it.

The purpose of this chapter is to synthesize the remarkable progress that has recently been made in understanding the role and response of the terrestrial biosphere in the face of climatic change. We first examine the biogeochemical and biophysical roles that the terrestrial biosphere has played in the context of past climate change, and what these roles mean for the future. Realistic biospheric dynamics appear to impart significant positive climate feedbacks, and thus will likely amplify future changes in temperature, precipitation, and other climate variables. The terrestrial biosphere has also played a significant role in modulating atmospheric trace-gas and aerosol concentrations through time. Here, the emphasis is on the Late Quaternary, where we have a rich, well-dated record of large-scale climatic change, but we also place biosphere dynamics in a longer-term context. Although our synthesis emphasizes vegetation change, particularly in the more heavily studied northern extratropics, we also consider past responses of animal populations. The implications of our synthesis are global.

The primary thesis of this chapter is that the terrestrial biosphere has evolved in concert with significant large-scale climatic change during the Quaternary. The magnitude, rate and "destination" of climate and atmospheric trace-gas composition change can define an "envelope of natural variability". By "destination", we mean the exact configuration of the climate system, including atmospheric composition, at a particular point in time. Over the last several hundred thousand years, the magnitude, rate and destinations of the climate system have, remarkably enough, stayed within some clear bounds. Thanks to recent advances in paleoceanography and ice-core paleoclimatology (see Chapters 2-4), we know many details of atmospheric composition, climate forcing, and climate response for the last 450,000 years (Figure 5.1). We also know that this period, encompassing four major glacial-interglacial cycles, is more or less representative of the last million years of earth history (Imbrie et al. 1993; Chapter 3). Although the spatial configuration of seasonal insolation and ice sheets varied through time with changes in the earth's orbit, both seasonal insolation and the total volume of land ice varied within well-defined limits. Glacial ice sheets grew repeatedly over large regions of North America and Eurasia, lowering global sea level substantially as they did so. However, the glacial ice sheets appear to have never grown more than it takes to lower global sea level by 130 m below today's level (Imbrie et al. 1993, Chapter 3). During the many intervening warm interglacial periods, global ice sheets almost always retreated to configurations (i.e. extent and height) similar to, or slightly smaller than, those of today. Melting of Greenland and Antarctic ice sheets at these times rarely produced more than about 5 meters of sea level rise above today's level (Cuffey and Marshall 2000, Lambeck and Chappell 2001). Similarly, everything that is known about atmospheric trace-gas concentrations suggests that they were tightly bounded within the natural envelope over the last several million years; for example, $CO_2$ varied naturally between about 180 and 300 ppmv (Figure 5.1, Petit et al. 1999), and probably stayed close to modern levels much farther back in time (i.e. even back to 25 Ma) (Zachos et al. 2001). Thus, extant life on earth (i.e. species and ecosystems) has evolved within a fairly well-defined envelope of climate variability

There are clear signs that anthropogenic climate forcing is driving the earth's climate to destinations well outside the natural envelope (e.g. unprecedented global warmth of the late 20th century) (Overpeck et al. 1997, Mann et al. 1998, Crowley 2000, Huang et al. 2000, Jones et al. 2001,) (cf. Chapter 6). For example, ambient levels of many atmospheric trace-gas constituents (e.g. 370 ppmv for $CO_2$ in AD 2000) already exceed those recorded in ice cores for the last 420,000 years, and are likely to continue their unprecedented growth (IPCC 2001). At local to regional scales, species and ecosystems in some parts of the world could experience a sustained period of climatic conditions unlike any experienced in the last million or more years (e.g. warm temperatures and increased potential evapotranspiration).

In the face of large future atmospheric composition and climate changes, aspects of the terrestrial biosphere will likely change in ways not seen in the Quaternary. Moreover, just as the future mean climate of many regions is projected to be well outside the natural envelope, it is possible that the variability shifts and rates of change associated with future climate change could also exceed the natural bounds of the last million years (Figure 5.2) (Jackson and Overpeck 2000). The attendant biotic response will take years to fully adjust, thus making it difficult to estimate biospheric feedbacks to the atmosphere in the coming decades. Projected future climate and atmospheric changes will also have significant deleterious impacts on biodiversity and basic ecosystem services, particularly when other anthropogenic stresses (e.g. land-use, invasive species, water depletion and pollution) are factored in. Moreover, the specter of unanticipated climate changes or "surprises" (Broecker 1987, Overpeck 1996, Overpeck and Webb 2000, National Research Council 2002) suggests that reliable prediction of regional climate, and thus of terrestrial biosphere change, will be difficult.

## 5.2 The roles of the terrestrial biosphere in the climate system

The terrestrial biosphere presently serves as a major net sink of atmospheric carbon, sequestering nearly 30% of total annual anthropogenic carbon emissions (ca. 8 Gt total) each year over the last decade (Watson et al. 2000). Given the enormous size of the terrestrial carbon pool (2500 Gt, distributed 20% in vegetation and 80% in soils and detritus; Schimel et al. 1996), change in the terrestrial biosphere has the potential to play an even larger role in global carbon cycle dynamics in the future. Assessing this role will require a better understanding of how the terrestrial biosphere will respond on local to global spatial scales in the face of significant climate change. However, the growing focus of attention on the terrestrial biosphere goes well beyond carbon dioxide, to the role of terrestrial vegetation in a range of climatically-sensitive biogeochemical and biophysical processes. The purpose of this section is to briefly discuss these processes from a paleoenvironmental perspective, and to highlight research needed to improve future assessments. Although the

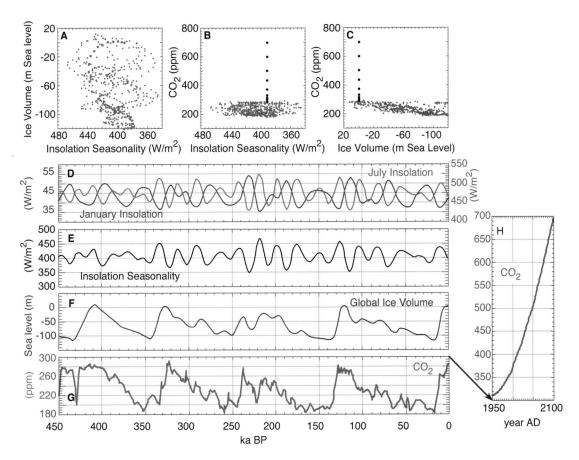

**Fig. 5.1.** Major climate forcing for the last 450,000 years (observed at 1000-year intervals to 1950), and for the period 1850 to 2100 (observed and projected at 25-year intervals). Scatter plots (A-C) and time series (D-H) of glacial ice volume (in meters of sea-level equivalent, see below), insolation at the top of the atmosphere (January, July and July minus January ("seasonality"), all calculated for 60°N (Berger and Loutre 1991), and atmospheric $CO_2$ concentration (Etheridge et al. 1996, IPCC 2001, Petit et al. 1999, Robertson et al. 2001). Note that $CO_2$ levels projected for the next century (red line in H, and black dots in A-C (IPCC 2001) are plotted versus years A.D. rather than B.P. Glacial ice volumes were obtained using the global average deepwater ('benthic') $\delta^{18}O$ (Imbrie et al. 1992) scaled to a 20,000 years B.P. glacial to present (interglacial) sea-level amplitude equal to the observed value of 120m (Fairbanks 1989). The trajectory (*sensu* Bartlein et al. 1997) of climate forcing over the last 21,000 years is displayed as green dots (A-C). Whereas atmospheric trace gas (e.g. $CO_2$) levels are expected to increase dramatically to unprecedented levels in the next century (A-C, H), both insolation and global sea level are not likely to change much relative to recent geologic variations. Figure from Jackson and Overpeck (2000).

last decade has seen great strides in identifying the roles played by the biosphere, the next decade will see a greater focus on quantifying these roles in the face of global climate change.

## 5.2.1 Biogeochemical roles

In the last decade, the history of atmospheric $CO_2$, $CH_4$ and (to a lesser extent) $N_2O$ concentrations over recent earth history (i.e. the last 450,000 years, Figure 5.1) has become well articulated (Blunier et al. 1995, Indermühle et al. 1999, Petit et al. 1999, Monnin et al. 2001) (Chapter 2). This work reveals that trace-gas concentrations have stayed within well-defined limits, and thus define a natural envelope of atmospheric composition variability within which the earth's present biosphere evolved. As chapters 2 and 4 detail, however, much remains to

be done in terms of understanding the mechanisms responsible for past trace-gas change, and in particular, the roles played by terrestrial plants and soils. On these long timescales, the terrestrial biota combined with oceanic processes to drive the large observed variability in atmospheric $CO_2$ and $N_2O$, whereas the observed changes in $CH_4$ were likely due almost entirely to changes on land with the possible exception of rare marine sediment methane hydrate instability (Kennett et al. 2000).

To simulate past trace-gas variability, we must first simulate the regional patterns of vegetation and soil response to past climate change in a realistic manner. The same holds true for simulating past changes in natural tropospheric aerosols. Vegetation serves as the primary stabilizer of soils and thus a key player in the modulation of natural "dust" aerosols (Pye 1987, Mahowald et al. 1999, Kohfeld and Harrison 2000, Mangan et al. in press). Similarly, aerosol and particulate emissions as a result of biomass burning play an important feedback role in the climate system (Levine 1991, Clark et al. 1996, Levine 1996a, 1996b, Watson et al. 2000). Whether derived from mineral or biomass sources, aerosols can have important biogeochemical, ecological and climatic impacts (Levine 1991, Penner et al. 1992, Levine 1996a, Levine 1996b, Overpeck et al. 1996, Avila and Penuelas 1999, Chadwick et al. 1999, Chapin et al. 2000, Shinn et al. 2000, Loreau et al. 2001). The important point here, however, is that it will be difficult to assess the mechanisms of past (or future) aerosol variability without first understanding how regional-scale vegetation and soils respond to climate change.

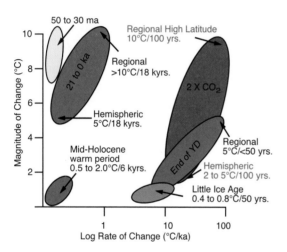

**Fig. 5.2.** Summary comparison of the rates and magnitudes of possible future climate change (estimated in terms of mean annual temperature) with those associated with several well-known periods of past change in regions that were vegetated. Rates of future regional temperature change could far exceed any widespread change in the late Quaternary (updated after Jackson and Overpeck 2000 using Zachos et al. 2001).

## 5.2.2 Biophysical roles

An understanding of the role of terrestrial vegetation in biogeochemical cycles is detailed in Chapters 2 and 4, but there is also reason to focus on the critical biophysical roles played by vegetation in the climate systems (i.e. in modulating energy, water, and momentum fluxes). Only 15 years ago, most studies focused on how vegetation change affects climate via changes in land-surface albedo. In contrast, the climate modeling community now considers a range of terrestrial feedbacks in addition to albedo, including evapotranspiration, surface

roughness, and snow masking (Hansen et al. 1984, Rind 1984, Hayden 1998, Eugster et al. 2000). Paleoenvironmental studies are the only source of long observational records of past vegetation, ocean, and climate change, and thus provide a unique opportunity to learn about the nature of these feedbacks, and also to evaluate how well we model them.

Biophysical climate feedbacks have long been a focus in the study of low-latitude climate variability. Deforestation in the Amazon and its associated climatic impacts are richly debated (e.g. Nobre et al. 1991, Dickinson and Kennedy 1992), just as are the 20th century biophysical (land-surface) feedback dynamics of the Sahel and North African climate (Charney 1975, Xue 1997, Zeng et al. 1999, Nicholson 2000, Wang and Elfatih 2000). Over the last decade, it has become clear that land-surface feedbacks have also been an important amplifier of climate sensitivity in the tropics and sub-tropics over the late Quaternary. Street-Perrot et al. (1990) highlighted the role of North African land-surface albedo change as a key positive feedback in the early Holocene, and this work has led to a host of studies focused on quantifying the role of this feedback in amplifying the radiative changes associated with Milankovitch forcing (Kutzbach et al. 1996, Coe and Bonan 1997, Broström et al. 1998). In a large community effort (the Paleoclimate Modeling Intercomparison Project, PMIP, Joussaume et al. 1999) a set of mid-Holocene simulations (made with 18 different global climate models) was compared. The conclusion of this comparison was that all of the models significantly underestimated the full magnitude of hydrologic changes needed to match the observed changes for that period – a result of the fact that all of the simulations lacked key feedbacks, including vegetation-climate interactions (Figure 5.3). More recent simulations with interactive vegetation and/or ocean dynamics have subsequently shown that these feedbacks do indeed help constrain North African climate sensitivity during the Last Glacial Maximum (ca. 20ka) and Holocene (Figure 5.3) though further improvements are clearly needed (Claussen and Gayler 1997, Kutzbach and Liu 1997, Texier et al. 1997).

The last decade has also seen significant work investigating land-surface (vegetation) biophysical feedbacks at high latitudes; not surprisingly, these feedbacks are dominated by the interaction of vegetation, snow, and temperature (Bonan et al. 1992, Eugster et al. 2000). Early paleoenvironmental investigations of these feedbacks focused on the positive boreal forest snow-masking feedback.

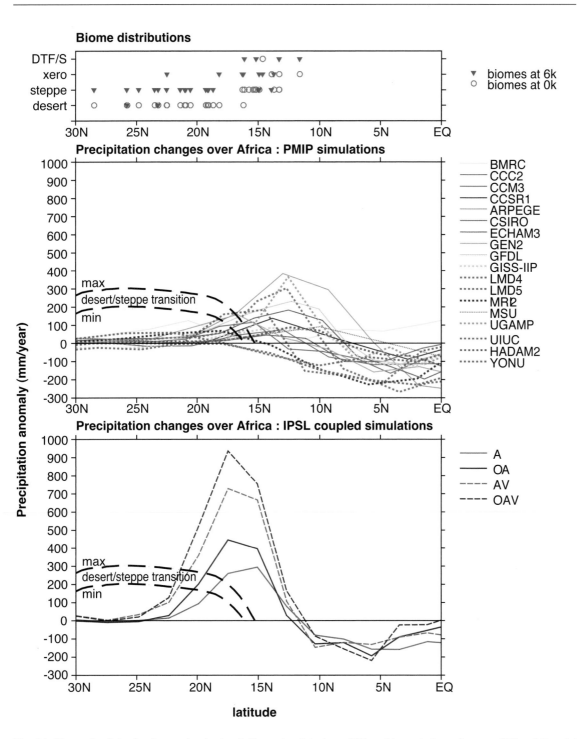

**Fig. 5.3.** Observed and simulated vegetation (top) and climate (precipitation, middle and bottom) change between 6000 yr B.P. and the present-day for northern Africa (20°W to 30°E) illustrating the importance of both ocean and vegetation feedbacks in estimating the correct sensitivity of the climate system to altered forcing (in this case astronomical forcing). Observations (top panel) indicate that the modern biomes (from north to south: steppe, xerophytic woodland/shrubland and tropical dry forest/savanna [DTF] shifted northward and displaced desert over much of North Africa in response to increased monsoon rainfall at 6000 yr B.P. Efforts to simulate these precipitation shifts using astronomical (Milankovitch) forcing alone (middle panel, no ocean or vegetation feedbacks, each line represents one of 18 atmospheric general circulation models) fail to generate the levels of precipitation increase (> 200 mm/year) needed to move the desert – steppe boundary as far north as indicated by the observations (Joussaume et al. 1999). In contrast, the addition (lower panel) of interactive ocean (OA), interactive vegetation (AV), and interactive ocean and vegetation (OAV) in a sequence of climate model experiments made with the same model (Braconnot et al. 1999) generates the simulation of incrementally greater precipitation. It is anticipated that more realistic ocean and land-surface feedbacks will eventually allow the northward expansion of steppe and wetter vegetation to be simulated correctly. Figure redrawn from Braconnot et al. 1999 and IPCC 2001.

However, subsequent work has illustrated that ocean feedbacks may have accounted for a large part (ca. 50%) of the circum-boreal region warming in the mid-Holocene that was previously ascribed to forest expansion and vegetation feedback alone (Figure 5.4; Hewitt and Mitchell 1998, Kerwin et al. 1999). These studies highlight the need to understand vegetation and ocean feedbacks more fully, and also how critical it is to tap the paleoenvironmental records of both climate forcing and response.

An understanding of vegetation feedbacks also should consider the influence of time-dependent patterns of vegetation change. Several studies have examined how land-use change may have altered regional to global climate change over the last several centuries. Chase et al. (2000) and Govindasamy et al. (2001) used the same land-cover change datasets, but different global climate models, and reached surprisingly different conclusions about the role of land-cover change in driving recent climate variations. These studies demonstrated that regional impacts were significant, yet even the direction of the change was difficult to estimate. One study (Chase et al. 2000) simulated virtually no net global temperature change, whereas the other experiment (Govindasamy et al. 2001) indicated that anthropogenic land-cover change was responsible for a significant amount (0.25°C) of global cooling over the last 1000 years (also in agreement with Brovkin et al. 1999). In yet another study, Claussen et al. (1999) showed how vegetation feedbacks may induce significant nonlinearities into the response of the climate system to gradual Milankovitch forcing.

The recent wealth of paleoenvironmental research on biophysical climate feedbacks is only the first step. We have yet to isolate the exact roles played by the terrestrial biosphere relative to competing influences, most notably ocean feedbacks. The next step is not only better climate system models, but also dynamic biosphere (e.g. vegetation) models that can simulate realistic responses of vegetation to changes in climate and atmospheric trace-gas concentrations. In addition, we need much improved observations of past environmental changes, including those of past vegetation, sea-surface state and aerosol levels (Kohfeld and Harrison 2000), all in the hope that we can improve our understanding of processes, while at the same time carrying out increasingly rigorous paleoenvironmentally-based evaluations of predictive models. The remainder of this synthesis focuses on what we know about the terrestrial biota's response to climate change, and thus what we must be able to simulate.

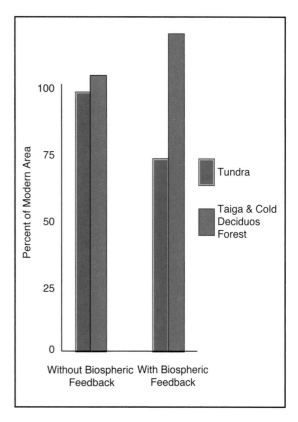

**Fig. 5.4.** Climate model simulations highlighting the importance of biosphere (e.g. albedo) feedbacks in estimating the correct temperature response (sensitivity) of the high northern latitudes to altered climate forcing. In this case (Foley et al. 1994; TEMPO Members, 1996), two identical climate model experiments were made, one with biospheric (albedo) feedbacks and one without. The experiment with the feedbacks generated 1.6°C greater warming at high latitudes on annual basis (>3°C in spring due to a greater snow masking effect), as well as a resulting greater northward movement of taiga and forest at the expense of tundra. This greater climate sensitivity and northward movement of biomes is in accord with paleoecological data, although more recent studies (Hewitt and Mitchell, 1998; Kerwin et al. 1999) suggest that ocean feedbacks may account for about half of the sensitivity ascribed to vegetation by TEMPO Members (1996).

## 5.3 Terrestrial biosphere changes in the past

Paleoecological and ecological studies reveal the hierarchy of effects of changing climate on the terrestrial biosphere that occur at different temporal, spatial and taxonomic scales. The consequences of high amplitude climatic change on Quaternary time scales are fundamentally different from those on shorter, ecological time scales over which the amplitude of climate change is relatively small. The nature of the biotic response to Quaternary climatic changes thus provides the best basis for understanding how ecosystems adjust in the face of substantial climate change. Such insights help define the envelope of natural biotic variability and serve as a benchmark for assessing biotic responses to projected future climate changes.

## 5.3.1 Response of the biosphere

Our understanding of the past indicates that the biosphere responds to climatic variations in several ways:

### Growth and/or death

There is an extensive literature focused on the response of individual plants and animals to climate variability, and the paleoenvironmental record provides key insights. Dendroclimatological studies provide long time series illustrating the strong role of temperature, moisture and other climatic variables in modulating plant growth (e.g. Cook and Cole 1991, Graumlich 1993a, Graumlich 1993b, Briffa et al. 1998, Cook et al. 1999, D'Arrigo et al. 1999, Hughes et al. 1999, Barber et al. 2000), as well as in driving tree mortality when environmental extremes exceed the ability of individuals to cope (e.g. Stine 1994, Arseneault and Payette 1997). More theoretical climate-growth relationships underlie whole classes of vegetation models (Botkin 1972, Solomon 1986), but the limited success of these models in simulating long records of past vegetation change indicates the complexity of climate-growth relationships. Moreover, dendroclimatological studies also illustrate how climate-growth relationships can change through time in as yet poorly understood ways (Briffa et al. 1998, Jacoby and D'Arrigo 1995).

### Species migration

The ability of species to shift their geographic location in response to climate change, referred to as species migration, is perhaps one of the strongest patterns in the Quaternary record. Species ranges are limited by features of the macroclimate (such as growing-season warmth, winter cold, potential evapotranspiration), and these parameters define a unique "climate space" for each taxon. Migration in response to changes in the geographic location of suitable "climate", or "environmental" space (Bartlein et al. 1986, Austin et al. 1990, Webb et al. 1993, Jackson and Overpeck 2000) is most evident on continental and sub-continental scales, where networks of fossil sites reveal large-scale biogeographic adjustments during glacial-interglacial cycles (Figure 5.5; Bernabo and Webb 1977, Webb 1981, Huntley and Birks 1983, Delcourt and Delcourt 1987, Huntley 1988, Webb 1988, Birks 1989, Prentice et al. 1991). Rapid range shifts have also been documented at smaller spatial and shorter temporal scales in response to abrupt climate changes (Ritchie and MacDonald 1986, Gear and Huntley 1991). Species migration has been best

documented for higher plants, because of the abundance of well-dated pollen sites, but the fossil records of mammals, molluscs, and insects show comparable responses (Elias 1994, FAUNMAP Working Group 1996, Ashworth 1997, Preece 1997).

The biogeographic changes evident in the last 21 kyr required rates of migration that would be considered unprecedented in modern times. For example, tree taxa in eastern North America and Europe moved at a rate of 200-1500 m/yr in response to climatic warming at the end of the last glaciation (Davis 1976, Huntley 1988, Webb 1988, Birks 1989). These rates are observed today only in the case of exotic species invasions, which are often assisted by human activities. Like the spread of weeds, rare long-distance dispersal events leading to the establishment of small, remote populations, followed by a quiescent phase of little discernible range change, and an active phase of explosive expansion were probably important components of Quaternary migrations (Pitelka and the Plant Migration Workshop Group 1997, Clark et al. 1998).

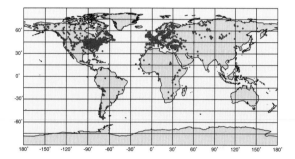

**Fig. 5.5.** An important component of our increased knowledge of terrestrial biosphere dynamics over the last decade has stemmed from the explosion of public-domain paleoclimatic and paleoecological data available at the World Data Center for Paleoclimatogy, Boulder: www.ngdc.noaa.gov/paleo/paleo.html. The map shows distribution of 1551 sites with fossil pollen data that are available globally. Other extensive paleobiological data holdings exist at the WDC, including thousands of sites in tree-ring and faunal databases, and hundreds of sites in plant macrofossil and insect databases. Collaborating data efforts of many scientists in many countries have been key to the success of PAGES data activities.

### Changes in community composition

Because individual organisms adjust to climate forcing independently, the composition and structure of biotic communities can change substantially. On glacial-interglacial time scales, communities have been subjected to wholesale reorganizations in composition as a result of species migration (Huntley and Webb 1988). In the course of such movement, the arrival of new species and the loss of others altered communities at particular locations. On shorter time scales, regional climate variations, often occurring abruptly, affected the composition and structure of communities by altering the natural disturbance regime and existing competitive interactions (e.g. Tinner and Lotter 2001). The paleoecological record indicates that communities can undergo dramatic changes in a matter of decades to centuries. These changes involve species movements, and thus represent individualistic responses to climate change, but the spatial scale of the change is limited. For example, drought events in this century have caused sufficient tree mortality to cause an expansion of steppe at the expense of forest in the American Southwest (Swetnam and Betancourt 1998). Similarly, deciduous forest and wooded steppe replaced steppe in southern Europe in less than two centuries in response to climate variations within the last glacial period (Allen et al. 1999, Allen et al. 2000). The movement of species was not great because areas of suitable climate space remained within the immediate topographically complex area.

### Changes through evolution

Little macroevolution is evident within the Quaternary, probably because of the rapid nature of environmental fluctuations relative to species lifespans (Bennett 1990, Bennett 1997). Nonetheless, morphological variations in some groups suggest that Quaternary climate change has resulted in continual selection for favorable ecotypes (Cronin 1985, Smith et al. 1995, Rousseau 1997, Davis and Shaw 2001). For example, morphological variation in fossil molluscs is evident throughout the Quaternary, although even extreme fossil morphotypes are found within the range of present-day forms (Rousseau 1997). Current genetic and morphological variation within western North American conifer species is also attributed to protracted periods of allopatry (geographic separation) during glacial periods, although isolation was apparently not long enough for speciation to occur (Critchfield 1984). Some groups of mammals (including humans) have also undergone substantial morphological change in the last 2 million years (Lister 1993).

### Extinction

The total loss of a species can result from an inability to migrate or adapt to environmental change. Extinction may occur with the loss of suitable environmental conditions, or else the loss of spatial or temporal contiguity of such conditions (Figure 5.6; Huntley 1999). Environmental change may also reduce or fragment a species range to the extent that population(s) lack the genetic variability to survive natural perturbations, such as extreme climatic events, an epidemic pathogen outbreak, or a wild-

fire. Species existing at the margin of suitable conditions are also rendered more susceptible to multiple stresses, such as human hunting, in the face of climate change.

The geologic record is punctuated by episodes of mass extinction as a result of rapid, sometimes catastrophic environmental change (Raup and Sepkoski 1984, Eldredge 1999, Kring 2000). The most recent prehistoric extinction event occurred in the late Pleistocene, when many large-bodied vertebrates (Stuart 1993, Lister and Sher 1995, Sher 1997), and at least one tree species (Jackson and Weng 1999), became extinct during the last glacial termination. At that time many other species were extirpated from parts of their overall range; for example, *Ovibos moschatus* (musk ox) was extirpated from Eurasia, although it persisted in the North American Arctic. Although human hunting pressures likely contributed to late-Pleistocene megafaunal extinctions (Alroy 1999, Martin and Steadman 1999), vertebrate populations were also stressed by environment changes imposed as a result of rapid global warming (Graham and Lundelius 1984, Graham and Mead 1987). Their failure to adjust to changing environments did not arise from the magnitude of the climate change alone, because other glacial terminations of the same general character were not accompanied by major extinction events. It is more likely that the rapid rate of environmental change at the end of the last glaciation led to spatial and temporal discontinuities in required environmental conditions. Landscape fragmentation and climate change in North America and South America, for example, coupled with the presence of human hunters, proved fatal.

The ability to adjust to environmental changes of different rates and magnitudes is a necessary strategy for species persistence. However, the capacity of the terrestrial biosphere to keep pace with climate change has varied across temporal scales (Prentice 1986, Webb 1986, Prentice 1992, Webb and Bartlein 1992, Webb et al. 1998). For example, short-term fluctuations in climate have often invoked little or no responses, especially in long-lived organisms. Intermediate rates of climate change have caused a complex nonequilibrium response, in which some parts of the system respond immediately and others lag behind. Slow variations in climate solicited a time-dependent equilibrium response. On millennial and longer time scales, the terrestrial biosphere has been in dynamic equilibrium with climate, and variations in the biosphere directly and indirectly contributed to changes in the climate system. On shorter time scales, lags in the response of organisms and ecosystems have led to periods of disequilibrium. The magnitude of the disequilibrium depends on the rate and location of changing potential range and the intrinsic characteristics of the taxa to migrate or adapt. Complex responses may arise from non-linear interactions occurring at different temporal scales. For example, gradual changes in climate may result in radical shifts in the equilibrium state (Ritchie and MacDonald 1986). In addition, more than one equilibrium state may occur because of feedbacks within the ecosystem. For example, the response to natural disturbance, such as fire events, can result in a variety of stable states in the vegetation, depending on the intensity and duration of the disturbance and the legacy of recent history. Consequently, changing climates may yield several outcomes in the response of the biosphere.

Spatial scales are also an important consideration in assessing biosphere response. At the global and hemispheric scale, large patterns in vegetation physiognomy are subject to climatic constraints, as evidenced by the fact that the distribution of biomes can be predicted by equilibrium models that assume a direct control by climate (Prentice et al. 1992, Haxeltine and Prentice 1996). Changes in vegetation at the global scale affect changes in total carbon storage in vegetation, peats, and soils, and these carbon pools are an important component of the climate system. At the regional scale, species have been shown to be in equilibrium with climate during some parts of the Quaternary and out of phase at other times. Thus, they are more limited by their intrinsic capacities at some times than at others. Vegetation changes at this scale have altered regional climate by changing evapotranspiration rates and albedo properties (Foley 1994, Kutzbach et al. 1996, Claussen and Gayler 1997, Broström et al. 1998, Farrera et al. 1999). At the landscape and local scales, biota may require a few hundred years to respond to climate change or the appearance of new species (Solomon et al. 1981, Davis and Botkin 1985, Overpeck et al. 1990).

### 5.3.2 The temporal hierarchy of climate change and biospheric response

The causes of prehistoric climate and biosphere variations are often described as hierarchical ranging from those operating at a global scale over millennia to those occurring at the local scale over seasons to decades (Table 5.1, Mitchell 1976, Delcourt et al. 1983, Bartlein 1988, Prentice 1992, Webb and Bartlein 1992).

*The "tectonic" frequency band (>1 million years)*
On the longest and largest scales are climate variations resulting from changes in the distribution of land, sea, and mountain ranges, as well as long-term

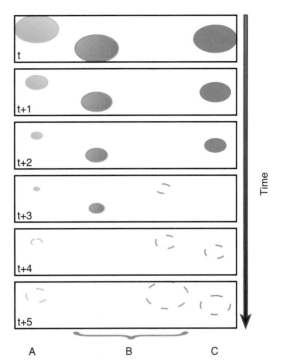

**Fig. 5.6.** Three alternative extinction mechanisms are illustrated by the sequence of panels. Time advances from the top panel downwards, as indicated by the arrow. As time advances the climate of the geographical area represented by the rectangle changes progressively. This climate change impacts upon the location and/or extent of the geographical area within which each species' climatic requirements/tolerances are met. These areas are represented by the ellipses that are shaded if the species is occupying them and empty with a dashed outline if the species is absent. Species A (green) experiences a progressive and severe reduction in its potential range from time *t* to time *t+3*, followed by an increase at times *t+4* and *t+5*. The extreme reduction of its range at time *t+3*, however, renders it extremely susceptible to stochastic extinction as a consequence of extreme environmental events or random population fluctuations; thus it has become extinct before the subsequent increase in its potential range at time *t+4*. Species B (red) experiences progressive but less severe range reduction from time *t* to time *t+3*. However, at time *t+3* a second discrete area of potential range becomes available in a different part of the overall geographical space. By time *t+4* the original component of the potential range has disappeared; the new component, however, has increased in extent and does so again at time *t+5*. The species nonetheless becomes extinct because it is unable to achieve the long-distance dispersal necessary to cross the spatial discontinuity between the two component parts of its potential range at time *t+3*. Species C (blue) experiences a progressive but moderate range reduction from time *t* to time *t+2*. At time *t+3*, however, no part of the geographical space offers climatic conditions that satisfy its requirements/tolerances and it thus has no potential range. Although suitable conditions are once again available at time *t+4* and the area of its potential range increases once again thereafter, it has become extinct at time *t+3* as a consequence of the temporal discontinuity in its potential range.

changes in the earth's energy balance due to changes in atmospheric composition (Ruddiman and William 1997, Zachos et al. 2001). These occur in the "tectonic" (or geologic) frequency band.

**Table 5.1**. Response of biota to climate variations on different time scales

| FREQUENCY BAND | SCALE OF VARIATION (YEARS) | KIND OF VARIATIONS | CAUSES | CHARACTERISTIC BIOTIC RESPONSES |
|---|---|---|---|---|
| **Tectonic** | $>10^6$ | Cenozoic cooling onset of glaciation | Tectonics, continental drift, atmosphere evolution | Creation of new biomes, speciation, major extinction events |
| **Orbital** | 10,000- 1,000,000 | Glacial- interglacial cycles | Seasonal cycle of insolation and trace gases, tectonics | Repeated formation and breakup of biomes, some extinction and speciation, selection at subspecies level |
| **Millennial** | 1000- 10,000 | Multi-millennial variations in climate within a glacial period | Ice sheets, insolation, trace gases, regional ocean-atmosphere dynamics | Species migration at the regional to continental scale, community reorganization, selection at the subspecies level |
| **Interdecadal-Centennial** | 10-100 | Decadal to centennial climate variations | Internal variations in the climate system, solar variability, impact of volcanic eruptions | Changes in community composition and structure through recruitment, mortality and natural succession |
| **Annual-Interannual** | <10 | Storms, droughts, ENSO events | Internal variations in the climate system, solar variability, volcanic eruptions | Adjustments in physiology, life history strategy, and natural succession following disturbance |

Climate variations on this long frequency band are responsible for the cooling trend in the late Cenozoic, as well as the onset of northern hemisphere glaciation about 2.5 million years ago (Maslin et al. 1998). Such changes, directly or indirectly, led to major extinction and speciation events, as well as the first appearance of new biomes, including deserts, grasslands, and tundra (Verba et al. 1995, Davis and Moutoux 1998). Tectonic-scale climate changes have also been tied to the evolution of ecosystems and hominids in Africa (De Menocal 1995).

### The "Orbital" frequency band (1 million to 10,000 years)

On glacial-interglacial time scales, or the "orbital" (Milankovitch) frequency band, climate change is attributed to variations in earth-sun orbital relations and their influence on insolation at different latitudes and during different seasons (Berger and Loutre 1991). In the Quaternary, the superposition of variations in the timing of perihelion with 19 and 23 ka periods, obliquity with a 41 ka period, and eccentricity with a 100 ka period has produced a sequence of changes in the latitudinal and seasonal distribution of insolation that are the "pacemaker of the ice ages". These variations resulted in changes in the size of continental ice sheets, ocean temperature and circulation, and atmospheric composition that constitute glacial/interglacial oscillations (Figures 5.1 and 5.7; Imbrie et al. 1993; Chapter 3, section 3.1). In terms of biotic responses, orbital variations on glacial-interglacial time scales resulted in repeated formation and break-up of biomes, including the appearance and disappearance of tropical rainforests, boreal forest, tundra and deserts (Huntley and Webb 1988, Overpeck et al. 1992, Markgraf et al. 1995, Colinvaux et al. 2000, Jackson et al. 2000, Jackson and Overpeck 2000, van der Hammen and Hooghiemstra 2000). Species responded individualistically in "dynamic equilibrium" with environmental forcing (Webb 1986, Prentice et al. 1991). Species migrations and biome shifts created opportunities for specialization of taxa and the elimination of less-fit or rare species through extinction. The biotic changes have resulted in continual genetic reshuffling as species ranges become alternately continuous and fragmented. Speciation events are rare in the Quaternary (Bennett 1997), probably because the direction of environmental change has been reversed over time scales too short for directional adaptive evolutionary responses to have played any significant role.

Biotic responses on orbital time frequencies are evident in long pollen records from different parts of the world (De Beaulieu and Reille 1984, De Beaulieu and Reille 1992, Reille and De Beaulieu 1995, Watts et al. 1996, Whitlock and Bartlein 1997, Davis 1989, Davis and Moutoux 1998, Allen et al. 1999, Allen et al. 2000, Whitlock et al. 2000). These records show shifts in vegetation types that match well changes in the seasonal cycle of insolation, global ice sheet size, and more regional changes in the ocean-climate system (Figure 5.7). Periods when orbital variations were different from those of the last 21 kyr resulted in vegetation assemblages with no late-Pleistocene or Holocene analogs. The close correspondence between changes in vegetation and variations in insolation and ice-sheet volume confirm that climate is the primary driver of regional vegetation change on these time scales.

Networks of paleoecological sites are available for the last 21 kyr and can be used to examine the spatial patterns of biotic change at orbital band frequencies within an interglacial period (World Data Center-A for Paleoclimatology: Global Pollen and Plant Macrofossils Databases). Because a primary Northern Hemisphere and tropical biotic response to these variations was species migration (Figure 5.8; see also Huntley 1988, Webb 1988, Prentice et al. 1991, Elias 1994, FAUNMAP Working Group 1996, Jackson et al. 2000, Webb et al. 1998), communities appear ephemeral, continually being formed and dismantled. Assemblages with no modern counterpart were particularly common prior to the Holocene when the combination of a large ice sheet, low summer insolation, cool sea-surface temperatures, and low greenhouse gases created unique climatic conditions in particular regions (Figure 5.8; Prentice et al. 1991, Overpeck et al. 1992, Williams et al. 2001). In contrast, the primary biotic response at higher latitudes in Southern Hemisphere was more a shrinking and re-expansion of biomes rather than differential species migration (Markgraf et al. 1995); climate change does not always result in climate or vegetation without modern analogs.

In mountainous areas, shifts in biogeographic range on orbital time scales were less dramatic, because areas of suitable conditions were available locally, often simply at a different altitude (Barnosky 1987, Thompson 1988, McGlone et al. 1993, McGlone 1997). Suitable microclimates to sustain populations during unfavorable periods, and sufficient habitat connectivity to allow population expansion during amelioration, have allowed montane species to respond rapidly to climate change (Figure 5.9). Allen et al. (1999) note that future warming could drive populations off the top of mountains, and hence overwhelm the natural resiliency of mountain environments. This history of repeated fragmentation and connection has led to considerable subspecific variation within extant montane taxa. This type of selection for favorable ecotypes is one consequence of climate changes on orbital frequency bands.

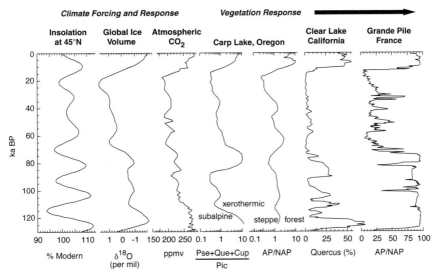

**Fig. 5.7.** Time series of climate forcing and large-scale climate response and vegetation change at two sites in North America (Carp and Clear Lakes, in Oregon and Northern California respectively) and Europe (Grande Pile, France). See Figure 1 for sources of insolation, ice volume and atmospheric CO₂ concentration data. The ratios of the sum of *Pseudotsuga-Larix* (Pse), *Quercus* (Que) and Cupressaceae (Cup) pollen percentages to *Picea* (Pic) pollen percentages, and of total arboreal pollen (AP) to nonarboreal pollen (NAP) provide an indication of vegetation type and openness at Carp Lake (Whitlock et al. 2000). *Quercus* pollen percentages at Clear Lake (Adam 1998).

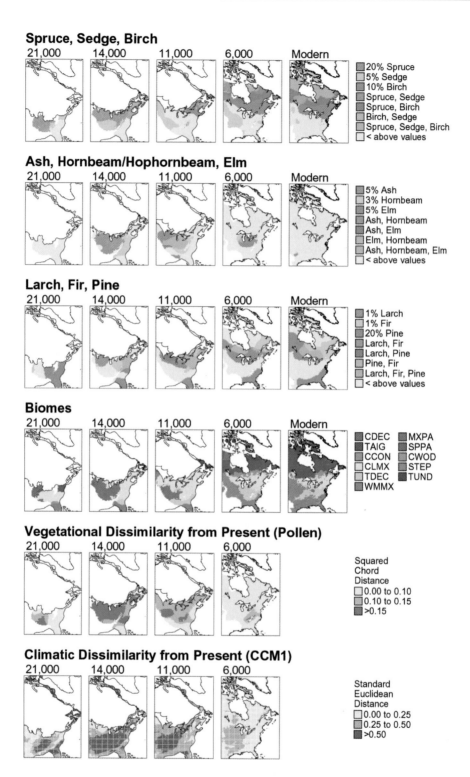

**Fig. 5.8.** Vegetation change and climate forcing over the last 21,000 years in eastern North America based on the analysis of 490 fossil pollen records and climate model results (after Williams et al. 2001). Late-Pleistocene and modern plant associations mapped (white areas = no data or ice sheet) as multi-taxon isopolls (top three rows, after Jacobson et al. 1987) and biomes. Differences in color between maps indicate a change in plant associations. Biome maps were created using the affinity score technique (Prentice et al. 1996) and the biome definitions of Williams et al. (2001), with spruce parkland and mixed parkland added to represent late-Pleistocene plant associations (Cold deciduous forest CDEC, Taiga TAIG, Cool Conifer forest CCON, Cool Mixed Forest CLMX, Temperate Decisuous Forest TDEC, Warm Mixed Forest WMMX, Mixed Parkland MXPA, Spruce Parkland SPPA, Conifer Woodland CWOD, Steppe STEP and Tundra TUND). The bottom two rows indicate the dissimilarity of the fossil pollen samples or climate simulations from their most similar modern counterpart.

## The millennial frequency band (10,000 to 1,000 years)

On shorter time scales, climate changes in the millennial frequency band include abrupt shifts in ocean-atmosphere-cryosphere (Fig 5.9; Chapter 3, section 3.2). Millennial-scale climate changes, including so-called Dansgaard-Oeschger (D-O) cycles, and Heinrich (H) events, are best known from marine and ice-core records from the North Atlantic region and Europe (Chapter 3). Heinrich event H0, the Younger Dryas (11.6-12.9 cal ka; Alley et al. 1993), is registered clearly in ice-core and ocean records in the North Atlantic and pollen data from Europe (e.g. Watts et al. 1996, Litt et al. 2001); and it is also described for other parts of the world (e.g. Hu et al. 1995, Peteet 1995). Evidence of anomalous conditions (not always cold and dry) corresponding with H1, H2, H3, and, in some cases, H4 through H6, is reported in Europe (e.g. Figure 5.9; Watts et al. 1996, Allen et al. 1999, Allen et al. 2000), eastern North America (Grimm et al. 1993), and the western U.S. (Benson et al. 1996, Benson et al. 1998, Hicock et al. 1999, Whitlock and Grigg 1999, Grigg et al. 2001). Although first described in Marine Isotope Stage (MIS) 2 and 3 (Bond et al. 1993), millennial-scale variations in climate have also been recorded in the Holocene, previous interglacials and early Pleistocene (Overpeck 1987, Bond et al. 1997, Raymo et al. 1998, Bond et al. 1999, Oppo et al. 2001). The ultimate cause of this millennial-scale variability is still not completely understood (Chapter 3), but it is responsible for changes in the strength and location of atmospheric features, and it accounts for some of the variations in climate and vegetation history on centennial to millennial time scales (Figure 5.9; Grimm et al. 1993, Watts et al. 1996, Allen et al. 1999, Whitlock and Grigg 1999, Allen et al. 2000, Tinner and Lotter 2001). The biotic response on these time scales includes regional shifts in species distributions and changes in the composition of plant communities (Figure 5.9). Heinrich events in southern North America, for example, shifted vegetation composition in both Florida (cool/dry versus warm/wet during H events) and Missouri (opposite phasing) (Grimm et al. 1993, Dorale et al. 1998, Grimm 2001) and D-O events in southern Europe caused alternations of forest and steppe (Allen et al. 1999). The rapid response on the part of forest taxa probably was accomplished by selection of cold and warm ecotypes within nearby species pools.

## Sub-millennial frequency bands (<1000 years)

Changes in climate occurring on time scales of centuries or less are ascribed to changes in volcanic activity, atmospheric greenhouse gas concentra-tions, solar output, and internal climate system dynamics (Rind and Overpeck 1993, Robertson et al. 2001; Chapter 3, section 3.3 and Chapter 6, section 6.10). Such changes are best documented in high-resolution records, including tree-rings, ice cores, lake sediments and corals (Overpeck et al. 1997, Mann et al. 1999, Briffa 2000, Crowley and Lowery 2000, Gagan et al. 2000, Urban et al. 2000, Jones et al. 2001). Likewise, variations at the inter-annual time scale, such as individual droughts, have been identified in the paleoclimate records where annually resolved records are available (Laird et al. 1996, Cole and Cook 1998, Stahle et al. 1998a, Stahle et al. 1998b, Woodhouse and Overpeck 1998, Touchan et al. 1999). At sub-regional scales, short-term climate fluctuations, including ENSO-related variations, helped to shape disturbance regimes and landscape-level vegetation patterns. Climatic extremes on these time scales have been responsible for dramatic changes in species mortality and recruitment patterns at the local to regional scale (Payette et al. 1996, Swetnam et al. 1999). In addition, ENSO-type variations play a role in life-history strategies that influence the pattern of reproduction and growth in particular species (Swetnam and Betancourt 1998, Finney et al. 2000).

### 5.3.3 The roles of changing disturbance regimes and atmospheric CO$_2$

#### Disturbance as an agent of change

Natural disturbances, such as fire, wind storms, disease, and extreme climatic events, are important triggers of rapid biotic response that can serve to reduce the response time to a given climatic change, and also to alter the composition of affected eco-systems (Davis and Botkin 1985, Overpeck et al. 1990, Clark et al. 1996, Swetnam and Betancourt 1998, Camill and Clark 2000). Long fire records suggest that many major vegetational changes during the Quaternary were associated with shifts in the fire regime (Bird and Cali 1998, Long et al. 1998, Tinner et al. 1999, Carcaillet and Richard 2000, Clark et al. 2001). On millennial time scales, variations in fire regime are attributed to climate changes caused by variations in the seasonal cycle of insolation, even in areas where the vegetation did not change (Millspaugh et al. 2000). At centennial and shorter time scales, climate/weather, vegetation composition, and ignition frequency all shape the fire regime, because vegetation determines fuel availability, and fires, in turn, create landscape patterns that influence vegetation recovery (Bergeron and Archambault 1993, Swetnam 1993, Clark et al. 2001). Changes in disturbance regime

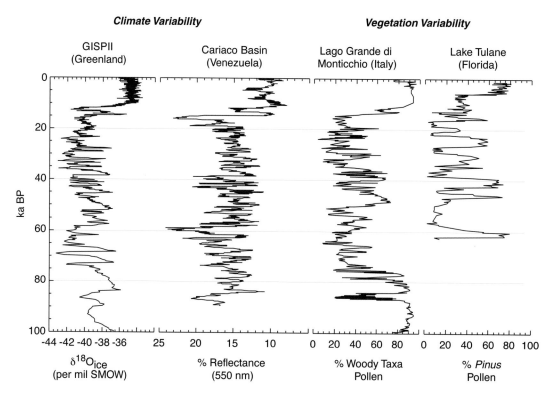

**Fig. 5.9.** Time series of millennial-scale climate variability in the circum-North Atlantic region (Peterson et al. 2000, Stuiver and Grootes 2000) and reconstructed vegetation variability for sites in eastern North America (Grimm et al. 1993) updated by (Grimm 2001) and Europe (Allen et al. 1999).

shift the competitive balance within communities, favoring species that are better adapted to the new conditions. It also creates opportunities for new species to invade and colonize, as evidenced by the role of fire in the migration of spruce across Scandinavia (Bradshaw and Zackrisson 1990, Bradshaw and Hannon 1992).

### The direct effects of changing atmospheric $CO_2$ concentrations

Changing concentrations of atmospheric $CO_2$ have the potential to affect the biosphere directly, just as they do indirectly though climate change. Altered atmospheric $CO_2$ has the potential to influence plant growth optima, canopy density, stomatal conductance, and water-use efficiency, and apparently did so at the Last Glacial Maximum (LGM – 21,000 yr B.P.), when $CO_2$ levels fell to about 200 ppmv (Solomon and Shugart 1984, Jolly and Haxeltine 1997, Street-Perrott et al. 1997, Collatz et al. 1998, Cowling 1999, Cowling and Sykes 1999, Levis and Foley 1999, Cowling et al. 2001, Huang et al. 2001). Indeed, it has been suggested that the ~200ppmv minimum seen in ice cores during glacial times represents a limit

that reflects reduced photosynthetic efficiency, below which the global biosphere is restricted in its ability to further lower atmospheric $CO_2$ levels (Falkowski et al. 2000). The direct impacts of lowered $CO_2$ at the LGM are hard to detect, however, highlighting the primacy of climate in driving much of the observed shifts in species and biome ranges. It is likely that LGM forest cover was more open (Cowling 1999, Farrera et al. 1999, Levis and Foley 1999, Cowling et al. 2001), and C4 species (e.g. grasses) were able to expand into many areas at the expense of C3 species as a result of the lower $CO_2$ levels (Collatz et al. 1998, Huang et al. 2001). Moreover, the biophysical impacts of these $CO_2$-driven changes on LGM climate may have been significant (Collatz et al. 1998, Cowling 1999).

## 5.4 Terrestrial biosphere change of the future: out of the envelope and into a world of disequilibrium

### 5.4.1 Future climate change

The latest assessment of future climate change (IPCC 2001) suggests that projected increases in greenhouse gases in the atmosphere will likely transform the climate in many regions beyond the historic envelope of variability observed in the last million years (Figs. 5.1 and 5.2). Even if global temperature changes by the end of the 21st century are at the low end of the current IPCC estimated range (1.4 to 5.8°C), the magnitude and direction of regional climate changes (e.g. in temperature and available moisture) could be without precedent in recent earth history, particularly when coupled with the large changes in atmospheric composition that will drive the climate change. Global and, in many cases, local rates of climate change are also likely to exceed any seen in the last million or more years (Figure 5.2), especially if the sensitivity of the climate system to doubled atmospheric trace gas concentrations is closer to the upper end of the estimated IPCC range. Lastly, there is no reason to believe that global warming will cease at the end of this century. Even if levels of atmospheric trace gases are stabilized, the earth's climate system will continue to react for thousands of years as a result of feedbacks among the oceans, atmosphere, cyrosphere and biosphere. Longer term impacts will include continued sea-level rise and continued slow warming (IPCC 2001).

### 5.4.2 Future biosphere change

The simplest assessments of the possible future state of the terrestrial biosphere are based on simulations of the equilibrium response of vegetation to some new atmospheric composition, for example a doubled atmospheric concentration of $CO_2$ (Figure 5.10). Such simulations have been made both for global vegetation, in terms of the extent and distributions of major units (biomes) (e.g. Woodward et al. 1998, Sykes et al. 1999); and for individual species at a variety of geographic scales from that of Great Britain (Hill et al. 1999), to that of North America (Solomon 1986, Overpeck et al. 1991, Bartlein et al. 1997, Shafer et al. 2001), and Europe (Huntley 1995, Huntley et al. 1995, Sykes et al. 1996, Sykes 1997, Hill et al. in press).

The key points to emerge from the equilibrium biome simulations are: first, simulations of past climate that do not incorporate terrestrial biospheric feedbacks underestimate important components of the climate response. Second, even without incorporation of these feedbacks, the climate simulated for the future is likely to result in substantial changes in the extent and location of biomes on the earth's surface, as well as in major shifts in the location of ecotones that are involved in important feedbacks (e.g., the forest–tundra

ecotone in northern Eurasia).

The potential for large ecosystem changes is further emphasized by simulated future climate-induced shifts in individual species ranges. The potential ranges of important tree species in North America (e.g. Shafer et al. 2001 and Figure 5.11) illustrate the magnitude of future biotic responses at the regional scale. They also emphasize (Figure 5.12; Bartlein et al. 1997) that in areas of topographic complexity, such as the western USA, the biogeographic response is not always simple poleward or upslope displacements. Although projected suitable conditions for some taxa are indeed located north of, or at higher elevations than, their present range, for other taxa they are found to the south or in the same general location (Bartlein et al. 1997). Such a wide range of possible responses becomes particularly critical for conservation strategies that aim to conserve biodiversity (i.e. threatened and endangered species) in the face of climate change.

Future changes in species ranges are not restricted to plants (Figure 5.13). Just as invertebrates and vertebrates exhibited large spatial responses to past climate changes (Graham 1992, FAUNMAP Working Group 1996, Ashworth 1997, Graham 1997, Preece 1997), large changes in animal distributions will be required to maintain equilibrium with projected future climate changes. Indeed, the paleoecological record, as well as model projections, suggest that species at all trophic levels are capable of large-scale biogeographic shifts in response to climate change, either directly or via climate-induced changes in their habitat.

### 5.4.3 The need to focus on the transient response

A major limitation of the simulations discussed above (section 4.2) is that they are portrayals of potential future equilibrium states that may never be realized, or at best, are likely to be achieved only following a substantial lag during which disequilibrium will prevail. The high probability of such lags arises principally from the unprecedented rate and large magnitude of predicted future climate changes (Figure 5.2). The most rapid recent global warming of large magnitude occurred at the last deglaciation. The transition from fully glacial to fully interglacial conditions spanned at least nine millennia (15,000 to 6,000 years ago) and represented a global warming of at least ca. 5°C (Webb et al. 1997, Pinot et al. 1999); this equates to an average rate of ~0.06°C per century, although warming was faster during parts

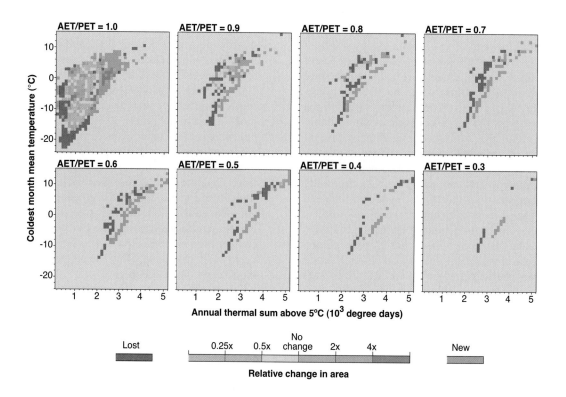

**Fig. 5.10.** Example illustrating potential shift in European multivariate climate space by the end of the 21st century given modest global warming (2.8°C). In this example, the multivariate climate space is defined by three key bioclimatic variables (coldest month mean temperature, annual thermal sum above 5°C, and water stress defined as the ratio of actual to potential evapotranspiration [AET/PET]). The changed climate scenario is derived from the climate simulated for a doubling of $CO_2$ by a global climate model (Schlesinger and Zhao 1989) close to the low end (2.8°C global mean warming for a doubling of atmospheric $CO_2$) of the IPCC (2001) range. Given continued present levels of greenhouse gas emissions, bioclimate will change dramatically throughout Europe by the end of the century, with large areas experiencing conditions without a present analog ('new') and many present combinations of conditions no longer represented ('lost'). Such changes will force substantial, although difficult to predict, changes in the terrestrial biosphere of the region.

of the deglaciation. In contrast, the most conservative estimates of future warming suggest that the global mean temperature will rise by ca. 1.5°C within the next century — a rate many times faster than during the deglacial transition. Both the paleoecological evidence (Davis 1976, Huntley 1991) and studies of recent species' range expansions (Thomas 1991, Hill et al. 1999) indicate that the maximum rates at which most species can expand their ranges are between 200 and 2000myr[-1]. The principal exceptions to such rates are exotic ruderals (species characterized by a short life cycle, rapid population growth and widely dispersed propagule (e.g. *Bromus tectorum*, cheatgrass, Mack 1981) or highly mobile vertebrates (e.g. muskrat and collared dove; Van den Bosch et al. 1992), which can spread even faster. Thus, most but not all, plant species will be unable to maintain equilibrium with the potential range shifts seen, for example, in Figure 5.13. If shifts of up to 1000 km are required to maintain equilibrium with the climate change simulated for the next century, as projected for Europe by the HADCM2 GCM (Mitchell et al. 1995), species ranges there would have to change at rates of up to 10 km yr[-1] – five times faster than the fastest rates in the past, and as much as 50 times faster than the rates achieved by many species.

Although it might be argued that some past climatic fluctuations, for example Dansgaard–Oeschger cycles, occurred at similarly rapid rates, these changes were of relatively modest global magnitude, and were soon reversed. Future warming is likely to be of relatively large global magnitude, unidirectional, and will continue for at least several centuries and/or to an eventual equilibrium with an atmospheric $CO_2$ several times the pre-industrial level or more (IPCC 2001). The lack of any modern or Quaternary analog insofar as climate is concerned suggests that species' ranges will significantly lag future climate changes and

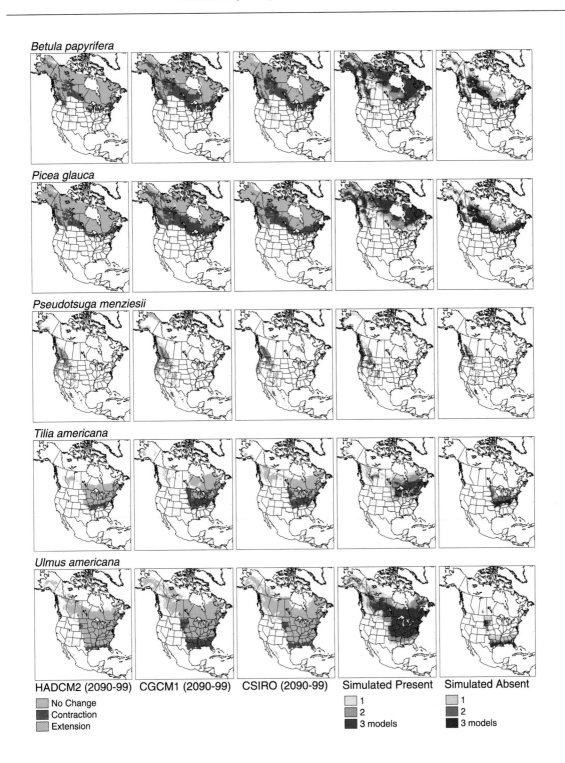

**Betula papyrifera**

**Picea glauca**

**Pseudotsuga menziesii**

**Tilia americana**

**Ulmus americana**

HADCM2 (2090-99)   CGCM1 (2090-99)   CSIRO (2090-99)   Simulated Present   Simulated Absent

No Change
Contraction
Extension

Simulated Present
☐ 1
☐ 2
■ 3 models

Simulated Absent
☐ 1
☐ 2
■ 3 models

**Fig. 5.11.** Comparison of observed distributions with future simulated distributions for fire tree species in North America using climate scenarios for 2090-2099 generated by three climate change simulations HADCM2, CGCM1, and CSIRO GCMs (Shafer et al. 2001, left three columns). "No change" indicates where the species is observed at present and is simulated to occur under future climate conditions; "contraction" indicates where the species is observed at present but is simulated to be absent under future climate conditions; and "extension" indicates where the species is not observed at present but is simulated to occur under future climate conditions. Agreement among the future distributions of each species as simulated by the three GCM scenarios is displayed by showing for each grid point the number of models that simulate a species to be present or absent (right two columns). Figure from Shafer et al. (2001).

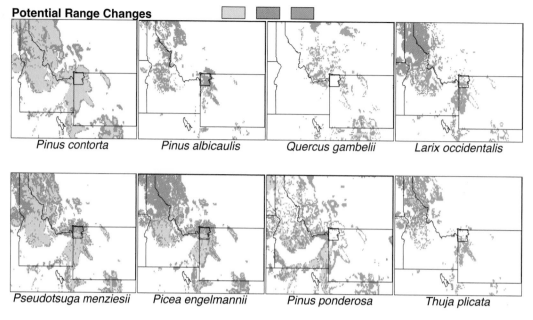

**Fig. 5.12.** Potential range changes for selected tree taxa in Yellowstone National Park under one global warming scenario. Green shading indicates grid points where a specific taxon occurs under both the present and $2 \times CO_2$ climate, red shading indicates grid points where a taxon occurs under the present climate, but does not occur under $2 \times CO_2$ climate, and blue shading indicates grid points where a taxon does not occur under the present climate, but does occur under the $2 \times CO_2$ climate. Figure from Bartlein et al. (1997).

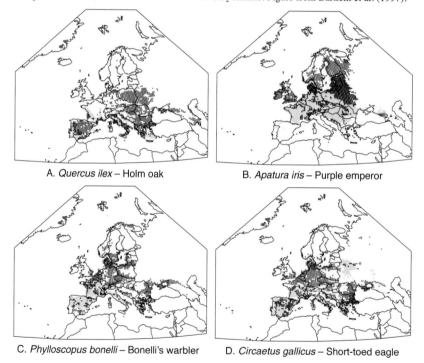

**Fig. 5.13.** The potential ranges of the species were simulated using a scenario derived from the last 30 years (2070–99) of the HADCM2 transient simulation using both greenhouse gases and sulphate aerosol forcing (Mitchell et al. 1995) applied to climate response surfaces fitted to the species' recorded European ranges and the climate corresponding to the period from which the records were derived (1931–60 from Leemans and Cramer (1991) for *Quercus ilex* and 1961–90 from New et al. (1999) for the remainder). As in previous studies (Hill et al. 1999; Huntley et al. 1995), three bioclimate variables were used to fit the response surfaces (coldest month mean temperature, annual temperature sum above 5°C, and an estimate of the ratio of actual to potential evapotranspiration). The data used to derive the models relate to a 50 km UTM grid. The models for all four species fitted well, as assessed using the kappa statistic (Monserud and Leemans 1992) $0.70 \leq \kappa < 0.85$ – 'very good' fit; $\kappa > 0.85$ – 'excellent' fit. The maps portray the present and potential future ranges

**(Fig. 5.13. continued)**
as simulated by the response surface model for each species. Blue dots signify grid cells simulated as potentially occupied under both the present and future climate; yellow dots indicate grid cells simulated as potentially occupied only for the present climate; red dots indicate grid cells simulated as potentially occupied under the future climate. The four species illustrated represent four trophic levels and are typical of many others that have been studied. **A.** *Quercus ilex* – Holm oak: an evergreen tree of southern and southwestern Europe ($\kappa = 0.78$) (distribution from Jalas and Suominen 1976). **B.** *Apatura iris* – Purple emperor: a widespread woodland butterfly whose larvae feed upon several *Salix* spp. (willows) ($\kappa = 0.87$) (distribution from Talman (1998) with the distribution used to fit the model truncated at 30°E because of the unreliability of the data from Russia). **C.** *Phylloscopus bonelli* – Bonelli's warbler: small insectivorous woodland bird of western and southwestern Europe ($\kappa = 0.78$) (breeding distribution from Hagemeijier and Blair 1997). **D.** *Circaetus gallicus* – Short-toed eagle: large raptor of partially-forested landscapes in southern Europe where it feeds primarily upon reptiles ($\kappa = 0.70$) (breeding distribution from Hagemeijier and Blair 1997).

possibly never achieve equilibrium with the new conditions because of their no-analog character.

Even with strenuous measures to reduce global trace-gas emissions, the future global warming driven by emissions to date is likely to be at least 1.5°C (IPCC 2001). As illustrated earlier (section 4.2), this warming alone will call for substantial range displacements by many species. Given the fragmentation of many species' habitats in the modern landscape, as well as the reduction in extent of available habitat, it is unlikely that many species will attain the migration rates observed during the late Quaternary (Overpeck et al. 1991, Bartlein et al. 1997, Collingham and Huntley 2000). In such circumstances, even a 1.5°C global mean temperature rise over the next century is likely to result in severe disequilibrium of the biosphere, with even the potentially most rapid migrants failing to adapt to climate changes. Were the climate change to cease at this level, centuries of stable conditions would likely then be required for many species and ecosystems to regain equilibrium with their environment.

Predicting the likely dynamic response of species is rendered more difficult by uncertainties about future disturbance regimes. Not only are the dynamic responses of species to climate change constrained by their dispersal abilities and by habitat availability, disturbance of the established vegetation is in many cases a prerequisite for the establishment of seedlings and hence for migration (Sykes and Prentice 1996, Pitelka and the Plant Migration Workshop Group 1997). Such disturbance may include changes in fire regime, pathogen outbreaks, and extreme climatic events, such as droughts, frosts or windstorms (Overpeck et al. 1990). Predicting future climate variability and extreme events is an area of current research in the climate modeling community, and the paleoenvironmental record can be used to help evaluate the performance of these models. However, predicting future "natural" disturbances will be more difficult than reconstructing past ones, as will be predicting future human disturbances.

## 5.4.4 More complicating factors

The foregoing sections make it clear that generating accurate model-based assessments of future biotic responses to climate change is not yet possible. It is thus not possible to obtain accurate assessments of how future changes in the terrestrial biosphere will feed back on the rest of the climate system, both in terms of biogeochemistry and biophysics. This line of reasoning also means that it will be difficult to anticipate future threats to biodiversity and ecosystem dynamics, even if we could know future climate change with perfect accuracy.

Even with the paleoenvironmental record as a guide, it is difficult to project the response of the terrestrial biosphere to future changes at all spatial and temporal scales. Hemispheric- to continental-scale climate predictions are available but are too generalized to be of use in regional or local assessments. Future average surface-air temperature is perhaps the easiest to estimate, whereas biosphere-relevant changes in the distribution and availability of moisture, the frequency of extreme weather events and natural and human-induced disturbances are difficult to predict, especially at the local-to-regional scales where societal impacts are greatest.

The paleoenvironmental record is rich with examples of "surprise" climate system behavior (Broecker 1987, Overpeck 1996, Overpeck and Webb 2000, National Research Council 2002; Chapter 3). Dansgaard-Oeschger and Heinrich events, for example, provide us with examples of abrupt climate system behavior. It is also clear that abrupt shifts in hydrologic variability (e.g. the frequency, amplitude and duration of droughts or floods) have occurred and could thus continue to occur, particularly if the mean state of the climate system is changing rapidly in response to elevated atmospheric greenhouse gas concentrations. Making accurate assessments of future biospheric change is simply not possible given the possibility of future "surprise" climate changes.

Although it is possible to use paleoenvironmental and 20th century records of biotic and environmental change to describe the environmental conditions required by plant and animal species, this knowledge may be of limited use in assessing future changes in many regions given the projected magnitude. Because future climate forcings are unprecedented, many local climates will be unique

(Crowley 1990, Webb 1992). Atmospheric composition will also be unlike that seen by any of the extant species or ecosystems. Disturbance regimes may be unlike those seen before, particularly in the face of growing anthropogenic impacts (e.g. Balmford et al. 2001). In addition, land-use, invasive species, pollution, surface-water use, groundwater depletion, and altered predation will add to the uniqueness of future earth environments. Unfortunately, the rich records of the 20th century and the millennia before, are unlikely to reveal all that is needed to make accurate assessments of future biosphere change.

## 5.5 Conclusions and future research needs

The paleoenvironmental record has provided invaluable estimates of past climate change (both rates and magnitudes, from local to global scales), as well as indications of how the terrestrial biosphere responded to those changes (Davis 1989, Davis 1990, Huntley 1991, Davis and Shaw 2001). Comparisons of these past changes with those projected for the next century reveal that the rate, magnitude and destination of future climate change in many regions lie beyond the historical range of variability. As a consequence, the pressures placed upon the terrestrial biosphere by these changes also will be unprecedented.

### 5.5.1 The problem with biosphere feedbacks and climate sensitivity

Anthropogenic greenhouse gas emissions have already committed the world to a mean global temperature increase of at least 1°C by the end of the 21st century (IPCC 2001). However, if emissions continue at current rates, atmospheric $CO_2$ will double pre-industrial levels before the end of the century, and cause mean global warming to at least 1.4°C, and perhaps significantly more (IPCC 2001). The paleoclimatic evidence discussed above, moreover, suggests that poorly constrained biospheric and oceanic feedbacks may limit the ability of state-of-the-art climate models to estimate the full sensitivity of the climate system to altered forcing. Future regional temperature and precipitation changes could thus be larger than presently anticipated. True regional climatic changes, especially at high latitudes where feedbacks are strongly positive, may be large, non-linear, and inherently difficult to project. In turn, these uncertainties render it difficult to make regional assessments of terrestrial biosphere change with any degree of certainty. Poorly constrained biospheric feedbacks

also result in major uncertainties with respect to other important parts of the earth's climate system. For example, even the relatively modest (~1°C or less) global warming during the last interglacial period led to substantially reduced ice volumes and corresponding increases in sea level of 5-6 m relative to present (Cuffey and Marshall 2000, Lambeck and Chappell 2001). Ice melting and sea level rises were even more substantial during earlier interglacials. These past changes suggest that climate sensitivity, or the magnitude of climatic response to a given (i.e., trace-gas) forcing, may indeed be larger than previously estimated.

### 5.5.2 Implications for future biodiversity conservation

The established global network of national parks and nature reserves is a key part of our global ecosystem and biodiversity conservation strategies. However, the future climatic conditions in many parks and reserves (e.g. Yellowstone Park, Bartlein et al. 1997) will differ substantially from those prevailing in the same locations today, and, as a result, the nature of the ecological communities in these parks and reserves is likely to change substantially. Consequently, parks and reserves may fail to protect the species and ecosystems they were set up to protect.

Yellowstone Park (Figure 5.12) is far from unique, and approximately half the world's protected areas are likely to experience climate-driven change of an extant biome (Leemans and Halpin 1992). Many such changes would result in almost complete species turnover, and hence likely threaten the rare or endangered species they were designed to protect. Moreover, many of the world's biodiversity hotspots (Myers et al. 2000) are also vulnerable to wholesale shifts in environment caused by abrupt climate change and projected sea level rise (Figure 5.14).

More generally, loss of biodiversity, particularly with respect to late-successional species, can be predicted as a consequence of the widespread disequilibrium between the terrestrial biosphere and climate. Ruderal species are likely to be the beneficiaries of this disequilibrium. Introduced species will also benefit and become even more widespread. One outcome of this climate-induced change is thus likely to be a tendency for communities to become dominated by early-successional and non-native species, and for vegetation to become increasingly homogeneous over broad geographical areas. Both at regional and global scales, biodiversity loss will occur principally among late-successional species characterized by Grime (1978)

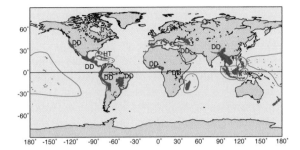

**Fig. 5.14.** Annotated map of biodiversity hotspots (after Myers et al. 2000) indicating those that are in regions known to have experienced decadal "megadrought" (DD – Cross et al. 2000; Heim et al. 1997, Hodell et al. 1995, Ledru et al 1998, Maley and Brenac 1998, Morrill et al. in press, Salgado-Labouriau et al. 1998, Sifeddine et al. 2001, Stine 1994, Verschuren et al. 2000) or abrupt shifts in tropical storm (i.e. hurricane or typhoon) landfall frequency (HT – Liu and Fearn 1993, Liu and Fearn 2000, Liu et al. 2001) in the past. Note that hotspots without annotation may also be susceptible to abrupt climate shifts, but detailed paleoclimatic data are lacking from these areas.

as 'stress-tolerators' (Thompson 1994). Projected climate changes also pose a substantial risk to genetic diversity (Huntley 1999, Davis and Shaw 2001). During the inevitable phase of climate-biosphere disequilibrium, species may occupy only a small fraction of their overall potential climatic range. If, as is often found to be the case, species exhibit genetically-determined clinal variation that is continuous in space in relation to major climatic variables (e.g. Mooney and Billings 1961), then the reduced population is likely to retain only a small part of the species overall genetic flexibility to climatic variation (Figure 5.15). Even if the climate subsequently stabilizes and the species has the opportunity to re-occupy the whole of its former potential climatic range, it may be unable to do so due to a loss in its former genetic variability.

The foregoing example serves to illustrate both one of the principal consequences of biodiversity loss – a loss of potential adaptability of the terrestrial biosphere to changed conditions – and also perhaps the principal value of biodiversity. Biodiversity at the species level, as well as at the level of intra-specific genetic diversity, renders the terrestrial biosphere adaptable to changing environments (Lawton 1999). Species that may, under one set of conditions, be relatively rare and/or play only a minor role in the biosphere, may under changed conditions become dominant (McGlone et al. 1993, McGlone 1997). Although rare species thus may appear "redundant" in relation to biospheric function under present conditions, they are better viewed in the same way as so-called "redundant systems" in an engineering context – they play a vital role when the species currently providing major biospheric functions are, as a result of changed conditions, no longer able to provide those

functions.

Given the importance of biodiversity for the future adaptability of the biosphere, and hence of its preservation, the potential problems faced by many species in the future must be addressed. A comparison with the paleoecological records indicates that many species ranges will be unable to immediately adjust to future climate change. It may thus be necessary to consider assisted migration or translocations of some species as part of an overall conservation strategy. In other cases, it may be necessary to plan to diversify landscapes and to incorporate more patches (and reserves) of wildlife habitat in order to facilitate species' migration through highly modified agricultural/urban landscapes (Huntley 1999). Deliberate manipulations of the disturbance regime may also be required in cases where species' migration is hindered by their inability to establish in a currently prevailing community; this is likely to be especially important in the case of many forest species and ecosystems.

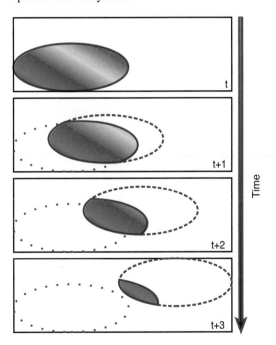

**Fig. 5.15.** Potential loss of genetic diversity as a consequence of rapid climatic change. Each rectangular panel represents the same geographical space, the four panels representing successive steps along a timeseries of climatic change that advances downwards, as indicated by the arrow. The ellipse in each panel represents the climatically determined potential range of the species in that geographical area at that time. As a consequence of climatic change the species' potential range shifts northeastwards with time. The shaded area in each panel indicates the area where the species is present at that time step, i.e., its realized range. At time $t$ the species fully occupies its potential climatically-determined range, as indicated by the shading of the entire ellipse. Thereafter the area occupied by the species progressively diminishes in size through time, because the rate at which the species' population is able to migrate is less than the rate required to maintain equilibrium with the rapidly changing

**(Fig. 5.15. continued)**

climate, whereas the species is unable to persist in areas of its former range that are no longer climatically suitable. The species' realized range thus becomes progressively more restricted to the southwestern part of its potential range. The spectrum of shading across the species' realized ranges represents clinal adaptive genetic variation in some character(s) related to the principal climatic gradient spanned by the species' realized range at time *t*. It is assumed that gene flow through the species' population is sufficiently rapid that there is no lag in gene 'migration' relative to the rapid climatic change. At time *t* the full range of genotypes (from red to blue) is present along the climatic gradient across the species' realized range. As the realized geographic range decreases in extent, however, it also decreases with respect to its range along the climatic gradient with respect to which the species exhibits clinal variation. Thus at time step *t+1* the species no longer occupies that part of the climatic gradient to which the red genotypes are adapted, with the result that they are lost from the population. By time step *t+3* only genotypes close to the blue extreme of the cline are able to persist within the species' realised geographic range. The extent to which such loss of diversity may be irreversible depends upon the underlying genetics; if alleles associated with the red genotype are not maintained in populations at the opposite end of the cline then the loss will be effectively irreversible. It should be noted that, were gene 'migration' to lag climate as much as does the migration of the species' population, then the genotypes able to occupy the 'realized range' at time *t+3* previously would have become extinct, and with them the species as a whole.

### 5.5.3 Principal uncertainties and research imperatives

Both the paleoenvironmental and biosphere dynamics research communities have accomplished a great deal over the last 10-15 years, but they have only set the stage in terms of gaining a predictive understanding of future climate and biosphere change. The importance of biosphere-climate feedbacks has been demonstrated, and the next decade needs to be devoted to quantifying the exact nature of these feedbacks through extensive monitoring, process-based studies, and earth-system modeling. In the next ten years, the focus in paleoclimate-biosphere research should include more focus on transient and site-specific responses, and how well these responses are simulated with predictive models. This shift in emphasis will be critical for assessing future climate changes, including the biogeochemical, biogeographic, ecosystem, and biodiversity consequences at regional and finer scales. At these spatial scales, the impacts of other stresses (e.g. disturbance, nutrients, atmospheric chemistry, pollution, land-use, and other human influences) will also have to be explicitly considered.

Assessments of future conditions without a strong paleoenvironmental component will not be successful. As made clear in this chapter, the centuries- to millennia-long records of past climate and biosphere change are the only information we have on the nature and consequences of large environmental changes. Future paleoenvironmental research

should improve our understanding of climate variability, as well as biospheric responses and feedbacks. It should also include an emphasis on high-resolution records that disclose the nature of inter-annual to century-scale climate and biosphere change, with a focus on short-term, as well as long-term responses of the biosphere to climate variability and change. Lastly, there should be more investigations that examine the ecological consequences of future climate change in a context of what actually happened in the past. This will provide much needed reality checks for simulations and assessments of future change, and will be most successful if carried out with greater interaction among disciplines that have traditionally worked in isolation, most notably the paleoenvironmental, ecological and land-use management, climate and climate modeling, and social science communities.

### 5.5.4 The final word

Meeting the research challenges in the preceding section will yield the understanding required to make realistic predictions of future regional climate and biosphere change, but probably only if future climate changes are at the low end of the IPCC 2001 range (<1-2°C; IPCC 2001). Paleoenvironmental observations and research will thus be a key component of what is needed. Gaining a much improved predictive understanding, coupled with improved strategies for the conservation of genetic, species and ecosystem biodiversity (e.g. redundant reserves, enhanced migration mechanisms, and control of non-climatic stresses) could help avert a major mass extinction.

However, a larger climate change (> ca. 1-2°C/century), or a lack of improved conservation strategies, will likely make a mass extinction inevitable. As this synthesis makes clear, it will probably be impossible to predict how regional climates, ecosystems and populations will be affected by changes in atmospheric composition and climate that are unprecedented, particularly when these changes are coupled with other anthropogenic stresses such as pollution, land-use, invasive species, predation, alteration to disturbance regimes, depletion of ground water, diversion of surface water, and elimination of migration routes and mechanisms. The inherent inability to predict how the biosphere will respond to future climate change also precludes accurate assessments of the biosphere's role in influencing future biogeochemical cycles (e.g. atmospheric composition) and climate. Because these influences are likely to be significant on regional scales, the likelihood of accurate regional climate predictions will also be further diminished.

The possibility of abrupt climate change also reduces the odds that extinctions can be avoided, unless new conservation strategies are implemented explicitly to reduce vulnerability to abrupt change. For example, many types of abrupt changes (e.g. widespread sustained drought, or shifts in the size and frequency of tropical cyclones) are, at present, not predictable. Anticipated large changes in climate forcing also increase the possibility that an unfavorable abrupt shift in climate variability could be triggered without warning (National Research Council 2002). Paleoenviromental research is at the heart of understanding the dynamics of abrupt climate change, and should also be central to understanding how the terrestrial biosphere responds to abrupt climate change.

However, just as the large and rapid climate change, coupled with human hunting activity, proved disastrous to large animals over the last deglaciation, the combined stresses of climate change and human activity could create an unprecedented ecological disaster. Given that it is entirely possible that a doubling of atmospheric $CO_2$ could drive a global temperature increase of 2°C or more, as well as larger regional changes (IPCC 2001), there is only one way to ensure that we do not trigger a mass extinction. First, we must limit future greenhouse-gas emissions to the atmosphere, and second, we must adopt a more sophisticated global conservation strategy that acknowledges the threats of climate change in the face of multiple additional anthropogenic stresses.

## 5. 6 Acknowledgements

Numerous colleagues and organizations have helped us and/or made available data used in this study: they include Pascale Braconnot, Wolfgang Cramer, Rhys Green, Ward Hagemeijer, Jane Hill, Sylvie Joussaume, Tapani Lahti, Phil Leduc and Chris Thomas, as well as the European Bird Census Council and the Committee for Mapping the Flora of Europe. Climate data for 1961–90 and for the HADCM2 scenario were made available to us through the UK DETR funded Climate Change Impacts LINK project. We had invaluable technical help from Ed Gille, John Keltner, Cristoph Kull and Shoshana Mayden. We also thank the authors and contributors of the other PAGES Synthesis chapters for their valued scientific input, as well as Vera Markgraf and Isabelle Larocque for valuable review and comments on the manuscript.

# The Climate of the Last Millennium

R.S. Bradley
Climate System Research Center, Department of Geosciences, University of Massachusetts, Amherst, MA, 01003, United States of America

K.R. Briffa
Climatic Research Unit, University of East Anglia, Norwich, NR4 7TJ, United Kingdom

J. Cole
Department of Geosciences, University Arizona, Tucson, AZ 85721, United States of America

M.K. Hughes
Lab. of Tree-Ring Research, University of Arizona, 105 W. Stadium, Bldg. 58, Tucson, AZ 85721, United States of America

T.J. Osborn
Climatic Research Unit, University of East Anglia, Norwich, NR4 7TJ, United Kingdom

## 6.1 Introduction

We are living in unusual times. Twentieth century climate was dominated by near universal warming with almost all parts of the globe experiencing temperatures at the end of the century that were significantly higher than when it began (Figure 6.1) (Parker et al. 1994, Jones et al. 1999). However the instrumental data provide only a limited temporal perspective on present climate. How unusual was the last century when placed in the longer-term context of climate in the centuries and millennia leading up to the $20^{th}$ century? Such a perspective encompasses the period before large-scale contamination of the global atmosphere by human activities and global-scale changes in land-surface conditions. By studying the records of climate variability and forcing mechanisms in the recent past, it is possible to establish how the climate system varied under "natural" conditions, before anthropogenic forcing became significant. Natural forcing mechanisms will continue to operate in the $21^{st}$ century, and will play a role in future climate variations, so regardless of how anthropogenic effects develop it is essential to understand the underlying background record of forcing and climate system response.

Sources of information on the climate of the last millennium include: historical documentary records, tree rings (width, density), ice cores (isotopes, melt layers, net accumulation, glaciochemistry), corals (isotopes and other geochemistry, growth rate), varved lake and marine sediments (varve thickness, sedimentology, geochemistry, biological content) and banded speleothems (isotopes). These are all paleoclimatic proxies that can provide continuous records with annual to decadal resolution (or even higher temporal resolution in the case of documentary records, which may include daily observations, e.g. Brázdil et al. 1999, Pfister et al. 1999a,b, van Engelen et al. 2001). Other information may be obtained from sources that are not continuous in time, and that have less rigorous chronological control. Such sources include geomorphological evidence (e.g. from former lake shorelines and glacier moraines) and macrofossils that indicate the range of plant or animal species in the recent past. In addition, ground temperature measurements in boreholes reflect the integrated history of surface temperatures, with temporal resolution decreasing with depth. These provide estimates of overall ground surface temperature changes from one century to the next (Pollack et al. 1998, Huang et al. 2000).

Proxies of past climate are natural archives that have, in some way, incorporated a strong climatic signal into their structure (Bradley 1999). For some biological proxies, such as tree ring density or coral band width, the main factor might be temperature – or more specifically, the temperature of a particular season (or even just part of a season). Ring density and width can also be influenced by antecedent climatic conditions, and by other non-climatic factors. Similar issues are important in other proxies, such as the timing of snowfall events that make up an ice core, or the rate and timing of sediment transport to a lake. Though we recognize that the details of such relationships are important, proxies are rarely interpreted directly in terms of such very

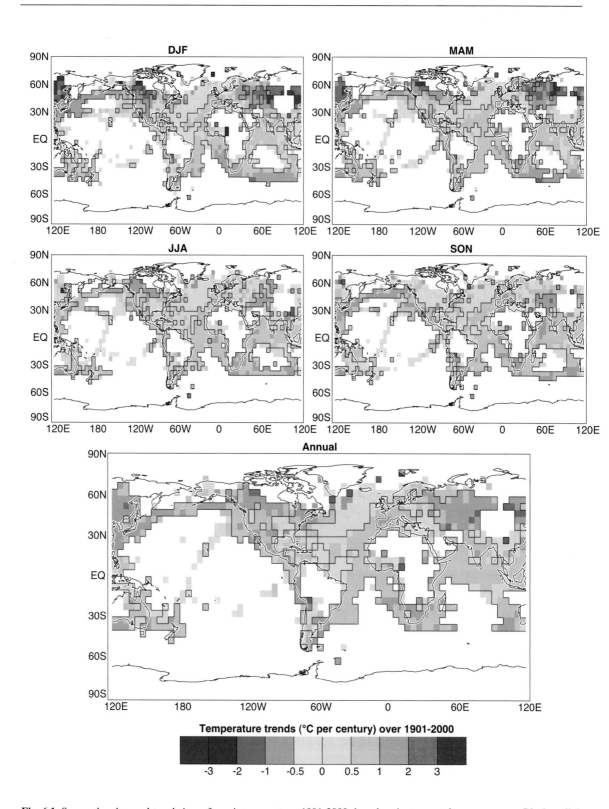

**Fig. 6.1.** Seasonal and annual trends in surface air temperature, 1901-2000, based on instrumental measurements. Black outlining surrounds those regions with statistically significant trends (at the 95% confidence level).

specific climatic controls, but rather in terms of temperature or precipitation over a particular season. In many cases the main climatic signal in a proxy record is not temperature alone. For example, evidence of a formerly high lake level may indicate higher rainfall amounts and/or a decrease in evaporation related to cooler temperatures. Such issues are grist to the paleoclimatologists' mill and are the subject of numerous studies. Suffice it to say that proxies that optimize a reconstruction of either temperature or precipitation are generally selected, and it is these studies that provide the basis for our review.

Changes in temperature have large-scale spatial coherence, making it easier to identify major variations with relatively few records. Spatially coherent precipitation changes are more local or regional in extent, but they often reflect circulation changes that may have large-scale significance (as, for example, in ENSO-related spring rainfall increases that commonly occur in the southwestern U.S. during strong El Niño events, Stahle et al. 1998). In this chapter, we focus mainly on temperature variations, but precipitation and hydrological variability are examined where there is good evidence for important changes at the regional scale. In particular, we ask two questions regarding each attribute:

- does the 20th century record indicate unique or unprecedented conditions?
- do 20th century instrumental data provide a reasonable estimate of the range of natural variability that could occur in the near future?

First we briefly summarise conditions during the Holocene epoch (the last 10,000 radiocarbon, or ~11,700 calendar years), as a background to climate variability over the last millennium. Then we examine the overall pattern of temperature change during the last 1000 years at the largest (hemispheric) scale, followed by a consideration of climate variability in several large regions. Finally we look at the forcing factors that may have played a role in the variations identified.

## 6.2 Holocene climate variability

The Holocene epoch follows the last major pulse of glaciation (the Younger Dryas interval) at the end of the last glacial period, and encompasses a period of time before there was any substantial anthropogenic forcing of climate. The Holocene has often been characterized as a period of relatively stable climate, yet there is much evidence to the contrary. In particular, the tropics and sub-tropics witnessed dramatic changes in hydrological conditions during this period. Early to mid-Holocene conditions in the northern deserts of Africa were significantly wetter

than in the late Holocene, as revealed by the evidence of extensive early Holocene lake sediments, and fossils of herbivores and aquatic reptiles in areas that are today utterly arid (Petit-Maire and Riser 1983). In fact, evidence for much drier conditions in the late Holocene (after ~4000 calendar years B.P.) is also found across central Asia and into Tibet (Gasse and van Campo 1994). Over much of this region, conditions today are the driest they have been throughout the Holocene. By contrast, lakes of inland drainage on the Altiplano of Peru and Bolivia have expanded and increased in depth from the mid-Holocene to the present. Lake Titicaca, for example, is currently close to its highest level in the Holocene. Similarly, in northern Chile, lacustrine and archeological evidence points to arid conditions from ~8000-3700 years B.P., followed by wetter conditions in the late Holocene (Grosjean et al. 1995). This is comparable to the situation in the western United States, especially Nevada and eastern California (Thompson 1992, Benson et al. 1996, Quade et al. 1998). Furthermore, low latitude hydrological changes in the Holocene were often abrupt (Gasse 2000, De Menocal et al. 2000) (e.g. at ~4200 calendar years B.P. in North Africa and the Middle East, when many freshwater lakes were reduced to swamps and arid lowlands within less than a century). As one might expect, such changes had significant impacts on ecosystems and the people living in those areas, in some cases resulting in complete societal collapse (Weiss et al. 1993, Dalfes et al. 1997, Weiss and Bradley 2001, De Menocal 2001).

A coherent picture is also emerging of a distinctly different pattern of El Niño/Southern Oscillation (ENSO) variability before ~5,000 years B.P. (Clement et al. 2000, Cole 2001). Most data and model results are consistent with a more La Niña-like background state and reduced inter-annual variability during this period (see Chapter 3, Section 3.31 for a complete discussion). For example, paleoclimatic observations indicate reduced incidence of heavy rains in Ecuador, absence of strong annual rainfall extremes in northern Australia, warmer SST along the northern Great Barrier Reef, and attenuated inter-annual variance in the ENSO-sensitive warm pool north of New Guinea (McGlone et al. 1992, Shulmeister and Lees 1995, Gagan et al. 1998, Tudhope et al. 2001). General circulation models forced with early Holocene orbital conditions simulate a cooler eastern/central tropical Pacific, due to intensified trades associated with a stronger Asian monsoon (Otto-Bliesner 1999, Bush 1999, Liu et al. 2000). The latter two references show warming in the westernmost tropical Pacific, much as La Niña brings today. Although these studies disagree on changes in the amplitude of inter-annual variability, a simpler model suggests that precessional forcing should result in weakened interannual ENSO strength. When radiation anomalies are strongly positive in boreal summer, as in the early to mid-Holocene, they result in stronger trade winds that cool the eastern Pacific during autumn, inhibiting the development of warm El Niño events. It is also of interest that Otto-Bliesner (1999) found that ENSO teleconnection patterns were very different at 6,000 B.P. compared to modern, this result cautions against inferring ENSO variability from sites not in close proximity to the tropical Pacific. For example, observations of mid-Holocene climate changes in continental South America and particularly in West Africa may reflect the effects of conditions in the Atlantic rather than in the PacificOcean.

There is a stark contrast between such a picture of major hydrological and circulation changes in the Tropics, with dramatic environmental consequences, and the record of relative stability seen in the well-known ice core accumulation and isotopic records from central Greenland (GRIP and GISP2 sites) (Dansgaard et al. 1993, Meese et al. 1994). However, ice core isotopic and summer melt layer data, from northern Greenland and smaller ice caps around the Arctic, do indicate a general cooling through most of the Holocene (i.e. a decrease in melt layers and lower $\delta^{18}O$), with warmest conditions in the first few millennia of the period (Koerner and Fisher 1990) (Figure 6.2). Furthermore, borehole temperatures at Summit, Greenland point to mean annual temperatures ~3°C warmer in the early Holocene compared to the last ~500 years (Dahl-Jensen et al. 1998). Thus, the GRIP/GISP2 isotope record appears to be anomalous.

There is also much evidence from other high latitude regions that temperatures generally declined during the Holocene (Figure 6.3). For example, diatom-based SST reconstructions for the Greenland and Norwegian Sea area indicate higher temperatures from ~9000 to ~4000 years B.P. (Koç et al. 1993). In central Sweden trees grew well above the modern altitudinal treeline from 9,000 to ~2,000 calendar years B.P. (Kullman 1989) and on the Kola Peninsula, Scots pine grew ~20 km north of the modern polar limit from ~7600 to ~4000 calendar years B.P. (MacDonald et al. 2000a). Similarly, trees grew north of the modern treeline in much of Siberia and in the Mackenzie River delta in early Holocene time (>8000 years ago) (Ritchie 1987, Burn 1997, Macdonald et al. 2000b). There is also evidence of extensive open water conditions in the Beaufort Sea and in the Canadian Arctic islands in the early Holocene, as documented by the skeletons of numerous bowhead whales and other marine mammals (that require relatively ice-free conditions) on raised beaches dating from that period (Dyke and Morris 1990, Dyke and Savelle 2001). At that time, driftwood was carried far into (seasonally ice-free) arctic fiords, whereas pervasive sea ice prevented such movement in the late Holocene. Furthermore, during the last few millennia of the Holocene land-fast ice shelves formed along the shores of the Arctic Ocean on Ellesmere Island where they are still found today (Bradley 1990). This Holocene cooling was not limited to high latitudes. Treeline in the White Mountains of eastern California (37°18'N) was 100 to 150 meters higher than the modern level from the sixth millennium B.C. until roughly 2200 B.C., declining most rapidly after AD 1000 (La Marche 1973). Although caution should be exercised in interpreting treeline movements in such arid regions (Lloyd and Graumlich 1997), the preponderance of evidence supports LaMarche's estimate of a 2°C cooling in warm-season temperatures.

Ice core deuterium ($\delta D$) records from Antarctica also indicate a general decline in temperature through the Holocene, with warmest conditions in the first few millennia (Ciais et al. 1992, Masson et al. 2001). Studies of deuterium excess in four Antarctic ice cores show an overall increase through the Holocene which is thought to be related to warmer early Holocene sea surface temperatures in the precipitation source regions, the low-latitude oceans of the southern hemisphere (Vimeux et al. 2001).

In the northern hemisphere, late Holocene expansion of glaciers (from mid or early Holocene minima) and the redevelopment of small ice caps accompanied late Holocene cooling (e.g. Nesje and Kvamme 1991, Nesje et al. 2001). It is arguable as to when this period began, but there is much evidence that the onset of this "neoglaciation" occurred ~4000-5000 years B.P. (Porter & Denton 1967, Grove 1988). A series of oscillations in ice extent in mountainous regions around the world characterises the last few thousand years, but the latest of these (within the last few hundred years) was generally the most extensive, indicating the overall severity of climate during this interval. The generic term "Little Ice Age" is commonly used to describe this episode, which is now considered to have occurred during the interval ~A.D. 1250-1880, but with the main phase, after ~A.D. 1550 (cf. Bradley and Jones 1992a, Grove 2001 a,b) (see further discussion below in Section 6.5).

These data all show that there was a period of relatively warm conditions in the first half of the Holocene, in many areas warmer than in the 20th century, after which temperatures generally

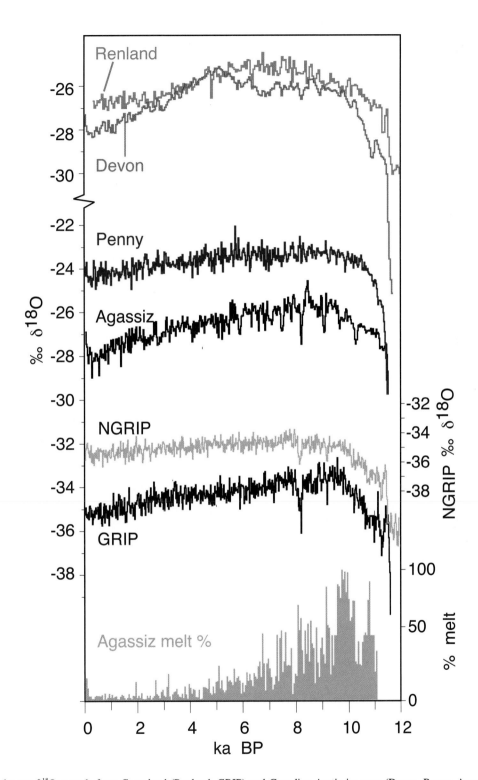

**Fig. 6.2.** Holocene $\delta^{18}O$ records from Greenland (Renland, GRIP) and Canadian Arctic ice caps (Devon, Penny, Agassiz). Also shown is the record of melt in cores from the Agassiz Ice Cap, Ellesmere Island, Arctic Canada (% of core sections showing evidence of melting and percolation of meltwater into the firn) (after Fisher and Koerner 2002).

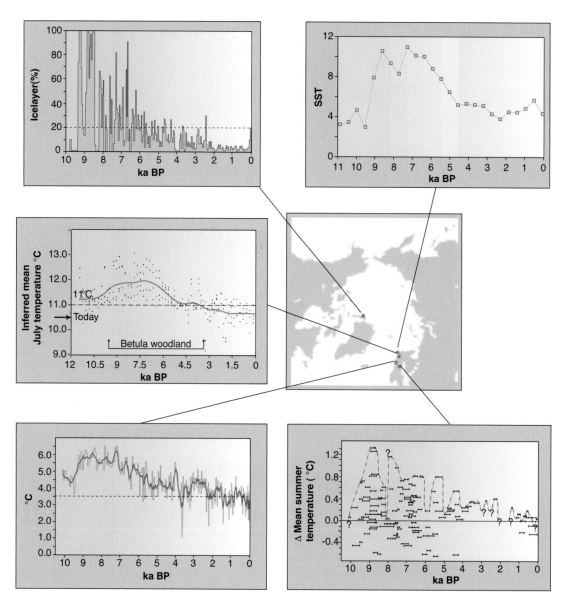

**Fig. 6.3.** Composite of records showing Holocene temperature changes. Top left: Melt record from Agassiz Ice cap (Koerner and Fisher 1990). Top right: Diatom-inferred SST (Koç et al. 1993). Middle left: Pollen-inferred mean July temperature (Seppä and Birks 2001). Bottom left: summer temepratures from oxygen isotopes (Lauritzen 1996). Bottom right: mean summer temperature from the upper limit of pines (Dahle and Nesje 1996)

declined. The decline was punctuated by centennial-scale warmer and colder episodes, with the most recent cold episode (~A.D. 1550-1850) being the coldest period of the entire Holocene, especially in arctic and sub-arctic regions (Bradley 2000).

## 6.3 Temperatures over the last millennium

Most high resolution paleoclimate records (i.e. those with annual resolution and a strong climate signal) do not extend back in time more than a few centuries. Consequently, while there are numerous

paleoclimate reconstructions covering the period from the 17th century to the present, the number of high resolution millennium scale records is very limited. Continuous records are restricted to ice cores and laminated lake sediments, where the climatic signal is often poorly calibrated, and to a few long tree ring records, generally from high latitudes. Inevitably, this leads to large uncertainties in long-term climate reconstructions that attempt to provide a global or hemispheric-scale perspective. Bearing this in mind, what do current reconstructions tell us about the last millennium?

Figure 6.4 shows a reconstruction of northern

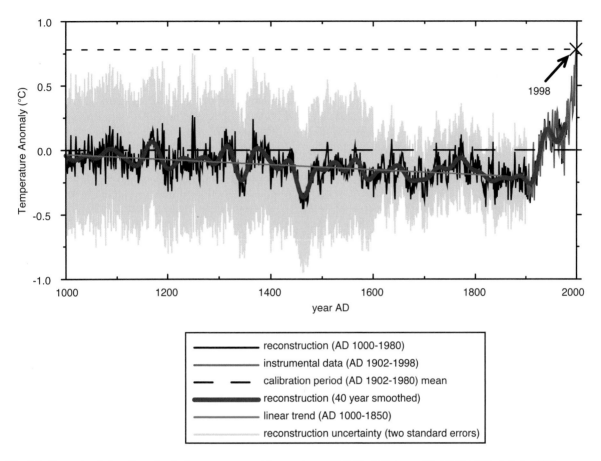

**Fig. 6.4.** Reconstructed northern hemisphere mean annual temperature with 2 standard error uncertainties (Mann et al. 1999).

hemisphere mean annual temperature for the last 1000 years. This is based on a network of well-distributed paleoclimatic records, the number of which decreases back in time. For the period since A.D. 1400, 397 records were used, but before A.D. 1400 this number drops to 14 (made up of 11 individual records, plus the first 3 principal components of tree ring width, representing a large set of trees in the western United States) (Mann et al. 1998, 1999). These paleoclimatic proxies were calibrated in terms of the main modes of temperature variations (eigenvectors) represented in the instrumental records for 1902-1980. Variations across the network of proxies, for the period before instrumental records, were then used to reconstruct how the main temperature patterns (i.e. their principal components) varied over time. By combining these patterns, regional and hemispheric mean temperature changes, as well as spatial patterns over time were reconstructed (Mann et al. 2000a). To accurately reproduce the *spatial* pattern requires that the proxy data network is extensive enough to capture several of the principal eigenvector patterns. With the data available, regional patterns of temperature variation could only be meaningfully reconstructed for 250

years, although the large-scale (hemispheric) mean temperature could be reconstructed for a longer period. This is possible because the proxy data network, even at its sparsest, exhibits a coherent response to temperature variability at the largest scale (Bradley and Jones 1993, Jones and Briffa 1998). Thus a reconstruction of hemispheric mean temperature back 1000 years is possible, using a quite limited network of data, albeit with ever-increasing uncertainty the further back in time one goes (Figure 6.4). This reconstruction shows an overall decline in temperature of ~0.2°C from A.D. 1000 until the early 1900s (-0.02°C/century) when temperatures rose sharply. Superimposed on this decline were periods of several decades in length when temperatures were warmer or colder than the overall trend. Mild episodes, lasting a few decades, occurred around the late 11th and mid-12th century and in the early and late 14th century, but there were no decades with mean temperatures comparable to those in the last half of the 20th century. Coldest conditions occurred in the 15th century, the late 17th century and in the entire 19th century. A critical question in any long-term reconstruction is: to what extent does the proxy adequately capture the true low-frequency nature of the climate record? Given that most of the long-term data used in all paleotemperature reconstructions are from tree rings, it is important to establish that the reconstructed temperature series are not affected by the manner in which biological growth trends in the trees are removed during data processing. This matter is especially critical when individual tree ring records, of differing record lengths (often limited to a few hundred years) are patched together to assess long-term climate changes. Briffa et al. (2001) have carefully evaluated this problem, using a maximum ring density data set that is largely independent of that used by Mann et al. (1998, 1999). By combining sets of tree ring density data grouped by the number of years since growth began in each tree, Briffa et al. provide a methodology that is designed to eliminate the biological growth function problem. They also estimate confidence limits through time (Figure 6.5).

The Briffa et al series shows similar temperature anomalies as Mann et al. in the 15th century (though no sharp decline in temperatures around A.D.1450) but markedly colder conditions from A.D. 1500 to ~A.D. 1800. The early 19th century is also colder in the Briffa et al. series. Their reconstruction thus describes a well-defined minimum in temperatures from ~A.D. 1550-1850 that conforms with the consensus view of a "Little Ice Age" (cf. Figures 6.4 and 6.5) (Bradley and Jones 1992 a,b). Though this period was not uniformly cold and temperature anomalies differed regionally, overall it was significantly below the 1881-1960 mean (by as much as 0.5 °C for most of the 17th century) in the regions studied. Independent reconstructions derived from borehole temperatures suggest that ground surface temperatures were even colder 400 to 500 years ago (~1 °C below levels in the 1980s) with temperatures subsequently rising at an increasingly rapid rate (Pollack et al. 1998, Huang et al. 2000). However, borehole-based temperature estimates have large geographical heterogeneity, resulting in a very small signal-to-noise ratio for mean hemispheric and global estimates. Indeed, a large number of borehole records do not capture the upward trend in 20th century air temperature in their respective regions (as discussed further below). Other attempts to assess northern hemisphere temperatures have taken a simpler approach than either Mann et al. (1998) or Briffa et al. (2001), by averaging together normalized paleo-data of various types (Bradley and Jones 1993, Jones et al. 1998) or by averaging data scaled to a similar range (Crowley and Lowery 2000). Such approaches do not provide an estimation of uncertainty, and indeed may lead to rather arbitrary combinations of very diverse data (often

having different temporal precision). Nevertheless, the resulting time series from all of these studies are similar, at least for the first 400-500 years of the last millennium. Thereafter, some series indicate especially cold conditions, from the late 16th century until the 19th century, but these estimates are all bracketed by the two standard error confidence limits of Mann et al. (1998) and Briffa et al. (2001) (Figure 6.6).

Although all of the reconstructions have much in common, they are clearly not identical. One explanation for the differences may lie in the geographical distribution of data used in each analysis. Each reconstruction represents a somewhat different spatial domain. In the Mann et al. studies, the "northern hemisphere mean" series is the same geographical domain as the gridded instrumental data set available for the calibration period (1902-80). This means that some regions within the northern hemisphere (in the central Pacific, central Eurasia and regions poleward of 70°N) were not represented. However, because *global* eigenvector patterns were employed, the northern hemisphere mean is influenced by data from low latitudes and parts of the southern hemisphere. By contrast, the study of Briffa et al. is largely based on records from the northern treeline (60-75°N) where temperatures were particularly low in the 17th century. As the other reconstructions generally do not include data from sub-tropical or tropical regions either, if higher latitudes were colder at that time compared to the Tropics, this may explain why the latter half of the millennium appears colder in those "hemisphere mean" series.

Another reason for the differences in Figure 6.6 may be because each reconstruction represents a somewhat different season. In the Mann et al. (1998, 1999) reconstruction, mean annual temperature data were used for calibration, since data from both hemispheres were used to constrain the eigenvector patterns and data from different regions may have had stronger signals in one season than in another. For example, some data from western Europe contain a strong NAO (winter) signal, whereas data from elsewhere carry a strong spring precipitation signal related to ENSO. Both data sets nevertheless help to define important modes of climate anomalies that themselves capture large-scale annual temperature patterns (Bradley et al. 2000). Other reconstructions are for boreal summer months (April-September), which may also explain some of the differences between the series. If summers were particularly cool in extra-tropical regions in the 17th-19th centuries, relative to temperature anomalies in low latitude regions (equatorward of ~30°N) this implies an increase in the northern

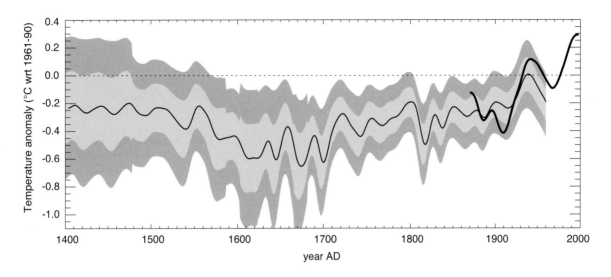

**Fig. 6.5.** Tree ring density reconstruction of warm-season (April to September) temperature from all land north of 20°N, with the ±1 and ±2 standard error ranges shaded. Units are °C anomalies with respect to the 1961-90 mean (dotted line) and the instrumental temperatures are shown by the thick line. Both series have been smoothed with a 30-year Gaussian-weighted filter (from Briffa et al. 2001).

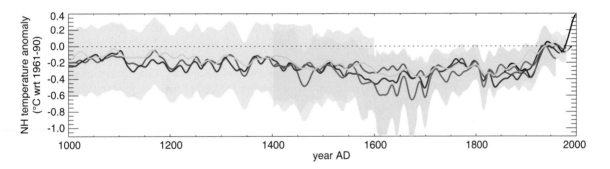

**Fig. 6.6.** Northern Hemisphere surface temperature anomalies (°C) referenced to the 1961–90 mean (dotted line). Annual-mean land and marine temperature from instrumental observations (black, 1856–1999), and as reconstructed by Mann et al. (red, 1000–1980, with ±2 standard errors shown by pink shading) and Crowley and Lowery (purple, 1000–1987). April-to-September mean temperature from land north of 20°N as reconstructed by Briffa et al. 2001 (green, 1402–1960, with ±2 standard errors shown by green shading), and by re-calibrating the Jones et al. estimate of summer northern hemisphere temperature (by simple linear regression) over the period 1881–1960 (yellow, 1000–1991). All series have been smoothed with a 30-year Gaussian-weighted filter.

hemisphere Equator-Pole temperature gradient during that period.

An independent assessment of temperature changes on the continents over the last few centuries is provided by a network of geothermal measurements in boreholes (Pollack et al. 1998, Huang et al. 2000). The depth profile of sub-surface temperature reflects a balance between heat loss from the surface, heat generated in the deep interior of the earth and the depth-dependent profile of heat diffusivity in the rock substrate. Changes in surface temperature propagate downward into sub-surface rocks, causing slight variations in the temperature depth profile. The depth to which such disturbances can be detected (by inversion of the sub-surface temperature profile) depends on the magnitude and duration of the surface temperature change, and variations of heat diffusivity in the rock, but whatever "signal" is transmitted from the surface, it is strongly attenuated with depth. This method thus provides a time-integrated perspective on paleotemperature with depth (Clow 1992, Beltrami and Mareschal 1995). It is not realistic to reconstruct a long-term annual, or even decadal temperature history from borehole data, but long-term trends or pre-instrumental mean temperatures can be assessed (Harris and Chapman, 2001). Figure 6.7 shows a comparison of century-long trends with data from co-located grid-boxes, derived by Mann et al. (1998). Here, (as in Huang et al. 2000) century-long

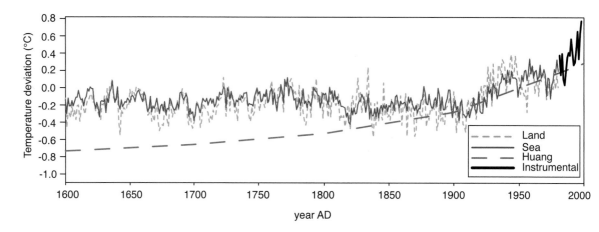

**Fig. 6.7.** A comparison of century-long ground temperature trends from boreholes with data from co-located grid-boxes, derived by Mann et al. (1998).

trends have been computed for each site, then averaged for all locations for each century and the means were concatenated back in time from the most recent 20th century trend. It is clear that the borehole data point to much greater cumulative warming since A.D. 1600 (and thus a much colder 17th century) than the multiproxy data set. However, borehole data are extremely noisy. The mean hemispheric signal of temperature change is not statistically significant between 1500 and 1900, only the change in the last 100 years emerges above the level of background noise (Mann et al. sub). Furthermore, a comparison of the data on a regional basis shows that whereas borehole data show continuous warming over the last 500 years in *all* areas, the Mann et al. data show overall negative (cooling) trends in the 16th century in Asia and North America, and cooling in all regions in the 19th century.

Why borehole temperatures increase at a rate greater than that indicated by proxy-based temperature reconstructions is not clear, but there are a number of possible reasons for the differences. Ground surface temperatures are not only affected by changes in air temperature (cf. Lewis 1998, Lewis and Wang 1998), but also by alteration of ground cover (e.g. due to land use changes), by changes in snow cover and by the amount of time it takes for near-surface soil moisture to become completely frozen in the winter (until all the moisture is frozen, further penetration of the winter cold wave into the ground is precluded). It is difficult to evaluate the importance of such effects on the diverse borehole data, but a study of data from northwestern North America suggests that ground surface temperatures significantly over-estimate air temperature changes in the 20th century (Skinner and Majorovicz 1999). Given that any major change in land use would likely affect both air and ground surface temperatures in the same way, it is hard to explain

all these discrepancies. Nevertheless, it is clear that borehole records commonly do not match observed warming trends in their respective regions in the 20th century, which indicates that using them as a simple proxy for air temperature is problematical. A site-by-site evaluation of the quality of ground temperature data is needed, together with land use histories and snow cover changes, to try and resolve the matter. Once this has been done, combining the valuable low frequency characteristics of borehole data with the higher frequency attributes of annually resolved proxy data should yield better overall assessments of long-term temperature changes (e.g. Beltrami et al. 1995).

## 6.4 Uncertainties in large-scale temperature reconstructions

All large-scale paleotemperature reconstructions suffer from a lack of data at low latitudes. In fact, most "northern hemisphere" reconstructions do not include data from the southern half of the region (i.e. areas south of 30°N). Furthermore, there are so few data sets from the southern hemisphere that it is not yet possible to reconstruct a meaningful "global" record of temperature variability beyond the period of instrumental records. For the northern hemisphere records, it must be recognized that the errors estimated for the reconstructions of Mann et al. (1999) and Briffa et al. (2001) are minimum estimates, based on the statistical uncertainties inherent in the methods used. These can be reduced by the use of additional data (with better spatial representation) that incorporate stronger temperature signals. However, there will always be additional uncertainties that relate to issues such as the constancy of the proxy-climate function over time, and the extent to which modern climate modes (i.e. those that occurred during the calibration interval) represent the full range of climate variability in the past. There is evidence that in recent decades some high latitude trees no longer capture low frequency variability as well as in earlier decades of the 20th century (as discussed below in Section 6.8) which leads to concerns over the extent to which this may have also been true in the more distant past. If this was a problem (and currently we are not certain of that) it could result in an inaccurate representation of low frequency temperature changes in the past. Similarly, if former climates were characterised by modes of variability not seen in the calibration period, it is unlikely that the methods now in use would reconstruct those intervals accurately. It may be possible to constrain these uncertainties through a range of regional studies (for example, to examine modes of past variability) and by calibration over different time intervals, but not all uncertainty can be eliminated and so current margins of error must be considered as minimum estimates.

## 6.5 The Medieval Warm Epoch and the Little Ice Age

Bearing in mind concerns expressed earlier about the uncertainties inherent in paleotemperature reconstructions, what evidence is there for a "Medieval Warm Epoch" (MWE) and a "Little Ice Age" (LIA) during the last millennium?

The original argument for a MWE was made by Lamb (1965) based largely on evidence from western Europe. Much of the evidence he cited was anecdotal and he suggested that temperatures between A.D. 1000 and 1200 were about 1-2°C "above present values" (probably meaning the 1931-60 average). In revisiting the concept of a MWE, Hughes and Diaz (1996) reviewed a wide range of paleoclimatic data, much of it reported since Lamb's classic work (Lamb 1965). They concluded that "*it is impossible at present to conclude from the evidence gathered here that there is anything more significant than the fact that in some areas of the globe, for some part of the year, relatively warm conditions may have prevailed*". Thus, they found no clear support for there having been a globally extensive warm epoch in the MWE, or indeed within a longer interval stretching from the 9th to the early 15th century. In fact, there is insufficient high resolution proxy evidence to be certain that *global or hemispheric* mean temperatures were higher during the MWE *sensu stricto* than in the 20th century, as data from different regions do not agree on the matter. Consequently, we cannot entirely rule out the possibility of a globally extensive warm episode (or episodes) for at least part of the period from A.D. 1000 to 1200, because of the

paucity of high-resolution records (especially from the oceans and the southern hemisphere) spanning that interval. It is interesting that Huang and Pollack (1997) find evidence (in a set of 6144 continental borehole heat flow measurements from around the world) that temperatures were in the range of 0.1-0.5°C warmer than "present" (the early 1980s -- the mean date of borehole logging) ~700 to 800 years ago. However, more definitive estimates (of timing and amplitude) can not be made and given the large variability of these data, such estimates may not reach statistical significance compared to temperatures in the late 20th century. High-precision borehole temperature measurements at the GRIP (Summit) site on the Greenland Ice Sheet (where the signal-to-noise ratio is relatively large) point to conditions that were 0.5-1°C above 1970 mean temperature at this one site around A.D. 1000 , but similar data from Law Dome, Antarctica show a temperature minimum at A.D. 1250, followed by warmer conditions in subsequent centuries (Dahl-Jensen et al. 1998, 1999).

High latitude tree ring data from some parts of the northern hemisphere also show evidence of temperatures in Medieval times well above the 20th century mean, at least in summer months (e.g. Briffa 2000), and so do some marine records from the North Atlantic, though the timing can not be precisely resolved (Keigwin 1996, Keigwin and Pickert 1999). There is also strong evidence from early European documentary records that winter temperatures in western Europe were quite mild during at least part of the period A.D. 750-1300 (Pfister et al. 1998). On the other hand, tree ring data from the southern hemisphere paint a different picture, with no clear evidence for a MWE, even when special care has been taken to conserve centennial scale variability (e.g. Villalba et al. 1996). Thus, whether there really were warm episodes of *global* extent in Medieval times and how these compare with late 20th century temperature levels (especially those in the last 20 years of the 20th century) remains an intriguing question that deserves further scrutiny. Until a more extensive set of data is available, the absence of evidence does not necessarily imply evidence of absence. Nevertheless, it must be stated that given the relatively limited evidence that does exist to support Lamb's original contention (Lamb 1965), the burden of proof must rest on demonstrating that his concept of a MWE has validity, rather than trying to show that it does not.

Perhaps of greater significance is that there definitely were significant precipitation anomalies during the period of the MWE, in particular, many areas experienced protracted drought episodes and these were far beyond the range of anything recorded within the period of instrumental records. For example, Stine (1994) describes compelling evidence that prolonged drought affected many parts of the western United States (especially eastern California and the western Great Basin) from (at least) A.D.910 to ~A.D.1110, and from (at least) A.D.1210 to ~A.D.1350. This led him to argue that a better term for the overall period was the "Medieval Climatic Anomaly" (MCA), which removes the emphasis on temperature as its defining characteristic (Stine 1998). The widespread nature of hydrological anomalies during the MCA suggests that changes in the frequency or persistence of certain circulation regimes may account for the unusual conditions during this period, and naturally this may have led to anomalous warmth in some (but not all) regions.

Numerous studies provide strong evidence that cooler conditions characterized the ensuing few centuries, and the term "Little Ice Age" is commonly applied to this period. Since there were regional variations to this cooling episode, it is difficult to define a universally applicable date for the "onset" and "end" of the period, but commonly ~A.D. 1550-1850 is used (Bradley and Jones 1992). However, there is evidence that cold episodes were experienced earlier, and glacier advances were common by the late 13$^{th}$ century in many alpine areas around the North Atlantic and in western Canada (Grove and Switsur 1994, Luckman 1994, 1996, 2000, Grove 2001a,b). The definitional problem is illustrated in Figure 6.4 which shows temperatures gradually declining over the first half of the last millennium, rather than there being a sudden "onset" of a "LIA". If this reconstruction is accurate, it might explain why different mountain areas registered the onset of this neoglacial episode at different times. As temperatures cooled the threshold temperature for positive mass balance and glacier advances may have been reached in some areas sooner than others, leading to seemingly heterogeneous regional responses. However, by the late 16$^{th}$ century almost all regions had registered glacier advances, and these conditions generally persisted until the mid- to late 19$^{th}$ century making the term "Little Ice Age" ubiquitous and meaningful for this interval. Nevertheless, even within the period 1550-1850 there was a great deal of temperature variation both in time and space (Pfister 1992). Some areas were warm at times when others were cold and *vice versa*, and some seasons may have been relatively warm while other seasons in the same region were anomalously cold. But whatever date one selects for the "onset" of the LIA, there is little doubt that it was firmly at an end by the be-

ginning of the 20$^{th}$ century. The reduction in ice masses accumulated over preceding centuries has continued to the present (in fact, the rate has accelerated) in almost all regions of the world (Dyurgerov and Meier 2000).

No doubt the complexity, or structure that we see in the climate of the LIA is a reflection of the (relative) wealth of information that paleoclimate archives have provided for this period. Having said that, when viewed over the long term this overall interval was undoubtedly one of the coldest in the entire Holocene. If we had similar data for the last 1000 years, our somewhat simplistic concepts of Medieval climatic conditions would certainly be revised and strong efforts are needed to produced a comprehensive paleoclimatic perspective on this time period. Only with such data will we be able to explain the likely causes for climate variations over the last millennium (see Sections 6.11 and 6.12).

## 6.6 20$^{th}$ century temperatures in perspective

One thing that all reconstructions shown in Figure 6.6 clearly agree on is that northern hemisphere mean temperature in the 20$^{th}$ century is unique, both in its overall average and in the rate of temperature increase. In particular the 1990s were exceptionally warm -- probably the warmest decade for at least 1000 years (even taking the estimated uncertainties of earlier years into account). A caveat to this conclusion is that the current proxy-based reconstructions do not extend to the end of the 20$^{th}$ century, but are patched on to the instrumental record of the last 2-3 decades. This is necessary because many paleo data sets were collected in the 1960s and 1970s, and have not been up-dated, so a direct proxy-based comparison of the 1990s with earlier periods is not yet possible. Nevertheless, confidence that an accurate reproduction of the recent instrumental record would be obtained if all the available paleoclimatic data were updated to the present is provided by Figure 6.8. This shows that a set of proxy data calibrated against the 1902-1980 period of instrumental data captured mean annual temperatures well both during this period and during the preceding 50 years for which an *independent* set of instrumental data is available. The excellent fit over the late 19$^{th}$ century test period provides confidence that an updated set of proxy data would also accurately reproduce recent changes. However, one cautionary note is needed: in the case of tree rings from some areas at high latitudes, the decadal timescale climatic relationships prevalent for most of this century appear to have changed in recent decades, possibly because increasing aridity &/or

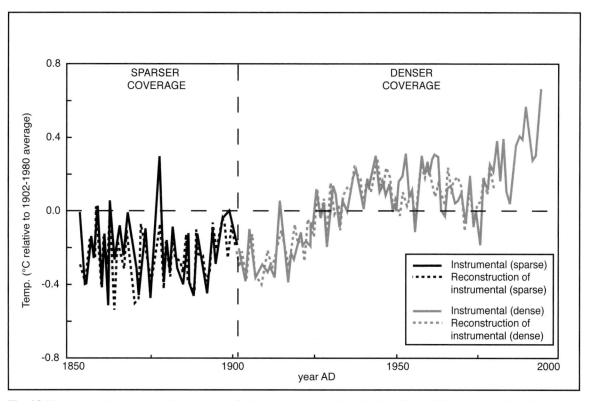

**Fig. 6.8.** Reconstructed mean annual temperatures for the northern hemisphere for the 19th and 20th centuries, from Mann et al. (2000) compared to the calibration data (1902-1980) and an independent period (1854-1901) for which instrumental data are available.

snowcover changes at high latitudes may have altered the ecological responses of trees to climate (cf. Jacoby and D'Arrigo 1995, Briffa et al. 1998). For example, near the northern tree limit in Siberia, this changing relationship can be accounted for by a century-long trend to greater winter snowfall. This has led to delayed snowmelt and thawing of the active layer in this region of extensive permafrost, resulting in later onset of the growing season (Vaganov et al. 1999). It is not yet known how widely this explanation might apply to the other regions where the partial decoupling has been observed, but regardless of the cause, it raises the question as to whether there might have been other periods in the past when the tree ring-climate response changed, and what impact such changes might have on paleotemperature reconstructions based largely on tree ring data.

In any case, the conclusion that temperatures in the 20th century rose at a rate that was unprecedented in the last millennium, reaching levels by the end of the century that had rarely, if ever, been exceeded in (at least) the preceding 900 years, seems to be an extremely robust result from all large-scale paleotemperature reconstructions, and it is confirmed by borehole heat flow data (cf. Pollack et al. 1998). The change in temperature has led to a major reduction in the mass of alpine glaciers in

almost all parts of the world (Dyurgerov and Meier 2000, Thompson et al. 1993, Brecher and Thompson 1998), an increase in permafrost thawing at high latitudes (Osterkamp and Romanovsky 1999, Osterkamp et al. 2000) and at high altitudes (Jin et al. 2000, Isaksen et al. 2001), a reduction in the extent and thickness of Arctic sea-ice (Rothrock et al. 1999, Vinnikov et al. 1999, Wadhams and Davis 2001), later freeze-up and earlier break-up dates of ice on rivers and lakes (Magnuson et al. 2000), an increase in the calving rate of Antarctic ice shelves (Scambos et al. 2000), shifts in the distribution of plant and animal species, latitudinally and altitudinally (Grabherr et al. 1994, Pauli et al. 1996), changes in the phenology of plant leafing and flowering (Myneni et al. 1997) and the storage of significant quantities of heat in the near-surface ocean (Levitus et al. 2000) as well as an overall rise in sea-level driven by the melting of ice on the continents and a steric change, due to the increase in ocean temperature (Warrick and Oerlemans 1990). Thus, regardless of arguments over instrumental versus satellite-based estimates of warming in recent decades (National Research Council 2000) there are multiple indicators of warming in the 20th century that paint a vivid picture of the global-scale environmental consequences of the temperature increase.

Figure 6.8 shows that the overall range in temperature over the last 1000 years has been quite small. For example, the range in 50-year means has only been ~0.5°C (from the coldest period in the 15th, 16th and 19th centuries, to the warmest period of the last 50 years). Within that narrow envelope of variability, all of the significant environmental changes associated with the onset and demise of the LIA took place. This puts into sharp perspective the magnitude of projected future changes resulting from greenhouse-gas increases and associated feedbacks (Figure 6.9). Even the low end of model estimates involving a scenario of minimal growth in future energy consumption suggests additional temperature increases on the order of 1-2°C by the end of the 21st century (Intergovernmental Panel on Climate Change, 2001) which would be far beyond the range of temperatures experienced over (at least) the last 1000 years.

The discussion so far has focused exclusively on the northern hemisphere record because there are insufficient data currently available to produce a very reliable series for the entire southern hemisphere. Data from the Mann et al. (1998) reconstruction (back to A.D. 1700) averaged for those parts of the southern hemisphere that were represented in the instrumental calibration period show a similar temporal pattern to that of the northern hemisphere, but generally warmer (less negative anomalies). However, much more work is needed on southern hemisphere proxy records to extend and verify this result. We now turn to selected regional studies where much information has been obtained about major climate systems (modes of climate variability) that affect extensive areas of the world.

## 6.7 The Tropical Indo-Pacific

The past decade of climate dynamics research has highlighted the important role that tropical climate systems play in orchestrating modern interannual climate variability (see discussion in Chapter 3, Section 3.3). With that recognition has come a substantial effort to understand the variability of those systems on longer time scales. Much of this effort has focused on developing continuous annual reconstructions of sea surface temperatures, or rainfall-induced changes in salinity using coral records from the tropical oceans, along with terrestrial records from adjacent continents that are impacted by ocean-atmosphere variability. Although continuous coral records generally span only the past few centuries, they are providing an emerging picture of coherent decadal and longer-term variability in regions where instrumental records are limited to the past few decades. Terrestrial records suggest

that teleconnections to ENSO evolve through time. Radiometrically dated fossil corals are revealing how ENSO responds to changes in background climate and forcing. Paleoclimate records indicate that ENSO is variable on many time scales, sensitive to certain background and forcing changes, and has a changing global signature. Century-scale records from the tropical Indo-Pacific support the global picture of a warming world, most coral records show a trend to unusually warm conditions in the late 20th century, with a few regional exceptions.

### 6.7.1 ENSO variability

Decadal variations in ENSO frequency, strength, and teleconnections are suggested by twentieth-century instrumental records (Trenberth and Shea 1987). In particular, instrumental records from the tropical Indo-Pacific show a recent shift towards warmer/wetter conditions in 1976. This shift appears as more "El Niño-like" conditions in the equatorial Pacific (Ebbesmeyer et al. 1991, Trenberth and Hurrell 1994, Graham et al. 1994) that set the stage for record El Niño events in 1982-3 and 1997-8. Unlike classic El Niño anomalies, the anomalies associated with this decadal shift reach into the mid-latitudes of both hemispheres (Zhang et al. 1997, Garreaud and Battisti 1999). In the tropical Indian Ocean, warming in 1976 (Terray 1995) coincides with a breakdown of previously established relationships between the Indian monsoon and ENSO and with changes in the relation of the monsoon to Indian Ocean SST anomaly fields (Kumar et al. 1999, Clark et al. 2000).

The change in ENSO in 1976 appears to be highly unusual, given the statistics of ENSO variability in the preceding 100 years of instrumental data. Time series (ARMA) modeling suggests that the 1976 shift should have a recurrence interval of about 2000 years, based on the pre-1976 variability in the Southern Oscillation and assuming statistical stationarity (Trenberth and Hoar 1996). A likely alternative to this "rare event" scenario is that the shift reflects a statistical non-stationarity introduced by a warming background climate (Trenberth and Hoar 1996). Other analyses have suggested, however, that the shift is somewhat less unusual (Rajagopalan et al. 1997) or even that it lacks statistical significance (Wunsch 1999). The ongoing debate over this issue argues strongly that longer records of ENSO are needed.

The 1976 shift appears as unprecedented in many ENSO-sensitive paleoclimate records. It occurs in

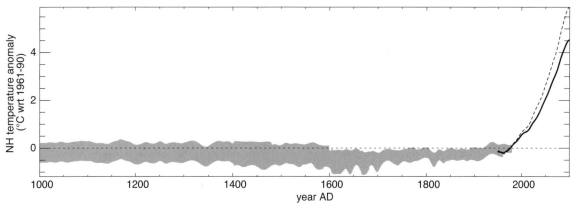

**Fig. 6.9.** Northern Hemisphere surface temperature (°C anomalies with respect to the 1961–90 mean) reconstructed by Mann et al. (1999) and Briffa et al. (2001), indicating the ±2 standard error ranges of the two reconstructions, compared to that simulated by the Hadley Centre's HadCM3 coupled climate model under increasing greenhouse gas and sulphate aerosol concentrations (SRES scenario A2) from 1950-2099. The dashed line is the model for latitudes higher than 20 degrees north. All data have been smoothed with a 30-year Gaussian-weighted filter.

all the ENSO-sensitive coral records, and in most of these, the post-1976 interval is the warmest/wettest period of the record. In the southwestern US, where El Niño events bring wetter and cooler conditions, the growth of high-elevation thousand-year-old trees shows an unprecedented increase beginning in the mid-1970s. This growth increase is attributed to a combination of increased cool-season precipitation and warmer growing season temperatures (Swetnam and Betancourt 1998). Global temperatures also reflect an upward shift in 1976 which, it is argued, indicates an intensification of the tropical hydrological cycle consistent with model simulations of a climate response to doubled $CO_2$ (Graham et al. 1994, Trenberth and Hoar 1996). Whether natural or anthropogenic, the mechanisms of this change have received intense scrutiny (Gu and Philander 1997, Zhang et al. 1998, Guilderson and Schrag 1998) and are the subject of ongoing analysis.

Many tropical Pacific records show a significant correlation with ENSO variability in their calibration intervals and can be used to assess how that system has changed through time. ENSO-sensitive records from the equatorial Pacific show clearly that the characteristic time scale of tropical Pacific variability changes over the past 1-4 centuries (Cole et al. 1993, Dunbar et al. 1994, Tudhope et al. 1995, Linsley et al. 2000a, Urban et al. 2000). In the Galapagos (1°N, 90°W), interannual and decadal modes of SST variability appear to change in strength simultaneously, implying a range of time scales for El Niño as well as linkages across these time scales (Dunbar et al. 1994). Shorter ENSO records from Tarawa (1°N, 173°E) and New Guinea (5°S, 146°) indicate shifts in variance among interannual periods that coincide with changes in the

strength of the annual cycle (Cole et al. 1993, Tudhope et al. 1995).

At Maiana Atoll (1°N, 172°E), a record that reaches to AD 1840 indicates a long-term trend from cooler/drier to warmer/wetter conditions that is associated with changes in ENSO variance through time (Urban et al. 2000, Figure 3.8). When background conditions are cooler/drier (the mid and late 19th century), variability occurs on a more decadal time scale, in contrast to the dominantly interannual variance of the 20th century. The decadal variance in this record offers a point of comparison for decadal variability in other tropical records, the Maiana record is coherent on decadal time scales with coral records from the equatorial western Indian Ocean and with other ENSO-sensitive records from the Pacific and on adjacent continents (Figure 6.10, Cole 2000). Other ENSO reconstructions that span this interval also note a weakening of interannual variance (Stahle et al. 1998) and stronger decadal variance (Mann et al. 2000b). This widespread signal is thus not a consequence of a changing spatial domain of ENSO impacting a single coral site, it clearly reflects a real change in the time-domain behavior of the system.

Land-based paleoclimate records from North America offer additional clues to past variations in the Pacific and its impacts. Stahle et al. (1998) developed a 272-year Southern Oscillation Index (SOI) reconstruction based on drought-sensitive tree ring chronologies from the southwestern US and northern Mexico. Their reconstruction indicates stronger interannual variance and a tendency for more cool events in the 20th century compared to earlier periods. A network of drought reconstructions over the continental US shows that, although the ENSO-drought link in the southwest is

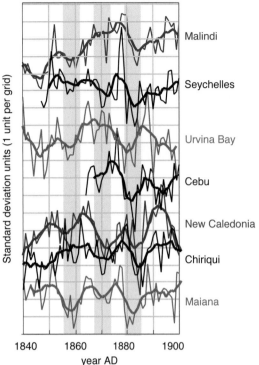

Malindi

Seychelles

Urvina Bay

Cebu

New Caledonia

Chiriqui

Maiana

Standard deviation units (1 unit per grid)

1840    1860    1880    1900
year AD

**Fig. 6.10.** Common patterns of decadal variability in tropical Indo-Pacific coral records during the 19th century. The records shown here all exhibit cool-dry events in the late 1850s, ~1870, and early 1880s (shaded bars). Small age offsets may be real, or may reflect age-model uncertainties. The top three and the bottom record correlate closely with ENSO in their calibration periods, but the remaining records are somewhat removed from ENSO centers of action or have competing climate influences on their $\delta^{18}O$, so do not correlate as strongly to interannual ENSO changes. The fact that they all reflect the decadal variance of the late 19th century suggests similarities with the 1976 shift, whose extent is latitudinally broader than typical ENSO variability.

relatively robust, drought in other regions shows a more variable connection to tropical Pacific variability (Cole and Cook 1998). In the early 20th century, El Niño events bring wetter conditions to the mid-Atlantic region, but by the mid-20th century, El Niño conditions are associated with drought. Cole and Cook (1998) attribute this change to the intensification of decadal variability in the North Pacific, with a downstream influence on mid-Atlantic regional moisture balance. Moore et al. (2001) suggest a similar reversal in the correlation of decadal anomalies in Mt Logan glacial accumulation and tropical Pacific variability.

### 6.7.2 Century-scale trends in the tropics

Looking beyond ENSO variability, coral isotope records from the tropical Indo-Pacific suggest a consistent pattern of declining $\delta^{18}O$ over the past few centuries, reflecting increasingly warm/wet

conditions (Figure 6.11). The rate of change observed between 1895-1989, for those records that span this interval, ranges from 0.3-2.1°C/century. The strongest recent trends are seen at off-equatorial sites (New Caledonia and the Philippines) and in precipitation-sensitive records where salinity changes likely make up a significant part of the coral isotopic trend (Nauru and Maiana). In two instances, the trend over this interval does not indicate warming. At Vanuatu, a strong warming trend that predates 1895 supports the general picture of a relatively warm 20th-century, despite the lack of such a trend between 1895-1989. A slight cooling at Galapagos is consistent with instrumental data from this region and may reflect an ocean dynamical response to greenhouse forcing, whereby warming in the western Pacific strengthens the trade winds, which enhances upwelling of cool water in the easternmost Pacific (Cane et al. 1997). Just as air temperature records from individual sites do not each reflect the global average air temperature change, we do not expect worldwide consistency in oceanic records, but the general pattern supports tropical warming in the 20th century. These records place in context the global analysis of instrumental ocean temperatures (Levitus et al. 2000, 2001) that shows an unequivocal accumulation of heat in the world ocean since 1955.

Records from tropical high-elevation ice caps provide some of the strongest evidence for tropical warming. Tropical glaciers worldwide are receding rapidly and many have already disappeared (Thompson et al. 1993, 2001, IPCC 2001). Isotopic records place these recent changes in a longer term context (Figure 6.12). Diaz and Graham (1996) have linked the recent ice losses to a systematic warming of the tropical troposphere driven by increasing tropical SST, time series of freezing level indicate a shift upwards in 1976, consistent with the shift in tropical SST (Diaz and Graham 1996, Vuille and Bradley 2000, Gaffen et al. 2000).

### 6.7.3 Tropical variability in the last millennium

Records that span the past millennium with high resolution are rare in the tropics, but three recently published datasets suggest common patterns at multidecadal-century time scales (Figure 6.13). The depth of Crescent Island Crater Lake, part of Lake Naivasha in Kenya, has fluctuated dramatically over the past millennium, with prolonged wet and dry phases coeval with cultural records of prosperity and hardship, respectively (Verschuren et al. 2000).

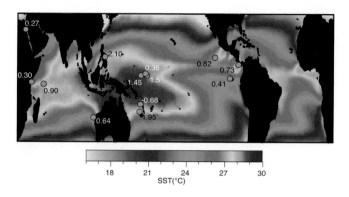

**Fig. 6.11.** Sites where annual coral isotope records span the interval 1895-1990. Numbers indicate the inferred SST trend in °C per 100 years, assuming all isotopic variability is due solely to SST changes and the slope of the SST-$\delta^{18}$O relationship is 0.22°C per 1‰. Site sensitivity issues (e.g. depth of coral, influence of rainfall) have not been taken into account in these calculations. Large central Pacific values are almost certainly due to the influence of rainfall on seawater isotopic content. Background colors indicate mean SST field.

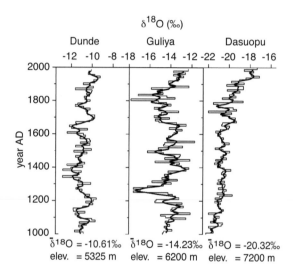

**Fig. 6.12.** Ice core records of recent isotopic changes in Tibet Unusually enriched values in recent decades at many locations may reflect exceptionally warm conditions.

On the other side of the globe, the strength of the trade winds at the Cariaco Basin (northwestern tropical Atlantic) has waxed and waned at the same time scale, recent decadal variability in this record correlates with both North Atlantic and eastern Pacific SST (Black et al. 1999). Lake sediment records from the Yucatan peninsula (Hodell et al. 2001) and inferred changes in riverine sediment input to the Cariaco Basin (Haug et al. 2001) also indicate marked century-scale oscillations in hydrologic balance. All of these studies point to solar variability as a possible cause for the multidecadal changes (Figure 6.13). High solar radiation corresponds to periods of Naivasha lowstand, weak Atlantic trades, and a drier Yucatan peninsula. The hydrologic responses are opposite those seen in an early Holocene speleothem record from Oman (Neff et al. 2001), which indicates monsoon intensi-

fication during high irradiance periods.

Model simulations of temperature change associated with solar forcing suggest some simple mechanisms that may help to explain these observations (Cubasch et al. 1997, Rind et al. 1999). Increased irradiance warms surface air temperature, and the correlation between solar forcing and temperature change is highest in the tropics and subtropics. However, the amplitude of the forced change is greatest at high latitudes, implying a reduced equator-pole temperature gradient when solar irradiance is high. Therefore, during solar irradiance maxima, weaker North Atlantic trades may result from the reduced latitudinal gradient, and the intensified monsoon in Oman could be driven by the enhanced land-ocean temperature gradient. Drier conditions in the Yucatan and Kenya during solar maxima may

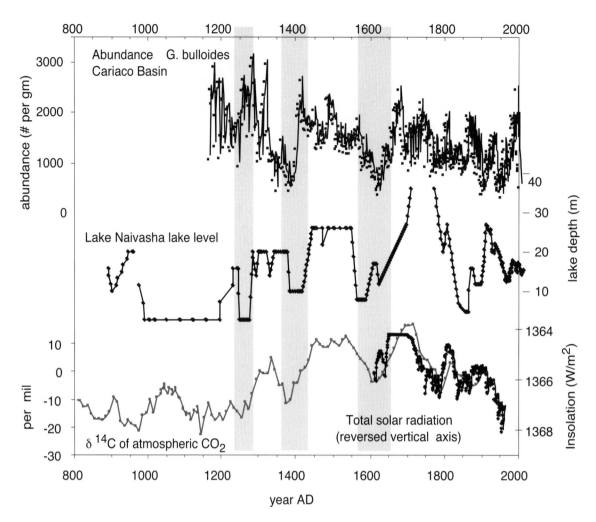

**Fig. 6.13.** Comparison of records of North Atlantic trade-wind strength (inferred from *G. bulloides* abundance at the Cariaco Basin, Black et al. 1999), Lake Naivasha level (inferred from sedimentological indicators, Verschuren et al. 2000), and solar radiation (inferred from the $\Delta\delta^{14}C$ of atmospheric $CO_2$ (Stuiver and Reimer 1993) and for the past 400 years from a reconstruction by Lean et al. (1995). Several of the multidecadal changes in these records are coincident (highlighted by grey bars), suggesting the possibility of a common response to radiative forcing on this time scale.

have resulted from the combination of a direct evaporative response to warming and potentially reduced moisture transports from weakened trade winds (both lie to the west of warm ocean basins under easterly trades). Additional well-dated, high-resolution records are needed to confirm and define the geographic dimensions of these relationships, and will undoubtedly add complexity to the preliminary interpretations we present here.

## 6.8 Hydroclimatic variability in western North America

There is an extraordinary wealth of natural archives of the last millennium's climate variability in western North America, including abundant ancient trees growing under climatic stress, geomorphological features associated with glacier activity

and lake level changes, lake and laminated marine sediments. This permits an unusual degree of cross-checking between completely independent natural archives. The annual resolution and century to millennial length of many of these records allows them to yield information on hydroclimatic fluctuations on all time scales from interannual to century and, in a few cases, millennial.

### 6.8.1 Tree ring networks

Fritts and his colleagues (La Marche and Fritts 1971, Fritts et al 1971, Fritts 1991) used a subcontinental network of tree ring width series to produce annual maps of seasonal climate features such as temperature, precipitation and sea-level pressure back to AD 1602. Their strategy was based on the predominance of moisture limitation as a limiting factor to plant growth in this mainly semi-arid region. These reconstructions then provided the basis for a detailed climatological study of the North American impacts of the El Niño-Southern Oscillation since AD 1602 (Lough and Fritts 1985), as well as of explosive volcanic eruptions (Lough and Fritts 1987), and an early cross-checking of natural archives and historical documents (Fritts et al. 1980). Other work, designed to reconstruct river flow on similar time spans (and focused even more strongly on moisture- sensitive trees), showed the early 20th century to have been anomalously wet on a 400-year time scale in the Upper Basin of the Colorado River (Stockton and Jacoby 1976), and provided rare evidence for a statistical association between the area of drought in the center of the continent and solar and lunar influences (Mitchell et al. 1979).

1746

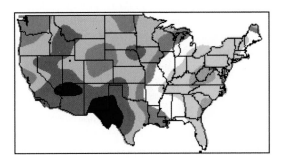

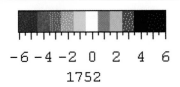

−6 −4 −2  0   2   4   6

1752

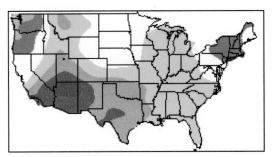

**Fig. 6.14.** Summer Palmer Drought Severity Index (PDSI) as reconstructed from a continental network of drought-sensitive tree ring width records (Cook et al. 1999). PDSI less than zero represents dry conditions. A.D. 1746 and 1752 were El Niño and La Niña years, respectively, as reconstructed by Stahle et al. (1998). These maps show that summer soil moisture conditions resembled those associated with the same phases of the ENSO fluctuation in the instrumental period.

A network of moisture-sensitive tree rings covering the whole of the lower 48 states of the USA (Meko et al. 1993, Cook et al. 1999) (Figure 6.14) has recently been used for a rigorous test of these findings (Cook et al. 1997). These last authors conclude that although they have not produced physical proof of a solar and/or lunar forcing of drought in the western United States, "the statistical evidence for these forcings appears to be strong enough to justify the continued search for a physical model...". They point out that, alternatively, the close-to-bidecadal drought area rhythm might be caused by unstable ocean- atmosphere interactions.

### 6.8.2 Interaction between time scales

In work focusing more narrowly on the region between the Rocky Mountains and the Pacific Ocean, from southern Canada to northern Mexico, Dettinger et al. (1998) used a dense network of moisture-sensitive tree ring records to show that the patterns of spatio-temporal variability in precipitation seen in the instrumental record are also evident in the previous two hundred years. They focused on variability in the 'ENSO' timescale, that is 3-7 years, and the 'interdecadal timescale', that is, more than 7 years. Similar patterns of intensity and distribution of precipitation were seen in the 18th and 19th centuries as in the 20th, on both time scales. They did, however, identify multi-decadal variations in regional precipitation that were present in the last 150 years, and before A.D. 1700, but not in the eighteenth and early nineteenth centuries (Dettinger et al. 1998). This feature is also seen in a reconstruction of precipitation in the Great Basin based on six very well replicated lower forest border bristlecone pine chronologies (Hughes and Funkhouser 1998), where fluctuations with a period of around 60 to 80 years appear clearly from AD 200 to 600, from 1300 to 1600, and then after 1850 (Figure 6.15). Minobe (1997) also detected similar oscillations over the past 400 years in tree ring reconstructions from western North America.

What might be the causes of such changes? It is of interest to note that the multi-decadal changes in this reconstruction in the late 19th and 20th centuries appear to correspond to major shifts in the Pacific Decadal Oscillation (PDO) identified by Mantua et al. (1997). The reconstruction of the PDO back to AD 1661 by Biondi et al. (2001), based on tree rings from southern California and northwestern Mexico, shows the same multi-decadal features since the late 19th century as the instrumental record (Figure 6.16). However, as in the work of Dettinger

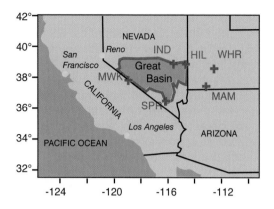

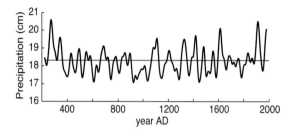

**Fig. 6.15.** Nevada Division 3 precipitation (July-June) reconstructed from a network of lower forest border stripbark bristlecone pine (after Hughes and Funkhouser 1998). The series has been smoothed with a 50-yr gaussian filter. 1 standard deviation unit equals 4.4cm, mean = 18.3cm. Map shows the location of tree ring sites (red + signs) and of Nevada Division 3 (green line).

et al. (1998), these are absent before the late 19th century. The first 200 years of the Biondi et al PDO reconstruction are marked by a strong bi-decadal fluctuation, uncorrelated with solar radiative forcing. This starts after the circa-1600 transition in frequency domain behavior seen in Figure 6.15. D'Arrigo et al. (subm), using tree ring width and density data from around the Gulf of Alaska, the northwestern conterminous U.S. and southwestern U.S. and northern Mexico to reconstruct the PDO since AD 1700, showed a different pattern. Their reconstruction showed a marked decadal mode of variation, that weakened in the mid-1800s, as also noted by Geladov and Smith (2001) and Villalba et al. (2001). These last authors saw this not only in tree ring chronologies in Northwestern North America but also in Southern South America. Villalba et al. (2001) also point out that the mid-1800s shift from a predominantly decadal to interannual pattern of variability means that "what we know of tropical Pacific variability through the analysis of instrumental records has been based largely on a period of predominant interannual variability". They suggest that there may have been a return to the pre-1850 decadal variability in the last two decades of the 20th century. The case for a strong

role for conditions in and over the equatorial Pacific in controlling hydroclimatic variability in western North America is further strengthened by the existence of decadal and multidecadal oscillatory modes common to the tree ring records from both hemispheres. The network of chronologies used by Villalba et al. (2001), which includes, *inter alia*, moisture sensitive records in North and South America, shares the same pattern of Pacific decadal sea surface variability over the past two centuries (Evans et al. 2001) as Sr /Ca ratios from coral at Raratonga (Linsley et al. 2000). This supports the proposition that "Pacific decadal variability is a basin-wide phenomenon originating in the tropics" (Evans et al. 2001). The predictability of the impacts of the ENSO phenomenon in extra-tropical America depends on the phase of the longer-term PDO variation (Gershunov and Barnett 1998), and thus great practical benefit would flow from a better understanding of Pacific decadal variability. Since its frequency domain behavior is not yet well known, it is likely that a more complete definition of the spectrum of variability will depend on natural archives such as those discussed here.

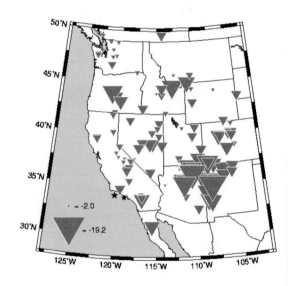

**Fig. 6.16.** Cumulative severity of A.D. 1561-1600 growth reduction in moisture-limited trees (from Biondi et al. 2000). The location of each symbol indicates the location of a tree ring width index chronology. These are expressed as dimensionless indexes with a mean of 1.0. The size of each symbol is proportional to the sum of all departures for the chronology over the period. The two symbols in the lower left of the map indicate the range of values on the map as percentages. Growth is reduced throughout this region in comparison to the long-term mean.

## 6.8.3 Extreme and persistent droughts and wet periods

Stahle et al. (2000) pointed out the extraordinary intensity and extent of drought in the middle and late 16th century, from central Mexico to northern Canada and as far east as the Atlantic coast. Grissino-Mayer (1996) identified the sustained drought at this time to be the most severe of the last 2000 years in the Four Corners region of the southwestern U.S. Biondi et al. (1997, 2000) identified major changes at the end of the 16th century in patterns of deposition of laminated sediments in the Santa Barbara Basin, off the central California coast, which are consistent with tree ring records that show a late 16th century dry period was followed by a series of unusually wet decades in the early 17th century (Figure 6.17). The post-1600 wet decades may well be the cause of the lacustrine event at 390 ±90 yr BP indicated by geomorphic data collected in the Mojave River Drainage Basin, southern California (Enzel et al. 1989). This probably coincides with a large flood in AD 1605 ±5 yr that is recorded as a silt layer in the sediments of the Santa Barbara Basin (Schimmelmann et al. 1998).

Giant sequoia tree rings from the western flanks of the Sierra Nevada show variations in the frequency of extreme single-year droughts over the last 2100 years (Hughes and Brown 1992, Brown et al. 1992, Hughes et al. 1996). The incidence of such droughts on a century time scale has varied more than threefold, with highest frequencies in the 3rd and 4th, 8th and 9th, and 15th and 16th centuries. The 20th century frequency was slightly below the 2100 yr mean. Not only tree rings show that there was a greater tendency for droughts to be intense and persistent between A.D. 400 and A.D. 1600 (Figure 6.15). Stine (1994) identified extreme low stands in Mono Lake, a closed basin on the California/Nevada border, that lasted from the early 10th century to the end of the 11th and from the beginning of the 13th to the middle of the 14th. These coincide with droughts seen in the rings of moisture-sensitive bristlecone pine from the neighboring White Mountains (La Marche 1974, Hughes and Graumlich, 1996) and in the Sierra Nevada (Graumlich 1993, Graybill and Funkhouser 1999). Similarly, the rate of change of $\delta^{18}O$ in sediments of Pyramid Lake (250 km north of Mono Lake) corresponds closely to a tree ring based reconstruction of streamflow in the mountains from which the lake is fed (Benson et al. 1999). These two records are, however, sometimes out of phase with those at Mono Lake., indicating a northward shift of storm tracks that resulted in periods of unusually high precipitation in the northern area at the same time as extended droughts prevailed around Mono Lake. Finally, there is evidence for a greater incidence of sustained, severe drought before approximately AD 1600 than since. Records that also support this observation are derived from sediments in Owens Lake, California (Li et al. in press), reconstructions of salinity in lakes on the northern Great Plains (Fritz et al. 2000, Laird et al. 1996), and sand-dune fields (Muhs et al. 1997).

In short, the paleoclimate record provides unequivocal evidence for droughts that far exceed anything in the 20th century, in terms of magnitude and persistence (Woodhouse and Overpeck 1998). A multidecadal drought today would have enormous economic and social impact, even in a well-off nation such as the U.S. The mechanisms by which these droughts are initiated and persist are not known, and therefore prediction is not yet possible. Intensified paleoclimate data collection, in conjunction with model experiments and modern process research, will be needed to document and understand this phenomenon.

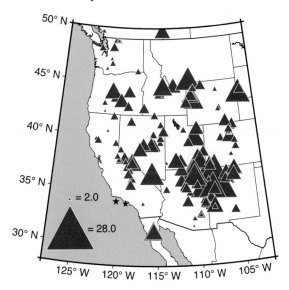

**Fig. 6.17.** Cumulative excess of A.D. 1601-1640 tree growth in moisture-limited trees (from Biondi et al. 2000). As Figure 6.16, except that growth is enhanced in comparison to the long-term mean.

## 6.8.4 Ecosystem impacts of climate variability

One unexpected by-product of the development of subcontinental networks of tree ring records has been the discovery of vastly accelerated growth rates in the last two or three decades of the 20th century of trees near timberline in the eastern parts of the American Southwest (Grissino-Mayer 1996, Swetnam and Betancourt 1998). These trees have been growing more rapidly recently than at any time in the last 1000 years, apparently as a result of an unprecedented period of springtime warmth coinciding with the wetter phase of interdecadal variability. The consequences of this acceleration on ecosystems are unknown.

Another complex set of climate-ecosystem interactions has been unraveled. Swetnam (1993) shows that long-term changes in fire frequency, revealed by dendrochronologically-dated fire scars in giant sequoia trees, may be related to century- scale temperature change, presumably through temperature effects on the composition and/or productivity of the forest (Figure 6.18a). There are also strong inter-annual fluctuations in fire frequency superimposed on this long-term variability. These are closely linked to precipitation and drought, which determine the quantity of fine fuels and the moisture content of the fuel in a particular fire season (Figure 6.18b). It would not have been possible to unravel these interactions without the development of high-resolution natural archives of climate and vegetation response. Such multi century histories of fire also have potential to be used themselves as natural archives of past climate (see Chapter 5).

## 6.9 North Atlantic region

The circum-North Atlantic region has yielded a unique combination of long instrumental climate observations, many documentary records and multiple sources of terrestrial and marine palaeoclimate information. A selection of these data is shown in Figure 6.19, together with two series that represent aspects of atmospheric and oceanic circulation variability. This figure provides a good perspective on the different types of evidence from which we can attempt to distill the spirit of how climate has varied across this region during the last 1000 years.

The Central England Temperature series (CET) spans the last 340 years and is the longest, continuous, homogeneous instrumental record in the world (Manley 1974, Parker et al. 1992, Jones and Hulme 1997). Mean annual temperature values are shown in Figure 6.19a. These are representative of temperature variability over much of western Europe (Jones et al. 2002). The CET can now be compared with a composite temperature record for the Benelux countries recently assembled from direct meteorological observations (for the period after 1700), weather diaries and journals, various account records (e.g. agricultural or river tolls), letters and early annals and chronicles (van Engelen et al. 2001). These data are plotted as a 'pseudo' annual series in Figure 6.19b, which is the average of the separate summer and winter index values that are

virtually complete from 1200 onwards.

The CET and Benelux records provide mutual confirmation of the unusual warmth of the 20th century, and particularly of the last 20 years, when viewed against the variability of the last 350 years. However, this warmth is most anomalous in the winter season, even when compared with evidence of unusual winter warmth in the Low Countries during the 11th century and again for a short period around 1200 (note that the earlier, more sparse data and the separate seasonal series are not shown here – but see van Engelen et al. 2001). The Benelux mean annual record and the early CET series clearly exhibit prolonged relative cold through much of the late 16th and the whole of the 17th centuries. This corresponds to what Lamb (1963) described as the 'pessimum period', or the worst part of the Little Ice Age. The early 1400s and early 1500s, the mid 1600s and most of the 18th century were periods of relative warmth. Figure 6.19c provides what is currently the best indication of the year-by-year variability of the winter North Atlantic Oscillation (NAO, see Chapter 3, Section 4.3) over the last 500 years. This was reconstructed on the basis of statistically defined associations between the NAO and a large number of predictors: mainly long surface pressure, temperature and precipitation records distributed over much of western Eurasia (Luterbacher et al. 2001, 2002). The reconstructed index is low during much of the late 16th and 17th centuries. It is high during the mid 19th and early 20th centuries, and in the early 18th and early 16th centuries, coinciding with distinct periods of warmth in England and the Benelux countries. While the NAO index is high also in recent decades, it has not reached the unprecedentedly high levels evident in the temperature records. The early decades of the 1400s were particularly warm in summer while in the mid 1700s the warmth was more in winter.

A long tree-ring-based reconstruction of individual summer temperatures for northwestern Sweden (Figure 6.19d – Briffa et al. 1992) again shows the 20th century to be generally warm, but no more so than the 11th, late 12th, the 15th and the early 16th centuries. The cold of the 17th century is striking in these data. It directly matches a period of high precipitation shown in a record of bog surface-wetness in western Ireland (Figure 6.19e), and one that is representative of general moisture trends over a wider area of the British Isles (Barber et al. 2000).

Moving to the other side of the Atlantic, Figure 6.19, graphs i-m illustrate a group of ice-core-derived proxies in Greenland and Eastern Canada: oxygen isotope ratios (Fisher et al. 1996) and the extent of snow melt that re-froze in the firn (Koerner and Fisher 1990, Fisher et al. 1995). Over the

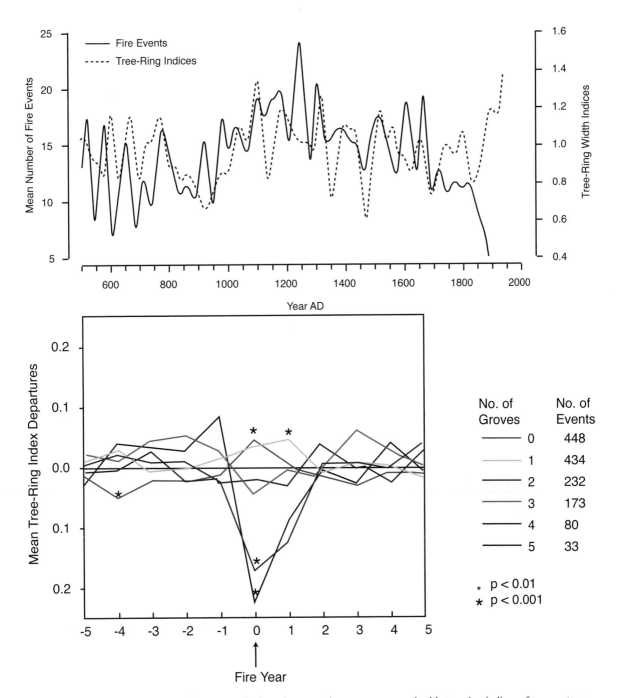

**Fig. 6.18. A.** 20-year running mean of fire events in five giant sequoia groves compared with tree ring indices of temperature-responsive bristlecone pine from near upper tree limit in the nearby White Mountains (from Swetnam 1993) **B.** Departures of tree ring width index (from A.D. 500-1850 mean) of precipitation-responsive bristlecone pine from near lowest forest border in the White Mountains, California. These are used as an index of regional drought, they are unaffected by fire in the nearby Sierra Nevada. They are shown for 5 years before and after fire in 0 to 5 giant sequoia groves in the Sierra Nevada. Small asterisks p<0.01, large asterisks p<0.001. The most extensive fires are clearly associated with drought years recorded by tree rings (from Swetnam 1993).

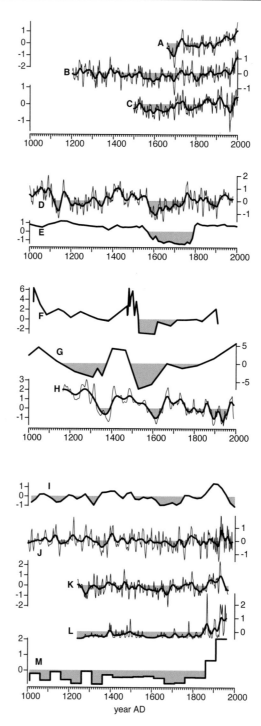

**Fig. 6.19.** Selected climate-related records for the last millennium around the region of the northern North Atlantic. All of the series are plotted as effective 10-year (thin line) and 50-year (thick line) smoothed and standardized values (with reference to the common base period 1659-1999). **A.** Central England mean annual temperatures (Manley, 1974) **B.** Pseudo annual temperatures for the Benelux countries (from van Engelen et al. 2001), **C.** Reconstructed winter North Atlantic Oscillation indices (Luterbacher et al. 2001), **D.** Warm season (A-S) temperatures reconstructed from tree-ring data (Briffa et al. 1992), **E.** Moisture index based on bog flora (Barber et al. 2000), **F.** SST from *Foraminifera* oxygen isotope composition (Keigwin 1996), **G.** An index of the speed of deep current flow, low values indicate a reduction in NADW production (Bianchi and McCave 1999), **H.** Foraminiferal abundance in the Cariaco Basin, indicative of trade wind intensity and possible changes in temperature in the North Atlantic (Black et al. 1999), **I.** Oxygen isotope data from the North Grip site ice core (Johnsen et al. 2001), **J.** Ice-core-derived oxygen isotopes from the GISP2 and GRIP sites (Johnsen et al. 2001), **K.** Several west Greenland ice-core oxygen isotope series (Fisher et al. 1996), **L.** High-resolution melt-layer data in an ice core from the Agassiz Ice Cap (Koerner and Fisher 1990, Fisher et al. 1995 ), **M.** A lower-resolution eastern Canadian ice-core melt record.

last millennium, these have a resolution and dating accuracy that is virtually annual, but individual core records provide indications of changing temperatures that are more reliable at decadal than yearly timescales. Appropriate averaging of data from different cores can reduce local non-climate 'noise' (Fisher and Koerner 1994) and give a stronger signal of local climate (White et al. 1997). The ice melt data provide a more direct indication of (summer) temperature than the isotopic data which have a complex association with regional temperatures, and also reflect changes in atmospheric circulation and shifts in the seasonality of snowfall over the core sites (Dansgaard et al. 1973).

A number of similarities can be discerned by comparing averaged or 'stacked' records from various west and central Greenland sites (Figure 6.19 k and j) (Fisher et al. 1996, Grootes et al. 1993) with a new single isotope record from northern Greenland (Figure 6.19i) (Johnsen et al. 2001) and melt data from the Agassiz ice cap in eastern Canada (Figure 6.19 l) (Koerner and Fisher 1990). There is general evidence of anomalous warmth in the last 100 years. There are indications of stronger and more recent warming in the melt data (the last 50 years) than in the isotope data, the latter show relative cooling in recent decades, as has been observed in Greenland instrumental records, especially in summer (Briffa and Jones 1993). The apparent magnitude of the early 20th century warmth is much greater in the northern Greenland and Canadian records than in the central or west Greenland data.

None of these records indicate decadal or multidecadal changes over the last 1000 years that exceed the mean temperature of the last century. However the data show that there was widespread warmth across eastern Canada and all of Greenland in the late 13th century and again in the late 19th and early 15th centuries. This accords with the warmth indicated by the west European documentary data and the Scandinavian tree-ring record. The middle 14th century was also cool on both sides of the Atlantic (see also Ogilvie and Farmer 1997). However, while the west Greenland evidence (Figure 6.19k) indicates cold conditions through most of the late 16th and the 17th centuries, in phase with western European evidence for the coldest part of the Little Ice Age, the central (and more northern) Greenland data show the cold phase continuing into the 18th century.

In trying to deduce the effect of the ocean on the patterns of temperature changes seen around the North Atlantic we are hampered by the lack of marine paleorecords that can be interpreted with the fineness of resolution that matches the terrestrial data over the last millennium. Figure 6.19g is a record of sediment grain size, a proxy for velocity of bottom-water flow, that can be used to infer the intensity of North Atlantic Deep Water (NADW) flow between Iceland and Scotland near 300 m water depth (Bianchi and McCave 1999). Although of low resolution, this record suggests that there was reduced NADW formation during the 13th century and in the 16th and perhaps early 17th centuries and enhanced production in the 11th and 15th centuries. These data also seem to indicate a strong trend towards enhanced NADW production over the last 200 years but the recent data are very sparse (see also Hansen et al. 2001).

The grain-size record can also be compared with a reconstructed low-resolution (~50 yr) sea-surface temperature history for the Bermuda Rise, at ~34°N 58°W in the Sargasso Sea (Figure 6.19f). This is based on statistically-calibrated counts of the abundance of temperature-sensitive *Foraminifera* in a sediment core (Keigwin 1996) and provides evidence of strong warmth at the beginning of the 11th century, and again in the 15th century, with cooler conditions in the 16th century. In these respects this sea-surface temperature history is similar to the inferred deepwater production record from the south of Iceland, despite the low resolution of both records, and it also shares the major characteristics of temperature change that are evident in the century-timescale features of the west European terrestrial data.

Keigwin and Pickert (1999) have suggested that sustained temperature anomalies such as the cold of the 18th century in Europe and the Sargasso sea region may have been times when the NAO remained in a predominantly negative phase (or the frequency of extreme negative years increased). Figure 6.20 illustrates the spatial association between the NAO and temperature and precipitation variability over the North Atlantic and adjacent land areas in the different warm and cold seasons. These patterns show that northeast Canada and southwest Greenland experience both colder and drier conditions, and northwest Europe is warmer and wetter, when the NAO index is very high, that is when the general westerly flow across the Atlantic is enhanced. At such times, winters are generally warmer and wetter in the southwest USA and over the western North Atlantic and noticeably drier on the Iberian peninsula and in the western Mediterranean.

In summer, though, associations are far less coherent. In high NAO years, there is no anomalous cold over eastern Canada and Greenland and the warm region over northern Europe covers much less of the area south of Scandinavia. Cold conditions over the Mediterranean are virtually absent, and the

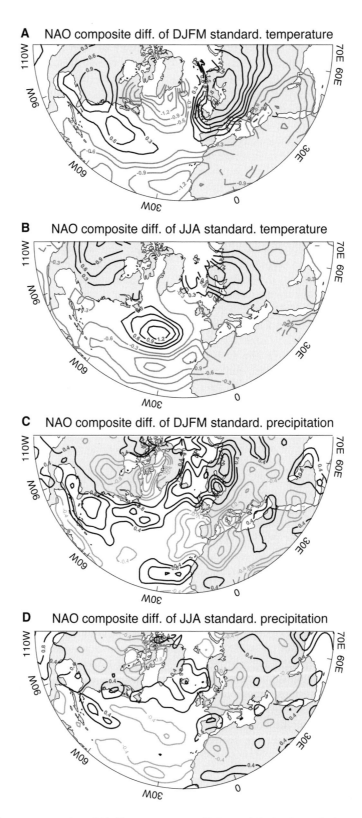

**Fig. 6.20.** Differences between composites of (**A**, **B**) temperature or (**C**, **D**) precipitation from (**A**, **C**) winters or (**B**, **D**) summers with positive North Atlantic Oscillation Index (NAOI) and those with negative NAOI. Seasonal temperature and precipitation were standardised to have zero mean and unit variance at each location prior to analysis. Thicker lines indicate positive anomalies.

warm band across the eastern United States and western Atlantic becomes two regions separated by an area with near normal temperatures. Moreover, although the pattern of precipitation associations in summer is similar to that in winter, the magnitudes of the correlations are so low as to be barely perceptible. This implies that the temperature evidence to support an NAO influence on major climate excursions would be found primarily in data that represent winter conditions, with variability that is out of phase on the two sides of the northern Atlantic.

The NAO index series (Figure 6.19c) shows a pronounced low phase in the late 1500s and 1600s that is coincident with cold in the Benelux record and in Fennoscandia (in summer), as well as with the sediment grain-size data and low Sargasso Sea surface-water temperatures.

The last record we shall discuss in this section is shown in Figure 6.19h. This shows high-resolution measurements of foraminiferal abundance in the anoxic, laminated sediments of the Cariaco Basin, off northeastern Venezuela (Black et al. 1999). These variations are interpreted as representing changes in upwelling intensity, which can be used as a proxy for trade wind strength. Stronger trade winds are associated with a southward shift in the ITCZ, which reduces precipitation in the southern Caribbean and enhances it in tropical South America. However, in analyses of recent instrumental data, it has been shown that stronger trades are also linked to cooler conditions in the North Atlantic, at least on an interannual timescale and probably (because of the absence of any lag) through an atmospheric rather than oceanic link. On the multi-decadal timescale it is hard to recognize any consistent associations between the proxy upwelling record and any of the northern temperature proxies. There is a generally positive correlation with the eastern records, and a negative correlation with the Greenland data in the post-1700 period, but these associations appears to be reversed before that date.

## 6.10 The Southern Hemisphere

Most of the evidence for climate variability in the southern hemisphere over the last millennium comes from tree rings, supplemented by a few ice cores and speleothems, as well as coral records that have already been discussed.

The major dendroclimatic resources of the Southern Hemisphere are found in southern South America, Tasmania and New Zealand. Sub-Saharan Africa has, so far, yielded only a few, relatively short, dendroclimatic records (Dunwiddie and LaMarche 1980, Stahle and Cleaveland 1997). LaMarche and

Pittock (1982) first demonstrated the potential of Tasmanian trees as climate records. Huon pine (*Lagarostrobus franklinii*) from a subalpine site in Tasmania has yielded a reconstruction of warm-season temperatures since 1600 BC (Cook et al. 1999b). As might be expected on an island exposed to such a vast expanse of ocean, the range of variability is smaller than in comparable Northern Hemisphere records. Although notable cool and warm periods of decade to century length were recorded during the first two millennia of this reconstruction, the last 1000 years were characterized by variability on shorter timescales. However, the last few decades of the 20th century have been marked by abrupt and major warming that is unique in the last millennium (Cook et al. 1992). The overall reconstruction contains consistent decadal to multi-decadal oscillations, which seem to be primarily associated with sea surface temperatures in a region extending west from Tasmania to South Africa, and in a region of the west Pacific north of the Equator. Extensive dendroclimatological work has been done in New Zealand (for example D'Arrigo et al. 1998, Norton and Palmer 1992, Ogden and Ahmed 1989, Salinger et al. 1994) and although there are species of great longevity, notably kauri (*Agathis australis*), millennial dendroclimatic reconstructions are still awaited.

Many dendroclimatic records, both temperature- and moisture- sensitive, have been developed for Argentina and Chile (Boninsegna 1992, Villalba et al. 1992). They include a 3620-year reconstruction of summer temperature from *Fitzroya cupressoides* at Lenca (Lara and Villalba 1993), which also shows reduced variability compared to Northern Hemisphere reconstructions. In this case it is clear that the reduced variability is in part the result of the standardization methods used to remove nonclimatic variation from the ring-width series. A new temperature reconstructiion, based on 17 millennia-length chronologies specially constructed to conserve low-frequency variability, shares many features with the Lenca series, but differs in showing a sustained period of cool summers between AD 1500 and 1650 (Lara et al. 1999). The long South American temperature reconstructions show decadal fluctuations rather similar to those in Tasmania (Villalba et al. 1996), but they do not show the sudden warming of recent decades. So far, however, the main connection that has been established between the South American and Australasian chronologies has been through reconstruction of a summer Trans-Polar Index (TPI) back to the mid-18th century, based on the pressure gradient between Hobart, Tasmania and Port Stanley (Falkland Islands/ Islas Malvinas) (Villalba et al. 1997). The anti-phase relationship between pressure at these two locations, or at least its interannual and high-frequency variability, was reconstructed as being stronger in the 20th than in the 19th century, and statistically associated with ENSO on 4-5 year time scales.

The establishment of relatively dense dendroclimatic networks in both North and South America has made possible a number of inter-hemispheric comparisons. Boninsegna and Hughes (2001), for example, showed a much clearer impact of explosive volcanic eruptions on temperatures in North America than South America, perhaps related to the "damping' effect of the extensive oceans of the Southern Hemisphere on such rapid changes. Elsewhere in this chapter we refer to the remarkable common patterns of decadal and multi-decadal oscillations between tree-ring chronologies from Northwestern North America and southern South America (Villalba et al. 2001), which raise fascinating questions about the mechanisms of inter-hemispheric climate linkage. These paleoclimatic records reveal robust teleconnections that can only be glimpsed using instrumental records, yet they have clear relevance to climate variability on time scales of social and economic interest.

Stalagmites from northeastern South Africa (Holmgren et al. 1999, 2001) provide a unique high resolution paleoenvironmental record from an area where few other records exist (cf. Tyson and Lindesay 1992). Variations in oxygen and carbon isotopes, and in coloration (reflecting humic acids washed into the deposit) all suggest that conditions were relatively cool and dry from ~A.D. 1500-1800. Warmer conditions prevailed from A.D. 900-1100 and ~A.D. 1400-1500. It has been estimated that the temperatures fell rather abruptly, by ~4°C, from the warmest to the coldest part of the last millennium (i.e., from ~1500 to the 1600s), with temperatures from A.D. 1500-1800 approximately 1°C below the mean for 1961-90 (Holmgren et al. 2001).

Ice core data from Peru (Quelccaya Ice Cap) clearly indicate a period of low oxygen isotope values (i.e. snow depleted in the heavier isotope, $\delta^{18}O$) from ~A.D. 1520-1880, interpreted as reflecting colder conditions (Thompson et al. 1985, Thompson 1992). It was relatively wet from A.D. 1520 until ~1720 (the wettest period in the interval A.D. 1450-1980), but dry thereafter, from ~1720-1860. Ice core records from Antarctica are quite variable (perhaps not surprisingly, given the area involved and the limited number of records) but generally lowest isotopic values were from the early 1600s until the early 1800s (Mosley-Thompson 1992, Peel 1992). At South Pole, the lowest values were in the 16th century, whereas at Law Dome,

nearer the coast in East Antarctica, the lowest values of the last 2000 years were in the late 1700s/early 1800s. Generally, these records fit with the far less resolved picture of glacier fluctuations in South America and the sub-Antarctic region, which show glacier advances during these colder centuries, with recession over the last ~150 years (Clapperton and Sugden 1988). By contrast, several ice cores from the Antarctic Peninsula show lowest oxygen isotope values in the 20th century, following a decline since the mid-19th century, in contrast to observed instrumental temperature observations which suggest warming over this interval (Peel 1992). This highlights the difficulties of using a simple isotopic/temperature interpretation in some locations where shifting moisture source regions, the presence of open water etc, make the climatic signal quite complex.

Overall, the limited data from the southern hemisphere suggest that conditions were generally colder during the interval A.D. 1500-1850, with different regions experiencing minimum temperatures at different times within this period. Not surprisingly in this highly buffered oceanic region, temperature changes appear to have been small, and in some cases changes in hydrological conditions related to shifts in atmospheric circulation may have been more significant than slight changes in temperature. Such changes are registered not only in terrestrial proxies, but also in the marine environment. For example, corals from the southwest Pacific indicate more saline conditions from ~A.D. 1700-1900, as a result of a stronger Hadley circulation and higher evaporation rates in the region (Hendy et al. 2002). This may be a reflection of stronger latitudinal temperature gradients during this interval of time.

## 6.11 Forcing factors: causes of temperature change in the last millennium

Temperature reconstructions for the last millennium (e.g. Figure 6.4) reveal three characteristic times-cales of variability: a long-term cooling trend, from A.D. 1000 to A.D. 1900 on which are superimposed multi-decadal and shorter-term (multi-year) anomalies. Here we consider some possible factors that may have been responsible for the observed changes, and also place those forcing factors in the context of the longer-term (Holocene) climate variations discussed earlier.

### 6.11.1 Orbital forcing

The pervasive influence of orbital variations (changes in precession, obliquity and eccentricity) undoubtedly played a dominant role in Holocene (and longer-term) climate variability (Figure 6.21). Overall insolation in the northern hemisphere summer season has declined over the last 10,000 years, by ~8% (at the top of the atmosphere) with a small increase in winter values. The effect of such insolation changes under the clear-sky conditions that commonly occur at high elevations, at high latitudes in the early summer (with 24 hour illumination) may explain why the melt record in Greenland and Ellesmere Island ice cores registers such markedly higher values in the early Holocene. The higher summer insolation values may also be responsible for enhanced monsoonal rainfall in continental Africa, though this effect has not yet been convincingly reproduced by GCMs with insolation forcing alone, surface water and soil moisture/vegetation feedbacks seem necessary to explain the magnitude of rainfall changes indicated by paleoclimatic data (e.g. Coe and Bonan 1997, Broström et al. 1998). Higher summer temperatures recorded by many proxy records from the early Holocene are a consequence of these higher radiation receipts. Even though the forcing mechanism for the higher temperatures was unrelated to greenhouse gases, there may be useful analogs in the environmental consequences of this warming that can illuminate how conditions may change in the near future (at least in areas relatively untouched by human activities) as a result of anthropogenic greenhouse gas increases.

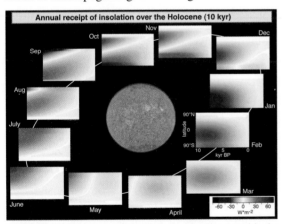

**Fig. 6.21.** Holocene orbital insolation anomalies (outside the atmosphere) versus month of year at 1000 year intervals. The key to each panel is illustrated in the February case.

On the timescale of the last millennium, orbital forcing is minor compared to the Holocene, but there has been a decline in July insolation of ~5 Wm⁻² within the Arctic circle over that time (Figure 6.22). In the Inter-Tropical zone, from January-June, insolation increased over the course of the millennium. Overall, the July insolation gradient from 30-60°N increased by ~3% over this interval

(which favors a slight equatorward displacement of the sub-tropical high pressure centers).

Orbital effects on insolation receipts were far less significant in the southern hemisphere, with the main features being an overall decline in insolation from August-November (especially in October) and a small increase in summer months (December-March).

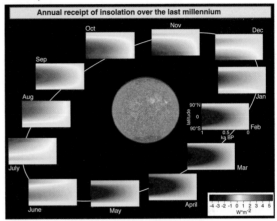

**Fig. 6.22.** Insolation anomalies (outside the atmosphere) for the past millennium, by latitude and month. The key to each panel is illustrated in the February case

### 6.11.2 Solar variability

Direct measurements of solar irradiance changes (TSI) from satellites only span the last two solar cycles but these data indicate a variation of ~ 0.1% in TSI over an 11 year solar cycle. Using long-term observations of the sunspot cycle, together with inferences based on the variability of sun-like stars, Lean et al (1992) estimated a change in TSI from the late 17th century/early 18th century solar minimum (Maunder Minimum) to the late 20th century of ~0.24%. To extend such a record further back in time requires the assumption that solar activity variations that affect cosmogenic isotope production in the upper atmosphere are diagnostic of (i.e. are proxies for) changes in TSI. If so, then the record of cosmogenic isotopes ($^{14}$C, $^{10}$Be) in trees and ice cores, respectively, can provide a long-term index of solar activity and TSI changes (Beer et al. 1996, Bard et al. 2000). Figure 6.23 shows the record of TSI as reconstructed by Lean et al. (1992) together with $^{14}$C and $^{10}$Be data for the past millennium. This suggests there was an overall decline in TSI from the early part of the last millennium to ~AD 1700, followed by an increase to higher levels (the highest of the last 1000 years) in the late 20th century. Analysis of northern hemisphere mean annual temperatures since A.D. 1600 and solar variations showed that the two records were strongly correlated over the period (Lean et al. 1995, Crowley and Kim 1996, Mann et al. 1998). Furthermore, numerous studies of regional climatic anomalies during the Maunder Minimum (~1675-1715) reveal the unusual nature of climate during that period, and its societal impacts (e.g. Borisenkov 1994, Wanner et al. 1995, Barriendos 1997, Alcoforado et al. 2000, Luterbacher et al. 2000, 2001, Xoplaki et al. 2001).

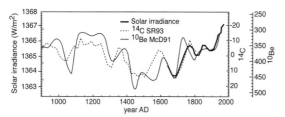

**Fig. 6.23.** The record of total solar irradiance variations as reconstructed by Lean et al. (1992) together with $^{14}$C and $^{10}$Be data for the past millennium.

The longer-term record of the $^{14}$C anomalies as recorded in the wood of precisely dated trees shows a number of periods with low solar activity, like the Maunder Minimum (Figure 6.24) which may be indicative of episodes of reduced TSI throughout the Holocene. Numerous studies have pointed out correlations between times of solar minima (so defined) and colder episodes, as recorded by a variety of proxies (e.g. Magny 1993, Stuiver & Braziunas 1993). Furthermore, there is growing evidence that solar forcing has played a role in modulating Holocene precipitation variability in the tropics, from northern South America and Yucatan (Black et al. 1999, Hodell et al. 2001, Haug et al. 2001) to east Africa and the Arabian Peninsula (Verschuren et al. 2000, Neff et al. 2001). Indeed, solar variability may also have played a role in mid-continental drought frequency on both short and long timescales (Cook et al. 1997, Yu and Ito 1999). Stuiver et al. (1991) have also noted the similarity between a strong 1470 year periodicity in $^{14}$C data and a similar periodicity in isotopic data from GISP2. This is close to the ~1500 year quasi-periodic record of ice-rafted debris in the North Atlantic which has led to speculation that such events are in some way driven by TSI variations affecting the climate system. However, no obvious mechanism for such linkages has been articulated, and oceanic circulation changes may have played a role in both the IRD variations and the $^{14}$C anomalies.

### 6.11.3 Volcanic forcing

Short-term changes in temperature are quite common in the time series of hemispheric mean temperature for the last millennium. There is compelling evidence that these result from gases and aerosols, ejected to high elevations during explosive

volcanic eruptions, reducing radiation receipts and leading to lower temperatures (Bradley 1988, Robock 2000). Several studies of tree ring records clearly demonstrate the effects of such changes on tree growth (Briffa et al. 1998, Boninsegna and Hughes 2001). Figure 6.25 shows reconstructed region-wide mean temperatures for North America and Europe since A.D. 1760. This period has sufficient proxy data to enable nine eigenvectors of temperature to be reconstructed, providing reliable information on the spatial patterns of climate variability at these regional scales (Mann et al. 2000). Of particular note is the sequence of years 1834-1838. 1834 was the warmest year in Europe over this period, with overall temperatures ~0.7°C above the 1902-80 mean. However, in the next four years, temperatures systematically fell to the coldest year of the entire record – 1838 (~0.82°C below the 1902-80 mean) (Figure 6.27). For marginal agricultural societies, this sequence of cold years was catastrophic. In the central mountains of Norway, for example, harvests were damaged by early frosts, leading to starvation (Nordli 2001) and in Japan there was widespread famine in 1836 (T. Mikami, personal communication). Similarly, dramatic cooling also affected North America (where 1837 was the coldest year of the last ~250 years). This rapid change was due to radiation and circulation anomalies associated with the explosive eruption of Coseguina in January 1835 (Bradley and Jones 1992). Such unpredictable events clearly have a major impact on temperatures over extensive parts of the globe, yet even the largest of these eruptions was small compared to numerous late Holocene eruptions recorded in Greenland ice cores (Zielinski et al. 1994, Zielinski 1995, 2000). If similar magnitude eruptions were to occur today, the rapidity of temperature change would likely have a devastating effect on society, even in a warmer "greenhouse world".

### 6.11.4 Internal "forcing" factors

All of the "external" forcings discussed above are superimposed on, or modulate "internal" variations of the climate system (Mann and Park 1996, Mann et al. 1995, Trenberth and Hurrell 1994, Gershunov and Barnett 1998, Delworth and Mann 2000). In previous sections, we have discussed the record of two such "internal" modes of climate variability – ENSO and the NAO. Here we consider variations in the North Atlantic thermohaline circulation.

### 6.11.5 Thermohaline circulation changes

Ice core and marine sediment records reveal that during the last glaciation there were rapid changes

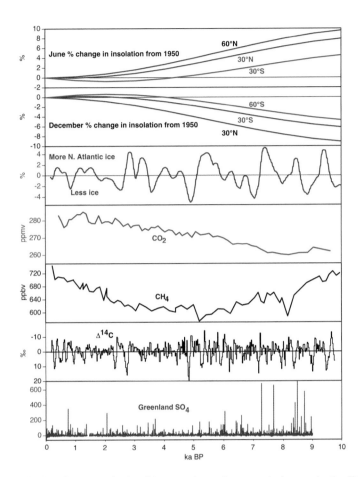

**Fig. 6.24.** An overview of forcing factors and related large-scale environmental changes in the Holocene. Top to bottom: Insolation anomalies at the top of the atmosphere (as percent departures from A.D. 1950 values) for mid-December and mid-June, for selected latitudes (from Berger and Loutre, 1991), an index of ice rafting in the North Atlantic, based on percentage of ice-rafted grains in ice cores from two sites (data stacked, smoothed and detrended) (from Bond et al. 2001), $CO_2$ in ice (ppmv) from Taylor Dome. Antarctica (from Indermühle et al. 1999), $CH_4$ in ice (ppmb) from the GRIP core site, Greenland (from Blunier et al. 1995), $^{14}C$ anomalies after removing low frequency geomagnetic field variations. Note data are plotted inversely, positive departures imply reduced solar activity and by inference, less irradiance (from Stuiver et al. 1991), volcanic sulfate in the GISP2 ice core from Summit, Greenland, expressed as residuals after removing low frequency variations with a spline function (from Zielinski et al. 1994). Many of these data sets were obtained from the NOAA World Data Centre for Paleoclimatology (Http://www.ngdc.noaa.gov/paleo/data.html)

in the climate of the North Atlantic region, and these have been linked to the cessation of deep-water formation and a concomitant reduction in the influx of warm sub-tropical water in the Gulf Stream and North Atlantic Drift (Chapter 3). Can such changes, albeit on a much reduced scale be responsible for the observed climatic changes in the late Holocene? Alley & Ágústsdóttir (1999), using GISP2 isotope data, argue that there has been a general decline in oceanic heat transport to the North Atlantic over the late Holocene. Their argument rests on an interpretation of seasonal temperature and accumulation trends and "expected" $\delta^{18}O$/temperature relationships. However, a reduction in North Atlantic heat transport would have led to colder and drier wintertime conditions,

yet Greenland accumulation data shows no Holocene trend (Meese et al. 1994).

Broecker et al (1999) also argue for late Holocene changes in deepwater formation, but focus attention on the Southern Ocean. Modern geochemical observations of oceanic phosphate, CFC-11 and $^{14}C$, appear to be incompatible with the notion of an invariant deep sea ventilation rate. These data can be reconciled if deepwater formation in the Southern Ocean was far higher for several centuries (up to ~800 years) prior to the 20th century. Since overall deepwater formation does not seem to have varied, this implies a parallel reduction in North Atlantic deepwater formation over the same interval, though this interpretation is not supported by WOCE (World Ocean Circulation Experiment) observational data (Ganachaud and Wunsch 2000). Nevertheless, Broecker et al. hypothesise that there may have been a reduction in North Atlantic deepwater formation (and a concomitant reduction of poleward heat transport by the ocean) during the LIA, with opposite conditions during the MWE resulting from an oscillation between deepwater production in the North Atlantic and the Southern Ocean. If such an oscillation has persisted over time, it might explain the ~1500 year periodicity in North Atlantic ice-rafted debris noted by Bond et al. (1997) and the LIA may have just been one of many oscillations within the Holocene, driven either by internal ocean system dynamics, or quasi-regular external forces not yet resolved

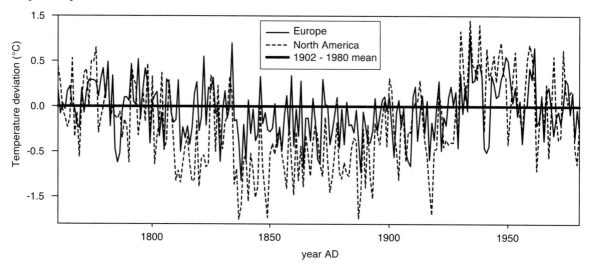

**Fig. 6.25.** Reconstructed mean annual temperatures for North America and Europe since A.D. 1760 (data from Mann et al. 2000).

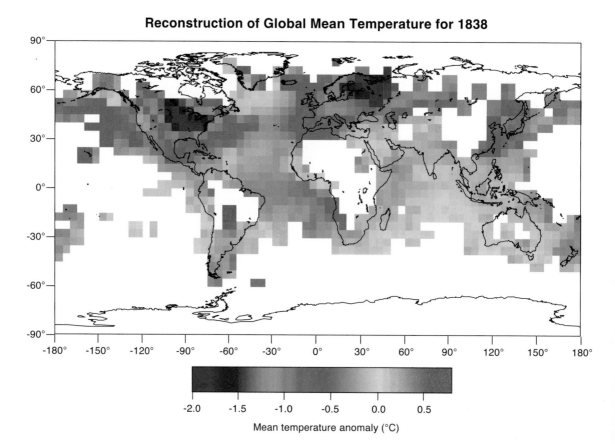

**Fig. 6.26.** Reconstructed mean annual temperatures in 1838 relative to the mean for 19202-1980 (data from Mann et al. 2000).

## 6.12 Anthropogenic and natural climate forcings over the past millennium: model results

Present and future global temperature reflects the combination of both anthropogenic forcings and natural ones such as solar irradiance and volcanic aerosols. Early work documented that solar forcing was insufficient to account for the temperature changes in the instrumental record (e.g. Kelly and Wigley 1992, Schlesinger and Ramankutty 1992). Later papers have refined this conclusion by using the paleoclimate record to assess climate sensitivity to radiative changes using correlation analysis (Lean et al. 1995, Bard et al. 2000). Climate model simulations of the recent past, driven by changing radiative forcings, yield temperature records that can be compared to paleo data and used to quantify the sensitivity of climate to radiative change. Refining existing estimates of sensitivity will require improvements on all sides of this question (models, data, and forcings). Model studies have proven very useful in attributing temperature change over the past 100-1000 years to specific forcings. The bottom line from virtually all such investigations is that although solar and volcanic forcing played an important role in pre-industrial climate variability, such natural forcings can explain at most about one-third of the warming of the 20th century, and hence anthropogenic forcings must have played a dominant role in the late 20th-century temperature rise. Here we describe recent modeling efforts to attribute temperature changes seen in paleoclimatic records to specific forcings, and to use these data-model comparisons in assessing climate sensitivity to changes in radiative forcing.

Energy-balance models (EBMs) provide a computationally efficient way to address the response of global temperature to radiative perturbations. EBM simulations of the temperature response to radiative forcings agree well with results from coupled GCMs on hemispheric and global scales (Gates et al. 1995, Raper and Cubasch 1996). However, the sensitivity of an EBM to radiative perturbations is adjustable, the results discussed here therefore can not be used to assess sensitivity, but they do give information on the relative roles of different radiative forcings.

A transient EBM experiment incorporating time-varying solar forcing over the past four centuries (Crowley and Kim 1996) found that solar variability correlates well with pre-industrial decadal-centennial temperature changes but cannot explain 20th century warming. A subsequent EBM analysis combined solar with volcanic and $CO_2$ radiative forcing, and explored the sensitivity of the exercise to different forcing time series (Crowley and Kim 1999). Using new multiproxy and tree ring temperature reconstructions for the past 600 years, they found that between 18-34% of observed variability was forced by the solar and volcanic changes, the residual variability was spectrally similar to that produced by long unforced GCM runs.

D'Arrigo et al. (1999) diagnosed the response of an EBM to various solar, volcanic and anthropogenic (trace gases plus aerosol) forcing scenarios over the 1671-1973 interval. They compared these results with a tree ring index of northern hemisphere temperature to identify the most successful forcing parameterizations and combinations. The simulations that explain the most temperature variability include anthropogenic forcings, the Dust-Veil Index of volcanic inputs, and any one of four solar reconstructions, these simulate 59-66% of the variance in temperatures derived from instrumental or tree ring data. The model generated a range of variability slightly greater than seen in the data -- cool intervals associated with extreme volcanic eruptions were stronger in the model output, and recent warming was overpredicted. A subsequent study (Free and Robock 1999) used the same model and a similar set of anthropogenic, solar, and volcanic forcings (the latter updated to reduce the impact of strong events) to explore in greater detail the relationship of these forcings to temperature reconstructions of the past 300-400 years. The updated volcanic indices improved the correlation between simulated and observed temperature, volcanism and solar variability had comparable influence on modeled temperatures in this study.

Crowley (2000) used an EBM forced with changing solar, volcanic, anthropogenic greenhouse gas, and tropospheric aerosol forcings, to simulate global temperature change of the past 1000 years. Solar and volcanic inputs explained 41-64% of pre-anthropogenic temperature variations, but anthropogenic radiative forcing had to be included to simulate the dramatic warming of the 20th century. Overall, this exercise suggests that 41-59% of decadal and longer-term temperature variability can be explained as a linear response to solar and volcanic forcings. This model also overestimated 20th-century warmth. The EBM sensitivity was set to 2°C for a doubling of $CO_2$, and the long-term temperature history agreed with reconstructed temperature histories (Mann et al. 1999, Crowley and Lowery 2000). Naturally, this agreement depends on the sensitivity prescribed – if a 4°C sensitivity had been prescribed, then the long-term temperature trend would be more in agreement with the larger values implied by borehole data (see discussion in Section 6.3).

General circulation models incorporate additional elements of the climate system and allow assessment of spatial patterns of climate change, as well as changes in climate other than temperature (precipitation, circulation, etc). An early GCM study adopted the strategy of imposing a fixed forcing (e.g. a different solar irradiance or volcanic aerosol loading) and assessing the equilibrium response to this forcing (Rind and Overpeck 1993). These authors found that climate would respond to radiative perturbations of reasonable (i.e. observed) magnitudes. Recent work with the HADCM2 model (The Hadley Centre's Second Generation Coupled Ocean-Atmosphere GCM) includes solar, volcanic, and anthropogenic (GHG and aerosol) forcing in assessing the causes of 20th century climate change. Spatial signatures of each forcing were developed and linear combinations of these were calculated to provide the best fit to observations, using optimal fingerprinting techniques (Tett et al. 1999, Stott et al. 2001) and linear correlation analyses (Johns et al. 2001). These studies disagree on the relative role of solar forcing in 20th century warming, depending on methodology and data source, it is either absent or minor. However, they concur that late 20th century warming is dominantly a result of anthropogenic forcing. A follow-on study with HADCM3 (Stott et al. 2000), using the same forcings in a transient mode, supports this conclusion and provides an estimate of the spatial pattern associated with both modeled and observed trends.

The 20th century is a relatively short period for developing confident assessments of the relationship between climate forcing and change, particularly as two of the leading forcings, greenhouse gases and solar irradiance, show positive trends, and solar variability contains substantial low-frequency variability. The combination of multi-century paleoclimatic reconstructions and forced multi-century GCM runs will be a powerful tool for attributing the causes of past change and diagnosing natural climate sensitivity. Such simulations are not yet commonplace, particularly with coupled ocean-atmosphere GCMs, but work in this direction is beginning. Cubasch et al. (1997) used the ECHAM3/LSG coupled ocean-atmosphere model to assess the climatic response to changes in solar forcing (from Hoyt and Schatten 1993) since AD 1700. They found that the 0.35% variations in the solar input time series (peak to trough) produced global mean temperature changes of about 0.5°C, with a pattern of stronger land-sea contrast resulting from increased irradiance. This direct response lagged the forcing by <10 years and was especially clear on the multi-decadal time scale. An inverse response of temperature to solar forcing was also noted in the North Atlantic temperature and thermohaline circulation, high irradiance led to weaker meridional circulation and cooler near-surface temperatures. They concluded that the warming simulated in response to solar variability contributed to 20th century warming, but could not explain it alone, and the pattern of solar warming differed from that generated by anthropogenic forcings in the same model.

Rind et al. (1999) used a slightly more conservative scenario of solar variability (Lean et al. 1995, extended to A.D. 1500 using cosmogenic isotope data) as input to transient simulations spanning the past 500 years. They used an atmosphere-only GCM coupled to a mixed layer ocean that included specified heat transports and diffusion of heat through the bottom of the variable-depth mixed layer. Analysis of the time-dependent temperature variability suggests that temperature response changes with latitude, that responses at different time scales have a preferred spatial expression, and that feedbacks are entrained at preferred time scales. They simulated a similar temperature change as the previous study (0.5°C warming from Maunder minimum to present), but the agreement was fortuitous: in Rind et al., the model was more sensitive, and the solar forcing less extreme. Their analyses indicated that even with a range of potential initial conditions, ocean heat uptakes, and initial solar conditions, the recent warming can not be explained by solar variability alone.

Robertson et al. (2001a) used the same model as Rind et al. (1999), but included a more complete set of radiative forcings. In addition to the extended Lean et al. (1995) solar record, they forced the model with a new, zonally resolved volcanic sulfate record (Robertson et al. 2001b) as well as increasing concentrations of $CO_2$, $CH_4$, and tropospheric aerosols (the latter scaled to greenhouse gases on a regional basis). The resulting global temperature history suggests that pre-industrial temperatures were 1.5°C cooler than present. This value agrees with the borehole temperature reconstruction (Huang et al. 2000) but is substantially greater than that indicated by various tree ring based and multiproxy reconstructions (Mann et al. 1999, Briffa et al. 1998, Jones et al. 1998, Crowley and Lowery 2000). On the other hand, the simulated interannual-decadal variability agrees closely with the Mann et al. multiproxy temperature reconstruction. Robertson et al. (2001) argued, however, that the paleoclimatic records used in Mann et al. (1999) may systematically underestimate low-frequency variability, even while capturing the high frequency changes accurately.

The GISS GCM has a sensitivity of 4.2°C for a

doubling of $CO_2$, which places it among the more sensitive of the models used in the IPCC assessment (Houghton et al. 1995, IPCC 2001). The agreement of the Robertson et al. (2001a) transient simulation and the Huang et al. (2000) data supports this high sensitivity. If, however, the multiproxy and tree ring records are correct in yielding a smaller estimated temperature difference, then the model would appear to be overly sensitive. To address the issue of climate sensitivity to increasing greenhouse gases alone, Robertson et al. (2001a) focused on the 1880-1930 interval, when greenhouse gases were rising but solar irradiance was stable. The degree of temperature change over this interval was consistent among the simulations and the observed datasets (including instrumental data) and thus supports the higher sensitivity of the GISS model. It should be noted, however, that the Hoyt and Schatten (1993) solar reconstruction shows an increasing trend over this interval which – if true – would reduce the greenhouse gas sensitivity estimated by Robertson et al. in this exercise. This study clearly highlights the need to improve the accuracy of both temperature reconstructions and forcings over the past several centuries.

Transient GCM studies suggest spatial and temporal fingerprints of solar forcing that can be useful in diagnosing the variability seen in paleoclimatic records. In both the Rind et al. and Cubasch et al. studies, the correlation of solar variability and temperature was strongest in the tropics with a lag of several (5-10) years, even though the amplitude may have been higher in the extra-tropics. The frequency-domain properties of the solar irradiance record vary through time, and those time scales were expressed in the temperature record differently at different latitudes. For example, in the Rind et al. study, 50-80 yr period variability was strongest in the tropics and 20 year variability persisted at all latitudes (even when it disappeared from the forcing!). Neither study found significant 11-year power in the simulated global temperature, likely because the modeled oceanic response time muted the response to forcing at this frequency. Oceanic regions showed a lower correlation to anthropogenic greenhouse forcing and a stronger correlation to solar forcing. These simulations thus provide clues as to where solar forcing may appear most strongly in paleoclimate records (such as the tropical oceans), as well as possible explanations for the absence of solar signals (e.g. the commonly sought-after 11-year periodicity).

## 6.13 Detecting twentieth century climate change

The detection of 'significant' trends, and being able to attribute them to particular climate forcings, requires knowledge of the expected climate signal and an estimate of natural climate variability (e.g. Santer et al. 1995, Tett et al. 1999). Almost all climate change signal detection and attribution studies to date assume a climate-model-based estimate of natural climate variability. This is a major and relatively untested assumption, and is a source of potential criticism that could be used to detract from all such studies (Bradley et al. 2000). Proxy-based reconstructions of past climate can be used to test this assumption, by providing information about pre-twentieth century variability which can be assumed to have been influenced relatively little by anthropogenic forcing.

There are two distinct approaches that can be followed. The first is to compare the model-based variability estimates (obtained from multi-century control simulations with constant external forcing, or from model integrations under natural forcing changes only) with the reconstructed past climate variability, in terms of levels of variability (standard deviation) or patterns of variability. This may be done at various time scales, encompassing the decadal to century time scales that are most important for climate change detection. This approach has been followed by, for example, Jones et al. (1998) who found different magnitudes and patterns of multi-decadal variability between climate model output and paleodata (but a similar lack of global-scale coherency in both data sets). On the other hand, Collins et al. (2000) found that one climate model simulation exhibited levels of interdecadal variability of summer temperatures across the Northern Hemisphere that were very similar to those reconstructed from tree ring density, unless the tree ring density data were standardised to maintain maximum multi-century variability (Briffa et al. 2001). Thus, so far such comparisons are somewhat inconclusive. Even when model-based variability appears to differ greatly from that derived from paleodata, uncertainty in the paleoclimate reconstructions may be so large that the difference cannot be considered significant. What do we then conclude about the detection studies that have used these model estimates of natural variability?

The second approach circumvents some of these problems. Rather than use the paleoclimate reconstruction to evaluate the veracity of the climate model simulation, it is possible to carry out the same detection study using either the model-based estimate of natural variability or the proxy-based estimate. If the climate change signal is detected significantly in both cases, then it does not matter if the model and proxy estimates are inconsistent.

An example of the second approach is given in

Figure 6.27, (based on Osborn and Hulme 2001) where recent warming trends are compared against the range of trends that have occurred naturally. We base our estimates of natural variability on the AD 1000 to 1900 section of the Mann et al. (1998, 1999) Northern Hemisphere (NH) mean annual temperature reconstruction, and the NH annual temperature simulated in a 1400-year control simulation of the UK Hadley Centre's climate model HadCM2 (Tett et al. 1997). These are compared with observed NH annual temperatures since A.D. 1901. In addition, we compare the observed warming with that simulated by the HadCM2 model, run with prescribed increases of greenhouse gas and sulphate aerosol concentrations (representative of the period from A.D.1901 to present). If we compare twentieth century mean temperature anomalies with the range of mean temperature anomalies occurring in the natural variability estimates, then we find the twentieth century to be very unusual (as already noted in Section 6.3). We do not compare *means* in this example, however, because that would require that the paleoclimate reconstruction is very reliable and homogeneous over multi-century to millennial time scales, a requirement that is both demanding and not easily verified. Instead, we consider in Figure 6.27 decadal to century *trends* of temperature, thus decreasing the reliance on obtaining accurate multi-century to millennial trends. For each of our two natural variability estimates, the distribution of all possible trends of a certain length (ranging from 10 to 100 years) is estimated, and Figure 6.27 shows the 95th percentile (°C/decade) of each distribution. The model and paleo-based natural variability estimates are in excellent agreement for trends longer than 40 years, though the model overestimates the strength of shorter trends relative to the paleoclimate record. This is not a like-with-like comparison, however, because (i) any natural forcings, e.g., volcanic or solar, are absent from the model simulations (and would probably increase the simulated trends at decadal and century time scales, respectively), while (ii) the paleoclimate record has lower variance than observed (because it is does not fully capture the variance of the observed record). If we assume that the residual variance in (ii) can be modeled as a white noise process (e.g. Collins et al. 2000), then this would raise the 95th percentile of the trends, principally at shorter time scales, thus bringing the paleo and model estimates into closer agreement. Mann et al. (1999) show that this white noise assumption is valid for their reconstruction after AD 1600, but before this, their reconstruction does less well at capturing low-frequency climate variability and the residuals are likely to be autocor-

related. Taking autocorrelated residual variance into account would raise the 95th percentile of the proxy-based trends for all time scales.

It is clear, therefore, that the model- and proxy-based estimates of natural temperature trends are quite similar, though some uncertainties need to be taken into account. But do the small differences matter? Figure 6.27 also shows the temperature trends, computed separately from the observed and simulated twentieth century data, over various length periods, all ending in 1999. With the exception of the decadal time scale, all trends lie above the 95th percentile of 'natural' climate variability – indicating a significantly unusual rate of recent warming, regardless of which estimate of natural variability is used.

This approach to detecting significant warming has been applied to other paleoclimate reconstructions, ranging in spatial scale from hemispheric to single site estimates (Hulme et al. 1999, Osborn and Hulme 2001). All hemispheric or quasi-hemispheric cases yield significant warming for recent trends of 30 years and longer, despite larger differences between model and proxy-based natural variability estimates in some cases. For localised sites, however, trends are typically similar to, or below, the 95th percentile of natural variability, thus precluding detection of unusual climate change at small spatial scales.

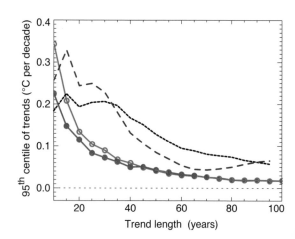

**Fig. 6.27.** Detection of significant 20th century temperature trends of varying length (all expressed as °C/decade). The purple line shows observed temperature trends from various length periods. The black line is the mean of an ensemble of four simulations from the HadCM2 coupled climate model forced by historical increases in greenhouse gas and sulphate aerosol concentrations. These can be compared against the estimates of the 95th centile of the various length trends that are possible due to natural climate variability. The green line is computed from a 1000-year control integration of HadCM2, with fixed external forcing. The red line with solid dots is computed from the pre-1900 portion of the 1000-year Mann et al. reconstruction.

## 6.14 Concluding Remarks

A wide range of proxies for past climate provide an invaluable long-term perspective on global climate variability, and proxies of past forcing allow natural factors affecting climate to be evaluated. Together, these records indicate that recent warming is both unusual and not explicable in terms of natural factors alone. By combining model simulations with paleoclimatic data, a better understanding of climate sensitivity and the climate system response to forcing is emerging. Nevertheless, many uncertainties remain. Paleoclimate research has had a strong northern hemisphere, extra-tropical focus (but even there the record is poorly known in many areas before the 17th century). There are very few high resolution paleoclimatic records from the tropics, or from the extra-tropical southern hemisphere, which leaves many questions (such as the nature of climate in Medieval times) unanswered. Furthermore, much remains to be learned about the mechanisms by which climate is recorded in proxies. Even so, proxy climate records have revealed important, hitherto unknown, features of climate variability, including regime-like fluctuations in the amplitude and frequency of the ENSO phenomenon and its extra-tropical teleconnections, interhemispheric connections over the Pacific Basin, extreme multidecadal droughts in mid-latitudes, and a confounding role for the North Atlantic Oscillation in temperature changes in that basin over the last century. Even so, variability of other major climate systems, such as the monsoons, is poorly documented. Our understanding of regional responses to forcing and how one part of the climate system may lead, or lag, another remains poor. Much more work on these topics remains to be done.

Climate variability over the last millennium provides the essential context for assessing future changes, even as anthropogenic effects become increasingly dominant. It is over the last millennium that modern societies have developed, coping with a wide range of climatic vicissitudes. Much of the world still lives at a subsistence level, very much affected by both inter-annual and inter-decadal climate variability. As we confront a world whose population is expected to increase from 6 billion people today to ~9-10 billion by 2060, paleoclimatic research can shed an important light on mechanisms of climate variability at these societally-relevant timescales.

# The Role of Human Activities in Past Environmental Change

F. Oldfield
University of Liverpool, PO Box 147, Liverpool L69 3BX, United Kingdom

J.A. Dearing
Department of Geography, University of Liverpool, PO Box 147, Liverpool L69 3BX, United Kingdom

## 7.1 Introduction

A crucial task for modern environmental science is to document and understand the ways in which human impacts on the earth system interact with other processes of global change. Such understanding is an essential prerequisite for establishing the consequences of further population growth and increased economic activity, with all their implications in terms of higher demands on energy, water and a wide range of resources, both renewable and non-renewable. Most of the research in this field draws on a combination of methodologies, for example remote sensing, environmental monitoring, experiments, large scale observation programs and modeling, all of which rely on a relatively short time span of empirical knowledge - usually a few years or decades at most. The purpose of the present chapter is not simply to consider past human impacts on the environment on longer time-scales but to explore the extent to which the longer time perspective contributes to an enhanced view of potential future changes and impacts.

Writings that highlight the potentially damaging effects of human activities on the environment have grown exponentially during the last decades, but have only relatively recently gained both a quantitative dimension and a historical perspective. We now realise that the time-span of human impact on the environment, at least at the regional level, ranges over millennia and not merely the last two centuries of industrialisation. The story of past human impacts, their interactions with climate variability and the human consequences of these interactions thus forms part of the essential context within which to evaluate present day trends and likely future consequences.

Much of the history of human impact on environmental systems discussed here is concerned with those aspects of global change that are, in terms of the distinctions made by Turner et al. (1990) 'cumulative' rather than 'systemic' (Table 7.1). Although this distinction is not entirely unambiguous, it implies that, in contrast to elevated atmospheric concentrations of $CO_2$ and $CH_4$ (Chapter 2), and their potentially systemic effects on global climate, most of the impacts discussed in this chapter achieve global significance by either the widespread nature of their effects or their cumulative magnitude. A familiar example is the eutrophication of freshwater ecosystems.

Table 7.1. Types of global environmental changes. Modified from Turner et al. (1990).

| Type | Characteristic | Examples |
|---|---|---|
| **Systemic** | *Direct impact on global system* | a) Industrial and land use emissions of greenhouse gases |
| | | b) Stratospheric ozone-depleting gases |
| | | c) Land cover induced changes in surface albedo |
| **Cumulative** | *Impact through worlwide distri- bution of change* | a) Groundwater pollution and depletion |
| | | b) Species depletion/ genetic alteration |
| | *Impact through magnitude of change* | a) Deforestation |
| | | b) Toxic pollutants |
| | | c) Soil depletion on agriculture lands |

## 7.2 Natural and human-induced processes of environmental change

For many ecosystems, hydrological regimes and biospheric processes, the problem of disentangling human from natural influences is dauntingly complex. There are still strongly divergent views regarding the extent to which human actions have influenced environmental systems in the past. The dilemma of interpretation arises largely from the inescapable fact that most of the environmental archives upon which reconstructions are based respond to both kinds of influence, especially in areas where human impacts have a long history. The most convincing insights have come from studies where deduction has been possible through retrospective research – as Deevey so neatly puts it: 'coaxing history to conduct experiments' (Deevey 1967).

Regional as well as global insights are important for understanding processes, for addressing contemporary problems at the human landscape scale and for contributing to future impact assessment. Global generalizations that overlook strong spatial diversity may be of only limited value. Among the most urgent regional needs are more high resolution records of climate variability and human impact from tropical regions where suitable environmental archives are scarce and much less critical research has been accomplished. The lack of such information often seriously compromises interpretation of contemporary ecological patterns, current trends and future implications (Fairhead and Leach 1998).

Seeking to establish the relative significance of natural and human drivers of past environmental changes is clearly important, but it is perhaps even more important to increase our understanding of the interactions between human and natural influences, especially in those situations where their effects are mutually reinforcing or where their combined impact is to drive systems over critical thresholds into modes of non-linear change. In this regard, the paleo-record is of considerable, though so far under-exploited, value. Even where such interactions can be modeled using data derived from experiments and observations, the validity of the models generated requires testing on timescales over which much of the essential evidence comes from paleo-environmental research based on proxy records. Proxy evidence for past human-environmental interactions makes it possible to extend the record by providing reconstructions over longer time-spans. For quantitative interpretation however, proxies require rigorous calibration against independent data derived both from present day observations and short term instrumental and documentary time-series.

An important issue is the potential vulnerability of human populations to human-climate interactions. Since, in reality, the effects of climate change on human societies are mediated by cultural factors, human activities may serve to amplify or moderate the impact of natural variability. The record of such interactions in the past contributes to our understanding of them and may well contain important information on, for example, the relationship between rates of change or persistence of stress on the one hand and human adaptability on the other. It is unrealistic to ignore the interactive nature of the relationship between environmental change and human societies just as it is inappropriate to conceptualize it as a simple one way causative linkage in either direction.

## 7.3 Past human impacts on the atmosphere

### 7.3.1 Greenhouse gases

Doubts no longer surround the conclusion that fossil fuel combustion over the last 200 years, and especially during the last decades of the twentieth century, is the main process responsible for elevating atmospheric $CO_2$ concentrations to levels that greatly exceed any recorded during the last 420 kyr (Chapter 2, Section 2.3). Forest clearance and increased biomass burning have made a significant additional contribution to this process, though these have not yet been fully quantified and remain the focus of ongoing research (see e.g. Foley et al. 1996). Increases in atmospheric methane concentrations have also been rapid and dramatic over the last few decades (Chapter 2, Figure 2.6). There can be little doubt that human activities underlie this trend, through expanding agriculture, especially paddy cultivation (Neue and Sass 1994) and growing livestock populations (Prather et al. 1995). Indeed, ascription of increasing atmospheric methane concentrations in the polar ice cores from the second half of the Holocene to tropical rather than high latitude sources (Chapter 2, Section 2.5) opens up the possibility that the effects of paddy cultivation on atmospheric composition may considerably predate the last two centuries.

The nature of these trends in past greenhouse gas concentrations in the atmosphere and their consequences for the functioning of the earth system are major themes in chapters 2 and 4 and are not considered further here.

## 7.3.2 Trace metals, other industrial contaminants and radioisotopes

Discernible widespread impacts of human activities on atmospheric chemistry begin with the early days of extensive metal smelting (Nriagu 1996). Although there are indications of these impacts from the Bronze Age onwards, they become much stronger during the time of the Greek and Roman Empires for which there are clear signs of enhanced atmospheric concentrations of lead, copper and other trace metals in ice cores from Greenland (Hong et al. 1994) and in European lake sediments (Renberg et al. 1994) and peats (Shotyk et al. 1996, 1998, Martínez-Cortizas et al. 1999) (Figure 7.1). Elegant lead isotope studies even permit ascription to particular sources (Rosman et al. 1993, 2000, Renberg et al. 2001) and are able to show that atmospheric concentrations of 'pollution' lead peak in Greenland and throughout Europe between 100BC and AD200. From early Medieval times onwards, metal burdens increase. Enhanced loadings to remote areas prior to the mid-19th century are recorded, but it is generally only close to industrial or urban sources that evidence for strong contamination is apparent (Brimblecombe 1987).

Within the period of widespread industrial and urban development over the last century and a half, evidence for atmospheric contamination became ubiquitous. Spatial patterns have changed, mirroring not only the process of industrial expansion but also trends in resource use as well as in production, abatement and dispersal technologies. Evidence for these changes comes from both documentary sources (e.g. Brimblecombe 1987) and paleoarchives such as lake sediments (e.g. Edgington and Robbins 1976, Galloway and Likens 1979, Rippey et al. 1982, Kober et al. 1999), ice cores (Rosman et al. 2000) and ombrotrophic (precipitation-dependent) peatlands (Aaby et al. 1979, Norton 1985, Clymo et al. 1990) all of which contain historical records of contamination by a wide range of compounds. Deciphering the record depends both on analysis of the growing range of industrially generated products, including distinctive particulates (Wik and Renberg 1991, Rose 1994) and on the development of dating techniques applicable to the period of industrialisation (Appleby et al. 1991, Appleby and Oldfield 1992). In developed economies, recent turning points detectable in the paleorecord are the rapid increase in power generation beginning in the late 1950's, the increased generation of often environmentally persistent organic compounds, for example polyaromatic hydrocarbons (PAH's) (Hites 1981) and organochlorine products (Wania and Mackay 1993), and the post 1970s trend towards reduced discharge limits, greater control and improved treatment. Following the first clear evidence for damage to ecosystems (e.g. Odén 1968), many contaminant emissions peaked in the 1970s and '80s. The record in environmental archives has confirmed the trend towards an amelioration of air quality in many parts of the world (Nriagu 1990, Boutron et al. 1991, Candelone et al. 1995, Schwikowski et al. 1999). Nevertheless, even in the richer developed countries, air quality problems persist (e.g. Blais 1998). Elsewhere, factors such as the legacy of previous political regimes, continuing dependence on fossil fuel as the main energy source, proliferation of vehicles, population growth and ongoing industrial development often combine and lead to a continued build-up in atmospheric pollution to the level where concern is increasing at both regional and global levels.

Significant radioactive contamination of the atmosphere on a global scale began with the fall-out from post-war nuclear weapons testing from 1953 onwards. Once more, paleoarchives preserve a vital record of the spatial patterns and temporal changes in atmospheric deposition resulting from weapons testing, nuclear power accidents such as the Chernobyl incident and discharges from nuclear installations (Figure 7.2). Not only do paleorecords complement and extend direct measurements, especially where these have been sparse, poorly organized, inconsistent or completely absent, they also serve as indicators of the subsequent behaviour and long term fate of radioactive species in the environment. This is especially important in the case of relatively long lived radioisotopes such as $^{137}$Cs (30 year half-life) and several isotopes of plutonium and americium. These same radioisotopes serve both as dating markers and as process tracers in studies of recent human impact on ecosystems (Walling and Quine 1990, Oldfield et al. 1993).

## 7.4 Paleoperspectives on acidification, eutrophication and the ecological status of lakes, coastal waters and peatlands

Just as environmental archives contribute records of past atmospheric contamination, many also allow us to identify the impact of contamination on ecosystems Nowhere has this been more clearly demonstrated than in the documented effects of industrial

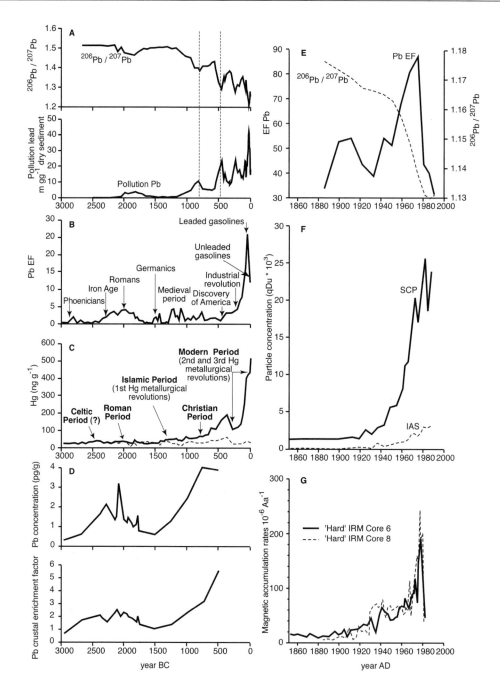

**Fig. 7.1.** Sedimentary histories of trace metal and industrially generated particulate deposition.

**A.** Pollution lead concentrations and [206]Pb/[207]Pb ratios as recorded in Koltjärn, a small lake in S Sweden (Renberg et al. 1994; 2000).

**B.** The total lead concentration record from Penido Vello, a peat profile in N Spain (Martinez-Cortizas et al. 1997) set against a series of historical events and cultural stages from 3000 BP onwards.

**C.** The total (solid line) and natural (dashed line) mercury (Hg) concentration record at the same site as **B** set against cultural and technological changes (Martinez-Cortizas et al. 1999).

**D.** Total lead concentrations and crustal enrichment values from Greenland ice for the period 3000 to 500BP.

**E.** A short-term record of total lead deposition and lead stable isotope ratios from La Tourbière des Genevez, an ombrotrophic (pre-cipitation-dependent) peat bog in Switzerland (Weiss et al. 1999). EF = the Pb Enrichment Factor. The 20th century pattern is comparable to that at Koltjärn, Fig. **A**.

**F.** Indicators of industrially generated atmospheric particulate deposition as recorded in the sediments of a small lake in NW Scotland, very remote from industrial sources (Rose et al. 1994). The inorganic ash spheres (IAS) are mostly fly-ash derived from coal-fired power stations, the spherical carbonaceous particles (SCP) are derived from both coal-fired and oil-fired power plants, but mainly the latter.

**G.** Magnetic measurements used as a proxy for industrial particulate at Big Moose Lake in NE USA.

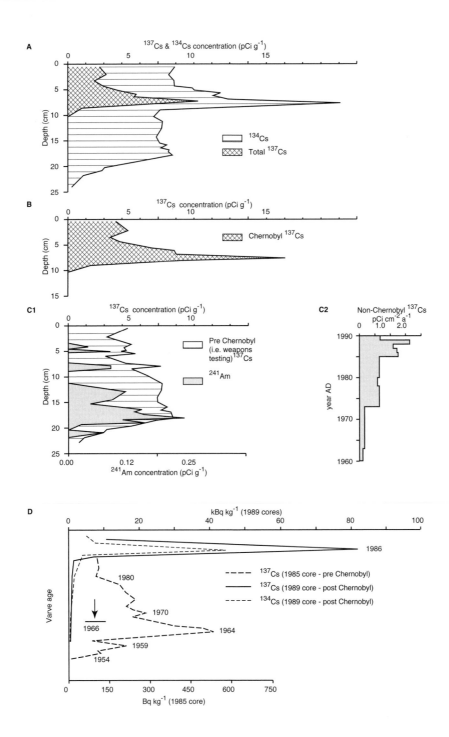

**Fig. 7.2.** Lake sediment records of direct and indirect atmospheric deposition of artificial radionuclides resulting from weapons testing and the Chernobyl accident. Plots **A** to **C** are from Blelham Tarn in the English Lake District (van der Post et al. 1997). Plot **D** is from Nylandssjön, Central Sweden (Crooks 1991). It is thus possible to confirm independently the integrity of the main features of the deposition record of $^{137}$Cs (derived from both weapons testing and Chernobyl), $^{134}$Cs (a marker for Chernobyl deposition) and $^{241}$Am, which was not dispersed by the Chernobyl accident. From the constant post-Chernobyl ratio of $^{134}$Cs to $^{137}$Cs. (**A**) It is possible to calculate the Chernobyl-derived $^{137}$Cs (**B**) and subtract this from the total trace in **A** to give a record of pre-Chernobyl $^{137}$Cs (**C**). This is consistent with independent dating and with the $^{241}$Am deposition history - **C(i)**. The record of weapons testing $^{137}$Cs calculated in this way can then be converted into a depositional flux, using the independent dating evidence from both varves and algae – **C (ii)**. Plot **D** shows Caesium traces from both immediately pre-Chernobyl (Feb. 1986) and Post-Chernobyl (1989) cores taken from the varved lake sediments of Nylandssjön, Central Sweden (Crooks 1991).

on surface water quality. Early evidence for the acidification of rain around industrial cities dates back over 100 years (Smith 1872) and from the 1950's onwards, Gorham (1958, 1975) and Odén (1968) were pointing to the likely impact of acidification on soils, vegetation and lake biota. Much of the acid was derived from sulfur-bearing coal burnt by the power generating industry, and more locally from the smelting of sulfide ores. Not until the 1980's were such processes seriously considered as potential contributors to observed changes in both surface pH and forest health in wide areas of Europe and eastern North America.

Attribution of the observed acidification to industrial processes required elimination of alternative explanations. These included declining upland agriculture and consequent catchment re-colonisation by soil-acidifying conifers or heathland vegetation; commercial afforestation; and natural long-term trends in soil development. By adopting a *post hoc* experimental strategy designed to test the validity of each of these, it was possible to eliminate all but industrial deposition as the *general* cause of acidification, though with two caveats. First, the added influence of the other factors was detectable in individual lakes. Second, the degree of chemical buffering in each lake-catchment system emerged as the most important single factor that determined vulnerability to acidification (Battarbee 1990). The following elements were crucial to the research strategy:

- paired lake-catchments were used to isolate and establish the role of each of the proposed causative mechanisms (e.g. Battarbee 1990, Whitehead et al. 1990, Birks et al. 1990a).
- robust, quantitative reconstructions of past lake water pH were developed using biological proxy records in the sediment that had been calibrated to measured values at the present day (Birks et al. 1990b)
- the capacity of sediment records to discern trends against the 'noise' of daily, seasonal and inter-annual variability was assessed; and
- detailed comparisons were made between recent trends and long term records (Figure 7.3).

Such elements have important wider implications for paleo-research, the more so as paleolimnological research results are increasingly being integrated into predictive models (Jenkins et al. 1990, Anderson et al. 1995, Jenkins et al. 1997) (Figure 7.4), the identification of critical loads (Battarbee 1997) and the tracking of aquatic ecosystem recovery (Dixit et

al. 1989, Allott et al. 1992, Anderson and Rippey 1994, Smol et al. 1998). Recent evidence from some impacted terrestrial catchments and the streams draining them suggests that recovery may be significantly delayed by the loss of calcium and magnesium from soils (Likens et al. 1996).

Parallel to research on the history of surface water acidification have been studies reconstructing the history of eutrophication. These too evolved from a need to establish the extent to which recent eutrophication was an expression of natural variability or a product of human interference with nutrient supplies to lake waters (e.g. Likens 1972, Battarbee 1978). There have now been many convincing demonstrations that accelerated eutrophication has been predominantly a consequence of human activities. These have included the routing of sewage effluents through integrated urban drainage systems from the late 19th century onwards, the discharge of industrial effluent to watercourses, the use of phosphate-rich detergents from the 1950's onwards and the application of artificial fertilisers to agricultural land (e.g. Lotter 1998). In many parts of the world, eutrophication is inexorably linked to population growth in regions where the treatment of effluent rich in human and animal waste is poor or nonexistent. Recent research has begun to provide increasingly quantitative bases for estimating past biologically available phosphate loading from sedimentary evidence (e.g. Bennion et al. 1996), thus extending the value of paleolimnological studies by providing additional information on the historical background to recent eutrophication.

Most of the research summarised above relies on adaptations of classical paleoecological approaches to fine resolution studies of recent time intervals. Less common are long term studies of past lake water pH and trophic status. Exceptions include the reconstruction of trends and variability in lake pH throughout the Holocene by (Renberg 1990), (Figure 7.3), as well as of the impacts of early agricultural societies on water quality in Sweden (Anderson et al. 1995) and of the Maya in Guatemala (Deevey et al. 1979, Scarborough 1993, 1994).

Acidification and eutrophication are not the only anthropogenic changes affecting water bodies. Even more dramatic in terms of human consequences in prehistory are the inferred increases in salinization linked to cultural demise, notably in Mesopotamia (e.g. Jacobsen et al. 1958).

By now, paleolimnology is not merely a tool in paleoreconstruction, it has become increasingly

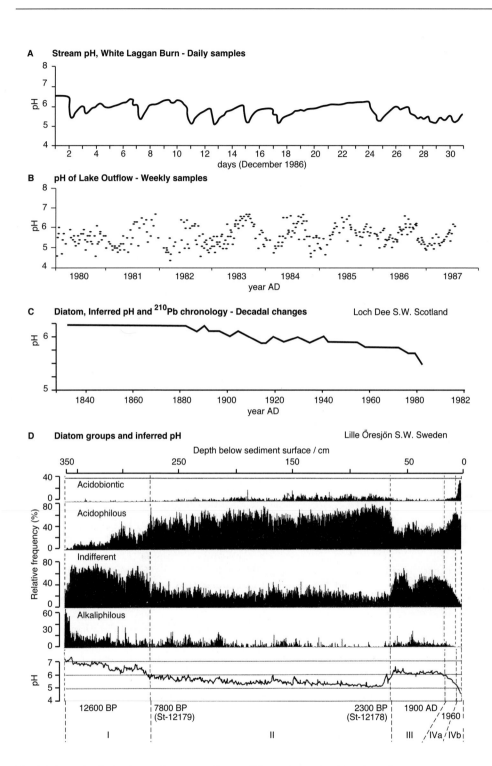

**A** Stream pH, White Laggan Burn - Daily samples

**B** pH of Lake Outflow - Weekly samples

**C** Diatom, Inferred pH and $^{210}$Pb chronology - Decadal changes    Loch Dee S.W. Scotland

**D** Diatom groups and inferred pH    Lille Öresjön S.W. Sweden

**Fig. 7.3.** Records of changing surface water pH over different timescales. Plots **A** and **B** show, respectively, daily and weekly direct measurements of pH at Loch Dee, SW Scotland. Plot **C** shows a diatom-based reconstruction of past pH from the analysis of recent sediments from Loch Dee dated by $^{210}$Pb (Battarbee 1998). The paleolimnological approach complements short term monitoring by allowing detection of the long-term decline in pH since the late 19th century, despite the fact that the diurnal and weekly variability exceeds the amplitude of change in the long term trend. In plot **D**, a similar century-long decline in pH at Lilla Oresjön (Sweden) is set in the context of a 12600 year long record at sub-decadal resolution. There are clear shifts in past lake water pH during the pre-industrial part of the Holocene, notable a slow, gradual decline in pH followed by a period during which settlement and farming around the lake led to enrichment and a reversal of the trend. The magnitude and pace of the 20th century decline resulting from acid deposition is seen to be unprecedented (Renberg et al. 1990). These examples show the value of the paleolimnological approach both in establishing trends against a background of short-term variability and in placing these trends in the context of long term natural changes.

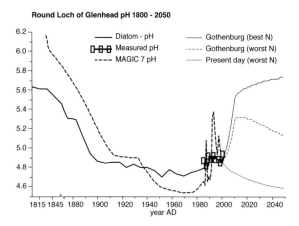

**Fig. 7.4.** The role of paleodata in substantiating models of trophic change in lake ecosystems. The left-hand side of the graph compares fossil diatom–inferred pH, modeled pH using MAGIC, and measured pH for the Round Loch of Glenhead, SW Scotland. The right-hand side of the graph shows alternative future scenarios developed from the MAGIC model under different atmospheric deposition scenarios (Battarbee 1998).

important as a diagnostic methodology for the assessment of ecological status and as a basis for setting remediation targets. Such targets are important in the design of management strategies (Dixit et al. 1996, Battarbee 1998).

An increasing number of recent studies have applied the types of 'paleolimnological' approaches outlined above to sediment records from near-shore marine environments. Andren et al. (1999, 2000) for example used silicieous microfossils (diatom frustules and chrysophyte cysts) to trace the impact of both climate change and human activities on marine ecosystems in the southern Baltic. They interpret a shift in the balance of diatom productivity from benthic to planktonic communities as a response to a thinning of the photic layer resulting from cultural eutrophication. At sites closest to the densely populated areas on the southern shores of the Baltic, the shift begins in the mid 19th century and parallels the evidence for cultural eutrophication in many North European lakes. Moving further north to the Gotland basin, the first clear evidence of an ecosystem response to increased nutrient supply takes place in the mid-20th century, coincident with lake sediment evidence also pointing to an increase in the rate of eutrophication as a result of use of artificial fertilizers and phosphate-rich detergents.

In the Adriatic (Figure 7.5A) from the Bronze Age onwards, changes in the benthic foraminiferal assemblage show a strong response to evidence for deforestation furnished by pollen data and, linked to this, accelerated delivery of terrigenous material to the sediments. Both the increase in *Valvulineria complanata* just before 4000 BP and its subsequent increase alongside *Bulimina marginata* coincide with the main periods of human impact on vegetation and soils (Oldfield et al. in press). These changes can best be interpreted as a response to a more stressed and oxygen-depleted benthic environment as a result of the increased deposition of terrigenous particulates, and, during the later period, organic matter resulting from higher autochthonous productivity (Asioli 1996).

Changes in organic carbon inputs, sedimentation rates, benthic foraminifer assemblages, diatom species distributions and sediment chemistry in Chesapeake Bay closely parallel the history of land clearance and erosion in the region that began on a broad scale in the 17th Century (Karlsen et al. in press). Anoxia in bottom waters of the Bay followed in the 20th Century as a consequence mainly of increased nutrient inputs (Cooper and Brush 1991, 1993, 1995). This response has been recorded by variations in the molybdenum concentration in the sediments (Adelson and Helz 2001).

The Baltic, Adriatic and Chesapeake Bay studies point to a strong link between human-induced changes in coastal regions and developments in marine ecosystems. In the case of the Adriatic, this linkage is evident from prehistoric times onwards. By contrast, the much wider scale development of anoxia recently observed in the northern Gulf of Mexico can be ascribed to increased nitrogen flux from the drainage basin of the Mississippi from the 1960's onwards (Goolsby 2000).

Comparable perspectives to those outlined above for lakes and near-shore marine environments are needed in order to assess the status of peatlands. Such systems are potentially highly sensitive to the combination of increased atmospheric deposition of anthropogenically linked contaminants and climate change, especially those that have developed to the stage where they are isolated from surface and ground water influences and are thus entirely dependent on atmospheric inputs (Clymo 1991). Although the stratigraphic evidence is not always easy to interpret, there are strong indications that sulphur deposition arising from past coal combustion, possibly in combination with climate variability, has led to changes in species composition, surface structure and erodibility (Lee 1998). When coupled with the role of peatlands in the global budget of atmospheric trace gases (Gorham 1990, 1991), such implied vulnerability makes the sustainability of such ecosystems a priority for future research.

In the text above, different processes and

influences have largely been treated in isolation, but it is important to increase our understanding of combined and synergistic effects and how these may interact with climate change in the future. Research devoted to this end is increasing, as witness studies of lakes where future threats are likely to arise from the combination of increased deposition of nitrogen species, from impacts of ozone depletion and increasing UV flux, from continuing inputs of acid and, in remote, high altitude and high latitude areas, from cold distillation processes that enhance deposition of organic contaminants from distant sources (e.g. Schindler 2000).

The paleorecords described here strongly indicate that research dealing with recent changes in nutrient and ecological status should set contemporary process studies against the longer term context provided by well-dated and quantitatively calibrated proxy records from paleo-archives. Only in this way can the true scope of human impacts be compared with variability associated with natural processes.

## 7.5 Past human impacts as a result of land-use and land-cover changes

Many of the impacts on lake systems outlined above are mediated by processes taking place on the land surface. Thus our next focus is on the many ways in which human activity has modified terrestrial ecosystems. Evidence for human impact on vegetation, derived largely from pollen analysis, dates back thousands of years in many parts of the world (Edwards and MacDonald 1991, Walker and Singh 1993) and scientific awareness of its significance in prehistoric times dates back for at least 60 years (Iversen 1941). For some widespread non-agricultural ecosystems, there is still uncertainty as to the extent to which what we see today is a response to prevailing climate and soil conditions, or a product of human interference and management. For example, it is likely that the extensive lowland heaths of Western Europe owe their origin in part to anthropogenically induced changes in plant cover, soil status and nutrient cycling beginning as much as 6000 years ago on the sandy Breckland soils of East Anglia (Godwin 1944) and during later stages of prehistory elsewhere (Bartley and Morgan 1990). In the British Isles, even biomes as apparently 'natural' as upland blanket bog include areas where Mesolithic artifacts dating from an even earlier period of hunting and gathering are associated with evidence for burning at the base of the peat profiles. This has encouraged the view that human activity

may have served as an important trigger to peat accumulation (e.g. Simmons 1969, Moore 1973, Casseldine and Hatton 1993). In each of these cases and others elsewhere in the world, the concept of decisive human intervention carries a range of connotations, from immediate environmental responses to longer term conditioning of the environment that may at some future time interact with climate in complex, non-linear and often unpredictable ways.

In many examples of ecosystem degradation through human activities, the key issue is a critical shift in the balance between the rate of depletion of key functional attributes and the rate of their renewal within a given system of production. Beyond a certain threshold in the shifting balance, a persistent state of lower productivity may develop that is difficult to reverse. In parts of the New Guinea Highlands for example, this type of transition has led to the conversion of a formerly productive forest-garden mosaic that reflected a cycle of forest clearance, subsistence horticulture and woody regeneration into a short grassland ecosystem. This *Themeda australis*-dominated grassland, with its low moisture retention, poor soil structure and low nutrient content yields a highly unproductive but persistent and quite extensive open landscape. Sediment-based evidence for changing catchment yields and erosion rates around small upland lakes in the highlands suggest that accelerated soil loss is associated with this process (Oldfield et al. 1980, 1985, Haberle 1994). Pollen evidence in the same region points to the spread of this type of ecosystem during at least the past 6000 years (Powell 1982, Haberle 1998), though too little is known to establish whether or not climate change also played a role in changing vegetation cover.

In areas where swidden (often termed 'slash and burn') agriculture is the main basis for food production and intensification of subsistence farming has led to a shortening of the regeneration cycle, the injudicious use of fire has often been cited as a crucial trigger in promoting rapid soil degradation that is difficult to reverse or repair (Pyne 1998, Redman 1999). Increasingly, it becomes apparent that in order to understand the present day status and future changes in contemporary systems that are undergoing this type of pressure it is necessary to study impact over relatively long time periods (Sandor and Gesper 1988, Sandor and Eash 1991).

Much uncertainty surrounds the degree of human impact on subtropical savannah landscapes in Africa (e.g. Mworia-Maitima 1997, Fairhead and Leach 1998), and on fire-adapted ecosystems in Australia (Bowman 1998), though there is growing evidence for human impact on the vegetation in African montane areas (Lamb et al. 1991, Taylor et al. 1999). Despite conflicting evidence, one conclusion becomes increasingly difficult to escape. Even in those countries where evidence for pre-colonial agriculture is sparse and where we tend to think of significant human impact as dating from 'recent' colonisation by Europeans, we should not minimise the impact of indigenous peoples on vegetation and soils (e.g. Bahn and Flenley 1992, Chepstow-Lusty et al. 1998, Dumont et al. 1998, Elliott et al. 1998, Behling 2000). Redman's summary of pre-conquest populations and likely environmental impacts in the Americas (Kohler 1992, Redman 1999, Chapter 1, Section 1.3) counsels strongly against either assuming negligible impact (cf. Fuller et al. 1998) or turning uncritically to reconstructions of pre-colonial conditions as templates for future management designed to reinstate 'natural' ecosystems. This last point is further reinforced once pre-Columbian impact is set alongside evidence for the ecological importance of past climate variability: pre-Columbian 'templates' used as goals in future management strategies may reflect climatic conditions that differed significantly from those experienced at present or predicted for the future (Millar and Woolfenden 1999, McIntosh et al. 2000). In those parts of the world with a long history of forest clearance and farming, the concept of a natural ecosystem has lost its meaning except insofar as modeling permits the postulation of some kind of potential vegetation or biome.

Of the long-settled areas of the world, the clearest evidence for the extended time span of strong human impact comes from Europe and the Middle East, largely because of the much greater attention devoted to these areas by Quaternary paleoecologists and archeologists. The degraded landscapes surrounding the Mediterranean have prompted numerous studies by archaeologists of the effects of human activity (e.g. Barker 1995). Geomorphologists (Vita-Finzi 1969), Quaternary stratigraphers (Van Andel et al. 1986, 1990) and palynologists (e.g. Atherden and Hall 1999, Ramrath et al. 2000) have contributed similar insights (Figure 7.5). In these cases, evidence for a dominant role for human intervention over millennia is overwhelming. Results from the Adriatic (Oldfield 1996, Oldfield et al. in press) also point to early human intervention but indicate that the first major phase of prehistoric clearance in the Bronze Age coincides with or quickly succeeds evidence for a change in sea surface temperatures as seen in the alkenone record (Figure 7.5). From this and other studies, the ques-

tion of the relative role of climate change and human impact in the evolution of Mediterranean landscapes remains open to debate (cf. Jalut et al. 2000).

In other areas of the world with early records of agriculture, paleo-ecological evidence for prehistoric human impact from sediment records is virtually universal and takes the form of characteristic changes in pollen flora, often with associated increases in sediment yields and fire incidence. In East Africa, there is strong evidence for deforestation, increased fire incidence and accelerated erosion dating to 2.2 kyr BP (Taylor 1990). Signs of human impact have been confidently inferred from ca. 4 kyr BP onwards in Central Mexico (Bradbury in press) and for over 2000 years in the Peruvian Andes (Chepstow-Lusty et al. 1998). Archaeological evidence from India gives similar indications, (Misra and Wadia 1999) whilst in the other densely settled areas of south east Asia, dates for the earliest clear signs of human impact on ecosystems are never later than 2 kyr BP and often much earlier (Maloney 1980, Maloney 1981, Stuijts 1993, Van der Kaars and Dam 1995, Kealhofer and Penny 1998). Ren et al. (1998) find strong evidence for forest clearance in the middle and lower Yellow River Valley region from 5000 BP onwards, with the date for the earliest signs of human impact getting progressively later to the north and west. Even on the islands of the South Pacific, dates of first inferred impact range from around 3 kyr BP in Fiji and 2.4 kyr BP in the Cook Islands to 1.2 kyr on Easter Island and 0.7 – 1.4 kyr BP in New Zealand (Flenley, in press).

Taking a broader perspective, there is evidence that in some environments, the cumulative impact of human activities over long periods may gradually transform ecosystems and that in others, initial impacts can be critical in switching ecosystems into different functional modes. In yet other cases, the pattern of ecosystem response is one of repeated recovery and apparent resilience over millennia.

At local to regional scale, the importance of land cover feedbacks to the climate system has often been demonstrated (e.g. Couzin 1999). Thus changes in ecosystems, whether naturally or anthropogenically generated, can have feedbacks via local climate that may reinforce their persistence. In addition, inferences derived from models of intermediate complexity that include such feedbacks suggest that their effects may even be significant at continental to hemispheric scale (Brovkin et al. 1999, Kleidon et al. 2000, Ganopolski and Rahmstorf 2001, Kabat et al. 2001). There is therefore a growing likelihood that changes in land cover

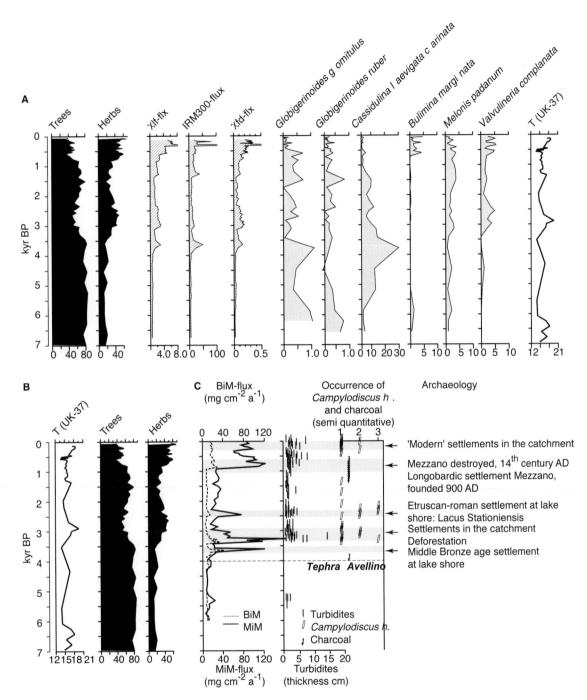

**Fig. 7.5.** Evidence for human impact and climate change in the Mediterranean region over the last 6000 to 7000 years. Part **A** shows pollen, rock-magnetic and benthic foraminifers records and Uk-37-based sea-surface temperature (SST) reconstructions from the mid-Adriatic (Core RF 93-30) plotted against a timescale derived from a wide range of chronological indicators (Oldfield 1996; Oldfield et al. in press). Pollen analytical evidence for forest clearance from ca. 3600 cal. BP onwards and from ca AD 1200 coincide with evidence for an acceleration in erosive input from the land surface (see especially the calculated fluxes of χlf χfd) as well as changes in the benthic foraminifers morpho-species assemblages indicative of increased stress as a result of higher sediment supply and organic enrichment. There are, simultaneously, major shifts in inferred SST and these, along with many other lines of evidence (eg. Jalut et al. 2000) point to climate changes taking place at roughly the same time. Parts **B** and **C** set the tree-herb pollen ratio and the Uk-37-based SST reconstructions from core RF 93-30 alongside several lines of evidence for human impact at the Lago di Mezzano site in C Italy (Ramrath et al. 1998). From the comparison, it can be seen that both of the main periods of forest clearance recorded in the Adriatic are strongly represented in the more site-specific record from Mezzano, which also correlates closely with the archaeological and historical record from the region. The data shown here are a small part of the large assemblage of data from the Mediterranean region that point to both climatic and human influences on late Holocene environmental change, but the nature of the balance and of the interactions between the two remains an open question.

may affect climate systemically. The possible contribution of land cover change to global warming in recent centuries is further considered in chapter 6.

It follows from the above that the climate-versus human antithesis represents not a simple dichotomy, but two complementary parts of a complex, interactive system. Distinguishing between directly climate-induced changes in vegetation and those reflecting feedbacks to anthropogenic impacts or delayed adjustments through long-term migration or succession is an area of palaeoecological research that has received little direct attention, but one that is vital to our understanding of natural ecosystem variability and the limitations of palaeoecological data as direct proxies of climate.

As the range of well dated climate reconstructions based on archives other than biotic response signatures increases, it should become possible to ascribe with greater confidence changes in extant ecosystems to dominantly climatic or anthropogenic forcing with greater confidence. This will increase our power both to generate models of past ecosystem transformation and to explore the nature of the interaction between natural and anthropogenic processes. The subtlety of such interactions is well demonstrated by the mystery of the dramatic and widespread European *Ulmus* (elm) decline some 6000 years ago. At least five hypotheses have been advanced to explain this feature that is common to virtually all Holocene pollen diagrams from W Europe. Peglar and Birks (1993) and Peglar (1993) use both high resolution pollen analysis and detailed charcoal counts to show that around Diss Mere in East Anglia, UK, the likeliest combination of causes is disease in forest stands that were already stressed by human impact. Their results suggest that multiple, including anthropogenic, threats to ecosystems are not new – they may date back thousands of years.

Fire plays a key role in ecosystem evolution, but like other variables, its frequency and impacts reflect both natural climate variability and ecosystem structure as well as human activities. Fire-scars on long-lived trees, sedimentary charcoal records (Patterson et al. 1987, Clark 1990, Lehtonen and Huttunen 1997), fire-related magnetic signatures (Gedye et al. 2000, and Figure 7.6) and geochemical markers provide a basis for reconstructing past fire frequencies. Quantitative reconstructions of the intensity and spatial extent of past fires are difficult to produce. However, recent studies (e.g. Whitlock and Millspaugh 1996, Clark et al. 1998, Tinner et al. 1998, Tinner et al. 1999) are improving the basis for interpretation through the analysis of sedimentary records of recent and well-documented fire events by combining charcoal counting with statistical and other techniques (e.g. Mworia-Maitima 1997). Paleoresearch should ultimately be able to provide bases for distinguishing between fire regimes of differing frequency and intensity. This will contribute to a better understanding of the role of fire in fire-stressed/fire-dependent ecosystems, of the impact on and ecological consequences of human modulation of fire regimes and of the likely long-term responses of present day ecosystems to different types of fire management. Failure of forest managers to understand until recently the vital role of fire in the regeneration and maintenance of widespread forest ecosystems in the western USA led to its injudicious suppression. The consequent build-up of combustible material has been one of the main factors exacerbating the incidence and effects of wildfires in recent times.

One of the most damaging impacts of human activity has been the destruction and degradation of wetlands as a result of reclamation drainage (Immirzi and Maltby 1992) and fire. Evidence for such activities date back to prehistoric times in established centres of civilisation. But, as discussed in the following section, and as Mitsch and Gosselink (1993) point out, it is only within the present century that the effects have attained global significance.

## 7.6 A paleo-perspective on human activity and biodiversity

One of the most controversial questions in Quaternary research is the possible role of prehistoric peoples in faunal extinction. Paul Martin's contention (Martin 1984) that human exploitation, through extensive hunting, was responsible for the extinction of the North American mega-fauna at the end of the last glacial period remains difficult either to prove or to refute, partly because the timing of the main prehistoric human expansion into the Americas coincides with incontrovertible evidence for major changes of climate. Nevertheless, a significant role for human exploitation pressure in changing ecosystems and faunal niches during the period of rapid natural environmental change remains a credible inference.

Many paleoecological studies lack the taxonomic resolution necessary for exploring questions of changing biodiversity at the level of species. Even where identification of higher plant remains can be made to the species level, it is rarely if ever possible to regard changing macrofossil assemblages as

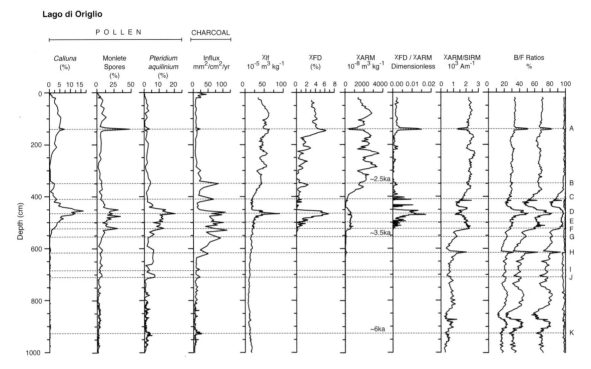

**Lago di Origlio**

**Fig. 7.6.** A sediment-based reconstruction of fire history at the Lago di Origlio, S Switzerland (Tinner et al. 1998, 1999). The figure illustrates a multiproxy approach using pollen and spore analysis, charcoal counts and magnetic measurements. The dashed and lettered horizontal lines are drawn at or close to the depth of peaks in the depositional flux of charcoal. The pollen and spore types plotted represent taxa that often respond positively to fire. The magnetic measurements plotted may be used to discriminate for fine–grained burnt iron oxides (Gedye et al. in 2000). Below 400 cm any fire impact on the magnetic measurements is set against a background detrital, magnetic influx dominated by hard remanence minerals. Thus the charcoal peak 'H' is reflected in the magnetic measurements by a sudden, brief 'softening' of the reverse field (BF) ratios. Above 400 cm, the background magnetic signature appears to be dominated by bacterial magnetite which gives rise to high values of χARM/SIRM. The fire 'spikes' from D upwards therefore also show up as peaks in the proportion and concentration of the finest grains which lead to high values for χFD and χFD%/χARM. The high concentrations of fire-related magnetic minerals are produced mainly by temperatures in excess of 500 – 600°C. Moreover, the minerals are much more likely to have originated within the catchment of the lake. The magnetic signature can thus give additional information on fire intensity and location. The pollen and spore record gives some indication of the impact, short- and long-term of individual fires and changing fire régimes. The fact that not all the indicators respond to each event and that, even when they do, the responses are not always in the same proportion, points the way toward a better understanding of the role of fire in the ecological history of the site.

being a reliable basis for tracing changes in biodiversity through time. Rarefaction indices derived from pollen or diatom records (for example Lotter 1998, Odgaard 1999), can, however, serve as proxies for biodiversity as noted below. Early paleobotanical literature is rich in studies that demonstrate the demise of plant taxa in Wurope with each successive glaciation. This serves as a reminder that what we see in many parts of the world are the remains of biota that survived many wrenching environmental shifts. This observation also applies on shorter timescales for many temperate mountain species now seen as threatened by future global warming. Evidence from temperate and sub-arctic environments points to a period in the early-to-mid Holocene when temperatures and tree lines were higher than they are now. What we currently see as threatened mountain biota survived this period. This does not constitute a reason for complacency in

current thinking on the preservation of mountain biodiversity but it does provide a compelling reason to understand the survival strategies and microhabitats that allowed the persistence of the present day biota through earlier periods of environmental stress.

The foregoing observations highlight the important role paleoresearch can play in reconstructing the history of landscape elements and ecological niches crucial for the survival of biodiversity. Interestingly, in those cases (mostly from N.W. Europe) where past biodiversity has been reconstructed from pollen diagrams using rarefaction indices, human impact has had the effect of increasing taxonomic diversity at landscape scale through the creation of a greater variety of habitat types (Berglund 1991, Lotter 1999, Odgaard 1999). Thus, many areas of high biodiversity or conservation value are not, as was previously thought, pristine ecosystems where high value has resulted from lack of human impact. This can be seen on a wide range of spatial scales. At Blelham Bog in northwest England, the habitats responsible for the tiny nature reserve's conservation interest are not a reflection of natural processes, as was originally believed, but are the product of human interference in the form of peat cutting and drainage (Oldfield 1969). On the larger scale, we may cite the evidence from the Bwindi Impenetrable Forest area of Uganda, one of the most species-rich areas of montane forest in Africa. The belief that this is a pristine, undisturbed ecosystem is seriously challenged by paleo-ecological evidence that includes signs of disturbance over the last 2000 years (Marchant et al. 1997, Marchant and Taylor 1998). Thus, any strategies designed to preserve biodiversity that ignore site, ecosystem or habitat history, past climate variability and human activities – all of which contribute crucially to contemporary patterns - seriously limit their chances of success. This is especially the case where zones of high biodiversity span steep physiographic gradients or ecotones, as for example, in many mountain areas.

The situation is different in the case of many aquatic microorganisms, for example diatoms, where the problem of taxonomic resolution is overcome because fossil remains allow identification to the species or even sub-species level. This provides a promising context within which to explore questions of endemicity as well as human impact in watersheds. The latter possibility is well illustrated by recent research at Baldeggersee, Switzerland, (Lotter 1998) in which the close links between eutrophication and diversity are resolved on a near-annual basis (Figure 7.7B).

## 7.7. Past human impacts on erosion rates, sediment yields and fluvial systems

Transformations of terrestrial ecosystems are expressed through shifts in rates of erosion (Dearing et al. 1990, Duck and McManus 1990, Van Andel et al. 1990, Higgitt et al. 1991, Zangger 1992, Foster 1995, Van der Post et al. 1997), sediment yield (Douglas 1967, Davis 1976, Macklin et al. 2000) and river channel change (Hooke 1977, Wasson et al. 1998, Fanning 1999) and are recorded in sequences of lake sediment, alluvium and colluvium (Dearing 1994) stretching back over thousands of years. Paleo-records are now able to provide a long term perspective for many contemporary fluvial and sediment systems, as well as valuable short term

perspectives where there is no other kind of record. The examples shown in Figure 7.8 illustrate the range of interactions between anthropogenic and climatic forcings and their impacts on river discharge and sediment transport during the Holocene. Where human activities have had significant and long-term impact, as in much of NW Europe, sediment records (Figure 7.9 c-e) often show linkages with the timing of settlement and agricultural changes (Dearing et al. 1990, Zolitschka 1998). Conversely, where human impact is considered to have been either recent or slight (Figure 7.9 a-b), the effects of climate variability may be seen, especially in alluvial sequences (e.g. Macklin 1999). The interaction between human impact and natural variability is particularly obvious in the incidence of extreme climatic-hydrological events such as floods or droughts (e.g. Eden and Page 1998, Thorndycraft et al. 1998, Foster et al. 2000). Not only land cover change but also modification of surface drainage through changes in soil structure, ditching and river channelization in combination modulate the expression of floods and accelerated sediment yield (Dearing and Jones in press, Foster et al. in press) and the human hazards to which they give rise. In many parts of the world, the short period of direct observation and monitoring is inadequate to capture the full interaction of these processes. This is clearly seen in the difficulties posed by effective

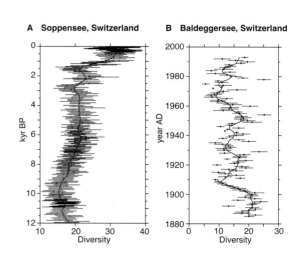

**Fig. 7.7.** Evidence for past changes in taxonomic diversity in both terrestrial and freshwater ecosystems. Both diagrams from lake sites in Switzerland use a rarefaction index (Birks et al. 1992) as a record of past changes in taxonomic diversity. Plot **A**, spanning the last 12000 years, illustrates the role of human activities in increasing diversity through the creation of a wider range of landscape and habitat types within the pollen source area of the site over the last 1500 years (Lotter 1999). Plot **B** spanning only the last 120 years of laminated lake sediment deposition shows how diatom species diversity declined with the onset of eutrophication around AD 1900 (Lotter 1998).

flood prediction within the framework of simple magnitude-frequency relationships as the boundary conditions of hydrological systems change through time (Knox 2000, Messerli et al. 2000).

New methods and techniques will improve our analysis of paleorecords. Adoption of a truly integrated lake-catchment framework where estimates of sediment yields based on lake deposits are compared with measurements of floodplain accretion, geomorphic evidence for slope instability and contemporary process monitoring (e.g. Dearing and Jones in press, Foster et al. in press) is allowing lake sediment properties to be directly calibrated to sediment sources and fluvial processes. Proxy records of sediment loads and flood intensities (eg. sedimentation rate, lamination thickness, particle-size) can be calibrated by comparison with monitored river discharges (e.g. Wohlfarth et al. 1998), rainfall records (e.g. Page et al. 1994) or documented records of flood events (e.g. Thorndycraft et al. 1998). Documented records of human and animal populations and land use provide independent evidence for the role of human activities (e.g. Higgitt et al. 1991) on fluvial and sediment processes (Figure 7.10). Such approaches when extended to paleorecords enable reconstruction and understanding of past sediment budgets and process-response mechanisms (e.g. Foster et al. 1988, Wasson et al. 1998, Fitzpatrick et al. 1999) that are comparable with contemporary studies of annual and seasonal changes, but over far longer timescales. As one example, long erosion records are already helping to define key properties of environmental change, such as resistance, resilience and lag-times between forcings and responses (Table 7.2). Contrasts in system trajectory, such as those shown in Figure 7.8, are providing a basis for defining the nature of human pressure, its interaction with natural variability and the sensitivity of the system under study.

**Table 7.2.** Lake sediment and model evidence for erosional responses to deforestation (Dearing and Jones, in press)

| Regions | Forcing | Response | yr |
|---|---|---|---|
| *Within $^{210}Pb$ timescale* | | | |
| Papua New Guinea[1] | 19[th]C clearance | x10 | >150 |
| New Zealand[2] | 19[th]C farming | - | 30 |
| Michigan, USA[3] | 19[th]C settlement | x10-70 | 10 |
| Vermont, USA[4] | 18[th]C settlement | x4 | 100 |
| Tanzania[5] | 19[th]C clearance | x4 | 10 |
| *Holocene* | | | |
| Germany (1050 AD)[6] | Clearance | x10-17 | 250 |
| Sweden (800 BC)[7] | Clearance | x4 | 200 |
| Pensylvania, USA[8] | Landscape evolution model | x40 | 50 |

[1]Oldfield et al. 1985, [2]Turner 1997, [3]Davis 1976, [4]Engström et al. 1985, [5]Eriksson and Sandgren 1999, [6]Zolitschka 1998, [7]Patterson et al. unpublished, [8]Tucker and Singerland 1997

Paleorecords, by improving our understanding of the fundamental dynamical behaviour of modern fluvial and sediment systems, will help form the basis for classifying their sensitivity to future impacts. Systems that have evolved into complex self-organized states under low levels of disturbance may be more susceptible to dramatic shifts in climate and land use than systems already conditioned by long histories of human impact (Dearing and Zolitschka 1999). Recent mathematical and cellular automaton models of long term erosion (Coulthard et al. 2000) suggest that sediment delivery over timescales of 10 to 100 years is a highly non-linear product of land cover change and high-magnitude rainfall events. Paleorecords can be used to test these models, thereby providing a framework of study across a very wide range of spatial (1-10$^6$ km$^2$) and temporal (1-10$^3$ yr) scales in many environments. In theory such approaches will overcome long-standing barriers to progress in hydro-geomorphological investigations and their application.

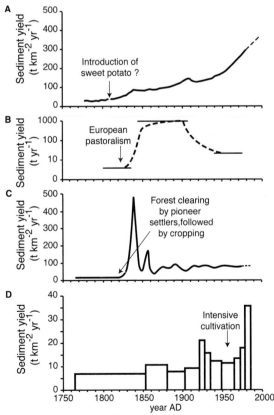

**Fig. 7.8.** Examples of sediment yield responses and subsequent trajectories of change associated with different kinds of major human impact over the past 250 years: **A.** Lake Egari, Papua New Guinea (Oldfield et al. 1980); **B.** southeastern Australia (Wasson et al. 1998); **C.** Frain's Lake, Michigan, USA (Davis 1976); **D.** Seeswood Pool, English Midlands (Dearing and Foster 1987).

## 7.8 Environmental sustainability and human vulnerability in the perspective of the paleorecord

The thrust of this chapter is towards demonstrating that in any evaluation of future sustainability or human vulnerability to environmental change, a deeper understanding of past interactions of environmental processes and human activities is essential. But there can be quite fundamental divergences in perspective across the divide between biological-physical and social sciences, and this divide must be bridged if the full potential of this area of research is to be realized.

The case of the demise of Anasazi in the semi-arid southwest USA provides a clear example. Dean et al. (1985) invoked persistent drought as a major causative agent in the decline of that society, while Kohler (1992) instead points to patterns of land use. These involved intensification and over-exploitation within a social context that favoured the continuation of such practices to the ultimate decimation of the resource base. The same contrast may be seen elsewhere, including the demise of Norse settlements in Greenland (Pringle 1997, Barlow et al. 1998), the collapse of classic pre-Columbian Mayan (Yucatán Peninsula) and Tiwanaku (Bolivian-Peruvian Altiplano) civilizations (Rice 1994, Hodell et al. 1995, Brenner et al. 2001, Nuñez et al. 2001), the evolution of food production in North Africa and the Middle East (Hassan 1994) and the collapse of Mesopotamian civilizations (Jacobsen et al. 1958, Weiss 1997, Redman 1999, Cullen et al. 2000). Even where the complexity of human-environment interactions is acknowledged, views on the nature of their interaction within the framework of multi-directional human-environment relationships differ widely. Most authors emphasize one or more of three components: damaging climate extremes such as droughts or cold (e.g. Hodell et al. 1995, Benedict 1999), human impact through non-sustainable resource use/environmental degradation (Redman 1999) and dysfunctional patterns of social organization not always directly related to resource use but nevertheless affecting it indirectly (Rappaport 1978, Crumley 1993). At one extreme, we have the view that maladaptive social systems are the proximal trigger, against a background of environmental degradation (Deleage and Hemery 1990) with no reference to the effects of climate variability. The opposing view is that climate extremes of exceptional magnitude or persistence are the triggers, particularly when they impact fragile environments and occur within the context of pat-

terns of social organization that increase the vulnerability of a society (Messerli et al. 2000). A balanced view requires a collaborative effort by scholars with complementary training and frames of reference. It also calls for studies with the best possible chronological control on all the lines of evidence. The value of this approach is self evident in several recent studies. Nials et al. (1989) linked the demise of the Hohokam culture in S. Arizona (Redman 1999) to an annually resolved record of past stream flow (Figure 7.11). A tephra-based link between the marine sediment record and archaeological evidence is cited by Cullen et al. (2000) as support for their view that climate change had a major role to play in the demise of the Akkadian Empire, while Macklin et al. (2000) used excellent chronological control to explore the relationship between environmental change and prehistoric settlement in Western Scotland. One of the most promising and indicative examples of synergy between environmental and cultural perspectives can be seen in the most recent study of the Anasazi by Dean et al. (1999), where a modeling approach was used within a context strongly constrained by both archaeological and palaeoenvironmental evidence, as well as tight chronological control. Transcending the biophysical-social divide is also well exemplified by Hassan's work on cultural and environmental change in Ancient Egypt (Hassan 1994).

Equally important, alongside case studies that embrace the complexity of human-environment interactions, are conceptual frameworks and models of change for uniting biophysical and cultural perspectives (Crumley 1994, Balee 1998). One of these is that of 'trajectories of vulnerability' developed by (Messerli et al. 2000), showing how detailed case studies of climate-human interactions in the past serve to highlight potential implications for the future.

Over the past century, technological responses to environmental variability have often succeeded in achieving protection against medium to high frequency, low to medium amplitude variations but, by both raising the thresholds of catastrophic impact and encouraging false confidence, they have increased vulnerability to high amplitude, low frequency events. If we view these trends in the context of past variability and human responses, sharp increases in vulnerability for many areas of human settlement appear unavoidable. Human responses to environmental stress and change that turn out to be inadequate or even destructive, in the long term run like a common theme through human history and prehistory right to the present day.

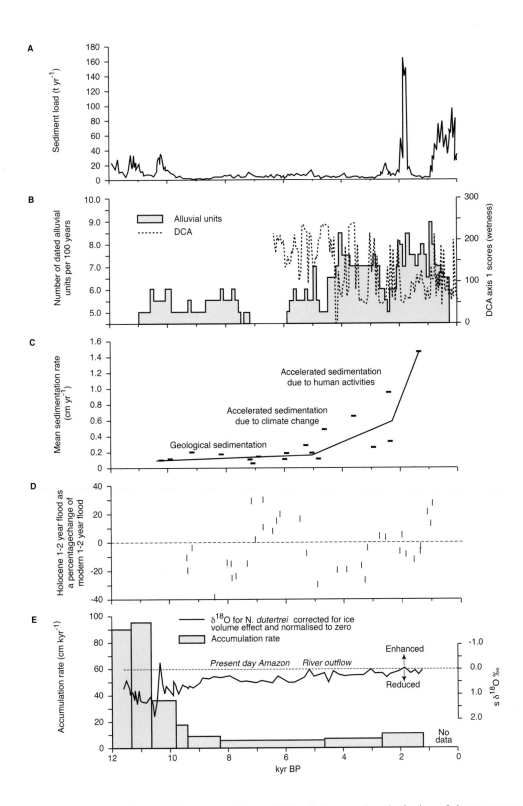

**Fig. 7.9.** Holocene proxy records of river discharge and sediment yield for landscapes where the forcings of change may range from solely climate to strongly anthropogenic: **A.** Lake sediment accumulation rates at Holzmaar, Germany (Zolitschka 1998); **B.** Frequency of dated alluvial units in Britain (Macklin 1999); **C.** Alluvial accumulation rates in the Yellow River basin, China (Xu 1998); **D.** Flood reconstructions derived from paleochannel cross-sections in southwestern Wisconsin, US (Knox, 2000); **E.** $\delta^{18}O$ measurements and accumulation rates from the Amazon Fan (from data in Maslin et al. 2000). All timescales are cal. [14]C yr B.P. (except **C**).

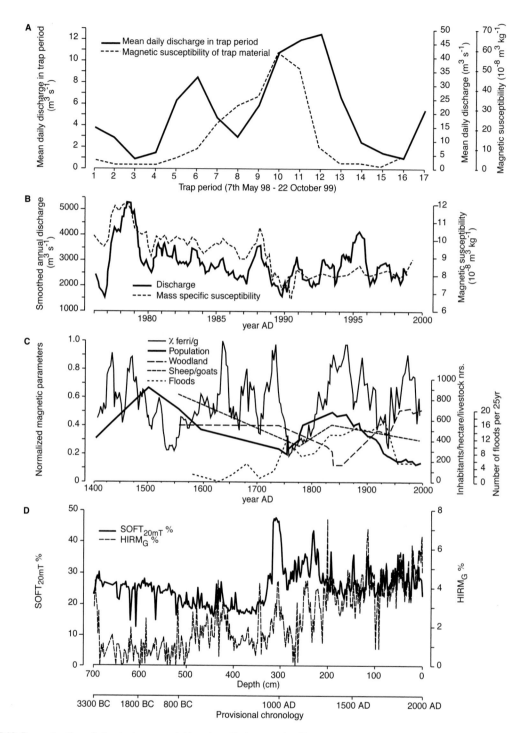

**Fig. 7.10.** Reconstruction of climate, human activities, river discharge and sediment transport at Lac d'Annecy based on calibrations and comparisons between sedimentary, instrumental and documentary archives over different timescales: **A.** Magnetic susceptibility ($\chi_{LF}$) of monthly trapped sediment and monitored river discharge (1998-1999) showing winter peak in susceptibility; **B.** Strong association between lake sediment $\chi_{LF}$ and river discharge record (1975-1999); **C.** Proxy records of river discharge (normalised $\chi_{ferri}$) compared with documentary records of major floods, human population, animal stocking levels and land use in the upland Montmin commune (1400-1999); **D.** Long records of lake sediment proxies for lowland surface soil (SOFT$_{20mT}$ %) and upland soil/unweathered substrates (HIRM$_G$ %) over the period 3300 BC – 1996 AD (Dearing et al. 2000; Dearing et al. 2001). Note the complex relationships between reconstructed discharge levels, historically recorded floods and land use changes (**C**), the major episodes of surface soil erosion following Cistercian clearances at ~1000 AD (**D**) and the trend of upland erosion, rising to the present day indicative of the long term conditioning of the montane zone by early clearance.

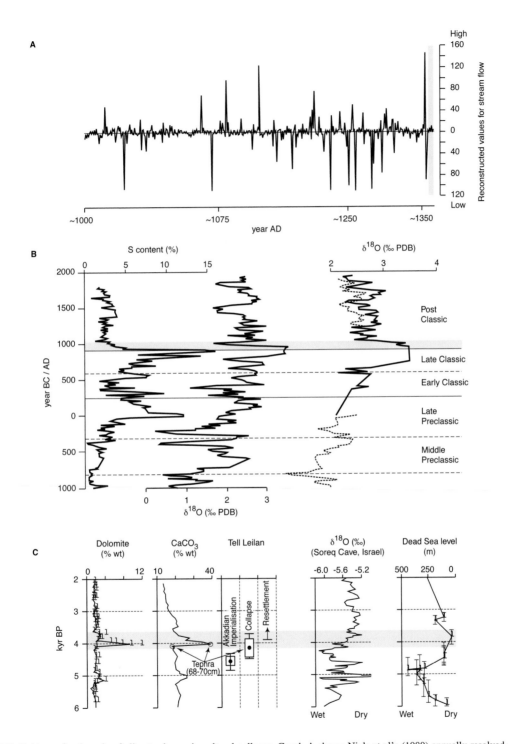

**Fig. 7.11.** Evidence for the role of climate change in cultural collapse. Graph **A** shows Nials et al's (1989) annually resolved record of past stream flow linked to the demise of the Hohokam culture in S. Arizona (Redman, 1999). The major swings in stream flow after AD 1350 reflect alternating severe flood and drought conditions thought to have been major contributors to sudden cultural change. Graph **B** shows sedimentary evidence from the sediments of Lake Chichancanab, Yucatan, Mexico, for a major and prolonged drought culminating during the period when Classic Mayan civilization collapsed (Hodell et al. 1995). Graph **C**, from Cullen et al. (2000) shows the close temporal correspondence between peak eolian mineral concentrations in the Gulf of Oman sediments and evidence for cultural collapse of the Akkadian Empire at the Tell Leilan site: note that a tephra layer permits precise synchronization of the marine sedimentary sequence and the archaeological record. Other lines of evidence for drought at the same time from speleothem stable isotope analyses and lake level changes are plotted alongside.

Among these is the tendency for a combination of population pressure, optimistic assumptions, short temporal perspectives and escalating technological fixes to raise the stakes in the interaction between nature and human populations. Setting these in the future context of likely major climate change and rapid population growth in the next century poses a daunting challenge.

## 7.9 Some future research priorities

Palaeo-environmental research has begun to shed crucial light on many aspect of present day ecology and ecosystem management for the future. This is confirmed by its role in identifying the origins and growing impacts of freshwater acidification and eutrophication (Battarbee 1978, Battarbee et al. 1990), its contribution to the definition of management targets in both aquatic (Anderson et al. 1995, Battarbee 1997, 1998) and terrestrial ecosystems (Millar 2000), its crucial partnership with modellers testing the long term performance of their simulations against the record from the past (Bradshaw et al. 2000, Bugmann and Pfister 2000), the essential contribution it makes to interpreting the dynamics of ecosystems under long term observation and monitoring (Fuller et al. 1998, Foster et al. 2000) and its essential role in deepening our awareness of the complex of processes that have combined, through time, to create the environmental goods and services which we value and upon which human life depends (Messerli et al. 2000). Common to all these examples is the demonstration that the palaeoenvironmental perspective can make an essential contribution to developing strategies for sustainable development.

Looking to the future, the consequences, both practical and academic, of discounting either human impacts on past environmental processes or climate influences on past human activities (no matter how strong and valid the antipathy to simplistic, old-style determinism may be) are equally damaging; a shared enterprise crossing the classic two-cultures divide is now an urgent requirement.

Putting this into practice will require:

- establishment of close links between the concepts of ecological dynamics (e.g. Lindblagh et al. 2000) and ecological modeling (Bugmann 1997, Bugmann and Pfister 2000);
- a continuing search for unifying frameworks of study, whether primarily biophysical (e.g. Oldfield 1977) or more conceptual (Dearing and Zolitschka 1999);
- better understanding of the nature of environ-

mental thresholds and nonlinear responses
- adoption of common research protocols and comparison of insights from diverse case studies (e.g. Wasson 1996) and from those with the range and depth exemplified by the cultural and environmental histories described for Southern Sweden (Berglund 1991);
- wider adoption of a *post-hoc* hypothesis testing approach (Deevey 1967) as illustrated in the acid-deposition section of the chapter, alongside model-data comparisons;
- greater reliance on quantitatively calibrated non-biological proxies of past climate change in order that biotic records may be interpreted as *responses* (Amman and Oldfield 2000)
- the use of all potential archives (e.g. Barber et al. 2000, Hughes et al. in press) and proxies (Oldfield and Alverson in press) for reconstructing climate change and human impact, including especially the work of environmental historians and historical ecologists who use documentary and archaeological sources (e.g. Egan and Howell 2001).
- The compilation of relevant global data bases (Klein-Goldewijk in press, Ramankutty and Foley in press) (Appendix B)
- Improvement of quantitative calibration of proxies in terms of environmental processes, conditions and patterns (e.g. Broström et al. 1998) and not only in terms of climate

The future research agenda that evolves from the above should promote a move beyond case-studies toward new theories, models and generalizations about future environmental responses linking together the full range of spatial and temporal dimensions (e.g. Dearing and Jones in press). The implications of such an endeavor for conservation (Birks 1996), landscape and ecosystem management (Heyerdahl and Card 2000) and environmental sustainability (Goodland 1995) are compelling.

## 7.10 Acknowledgements

We are particularly grateful to Rick Battarbee, John Dodson, John Flenley, Geoff Hope, Guoyu Ren, David Taylor, Peter Crooks, Sharon Gedye and Richard Jones for access to unpublished data and for the benefit of their help with sections of the text where our own experience was least adequate and to Keith Alverson, Ray Bradley, Carole Crumley, Vera Markgraf, Bruno Messerli, Tom Pedersen, Bob Wasson and Cathy Whitlock for essential constructive criticism. We also thank Sandra Mather for producing most of the figures.

# Challenges of a Changing Earth: Past Perspectives, Future Concerns

R.S. Bradley
Climate System Research Center, Department of Geosciences, University of Massachusetts, Amherst, MA, 01003-9297, United States of America

K. Alverson
PAGES International Project Office, Bärenplatz 2, CH 3011 Bern, Switzerland

T.F. Pedersen
Earth and Ocean Sciences, 6270 Univ. Blv, University of British Columbia, Vancouver, Canada V6T 1Z4

## 8.1 Introduction

Paleoclimatic research has revealed an astonishing picture of past changes in the earth system. Over the last 2 million years, climate has varied widely from ice ages to warm interglacials. Between these extremes global sea-level varied by up to 130 m, alternately exposing and submerging land bridges that provided gateways between continents for the migration of early humans. Ice sheets grew to such an enormous size, they depressed the earth's crust under their weight and forced airmasses around them. So much water was sequestered on the continents that the chemical composition of the oceans was altered, affecting all the organisms that lived in the seas. As the ice sheets grew, the world's terrestrial and marine ecosystems changed, in both distribution and composition. Continental interiors became drier, wind speeds increased and large areas of wind-blown loess accumulated. As ecosystems were altered, the composition of the atmosphere changed repeatedly, with low levels of important greenhouse gases during glacial episodes and higher levels in interglacial periods. The evidence for these changes has been carefully extracted from ice cores and sedimentary records on land and in the ocean. Advances in technology have enabled longer and better cores to be recovered, and more detailed and more precise analytical techniques to be applied to the records. Better chronological control has allowed more precise estimates of the timing of changes in different regions to be made, enabling leads and lags in the system to be established and rates of change to be examined. In addition to reconstructions of climate system parameters, paleodata have also yielded information about the *causes* of past changes, focusing attention on mechanisms within the climate system that translate the forcings into responses that may vary from one region to others.

Computer models of the climate system have elaborated on these mechanisms and helped to quantify the nature of interactions between different regions and sub-systems.

Research on glacial-interglacial timescales has thus revealed the dramatic changes that took place as man evolved from *Homo erectus* to *Homo sapiens*. Changes in climate over more recent intervals of time have been no less significant for human society. The origins of agriculture were associated with abrupt changes in climate in the Middle East during the pre-Holocene/Younger Dryas interval. Archeological research provides numerous examples of societal disruptions related to climate variability during the Holocene, reminding us that agriculturally-based societies remain vulnerable to abrupt and persistent climatic anomalies, even today. As we focus on climatic changes on interannual and inter-decadal timescales over recent centuries, it is increasingly clear that important modes of climate variability, such as ENSO and the NAO, have not been an unchanging feature of the climate system over time, and there is no reason to expect they will remain constant in frequency or amplitude in the future.

All of these studies remind us that our modern perspective on the climate system, derived from detailed instrumental observations over (at best) the last century or two, provides a totally inadequate perspective on its real variability. *The message from the paleorecords is that change is normal and the unexpected can happen.*

## 8.2 Understanding earth system variability

To understand the full variability of the climate system requires a comprehensive network of well-dated, quantitative paleoclimate data. Compared to contemporary environmental observational networks, the network of paleorecords is still sparse and in many cases poorly resolved in time. Where high resolution records have been recovered, significant advances in our understanding of how the climate system operates have often been made possible. It is essential that more such records be recovered, for the task of deciphering the past is far from over. In particular, a comprehensive understanding of variability on societally-relevant timescales (years to several decades) will require an enormous expansion in paleoresearch. Prediction of ENSO activity, for example, will not be fully achieved as long as the full spectrum of ENSO variability remains unknown. So far, the paleorecord provides hints that the modal frequency of ENSO variability (determined by instrumental observations) that the world has experienced over the last century has not been stable. Rather, the ENSO system may have operated at a lower frequency in the recent past (Figure 3.7). There is also evidence that there were periods in the past when ENSO variability may have essentially disappeared, or at least changed in character to a state quite different from that which we now take as its normal mode. Furthermore, it is clear that ENSO teleconnection patterns, that affect rainfall and temperature around the world, have changed over time, yet we have only a rudimentary understanding of how or why such changes have occurred. Predictive models based on limited instrumental data may be able to generate scenarios of unusual ENSO behaviour in the future, but confidence in such predictions will be limited unless there is sound evidence that such conditions are plausible, as demonstrated by records from the past.

### 8.2.1 Paleoclimate data and models

Models are useful tools to understand the complexities of past and future climate variations, but they are no substitute for reality. Unlike models of the past, paleorecords are integrators of forcings and feedbacks, incorporating all the processes operating on them. Global climate models can never be relied upon to give a definitive description of the natural variability of the true climate system since models cannot estimate variability associated with processes or feedbacks that are not present in the models themselves. The potential non-linearities of small-scale processes that are not resolved, dynamical feedbacks that are not accurately represented, and numerous potential surprises in the global climate system will forever limit the ability of models to accurately represent the true variability of the climate system. In fact, paleodata offer the best prospect of validating the low-frequency behavior of models. They add a critical element to the climate change detection and attribution problem that is not

possible from a comparison of 20th century model simulations and instrumental observations alone. Paleodata allow us to look at potential forcing mechanisms prior to the era of anthropogenic forcing, potentially validating or testing model-based estimates of climate sensitivity. For example, estimates of the sensitivity of climate to solar irradiance forcing, estimated empirically from paleodata have provided validation of model-based estimates. In this way, the magnitude of "natural" forcing, upon which anthropogenic forcing is superimposed, can be assessed. Furthermore, "natural" forcings may produce a spatial response in the climate system similar to the signal of anthropogenic (greenhouse gas) forcing, confounding the detection of significant anthropogenic climate effects. Understanding these responses is thus necessary in anthropogenic signal detection.

## 8.3 Nonlinear dynamics in the earth system

Throughout this volume, we have seen examples of how the paleorecord and associated modeling give insight into processes operating within and between earth system components, and into relationships between forcing mechanisms and system responses. An important theme that appears common to many different temporal and spatial scales is the nonlinear nature of climatic and environmental changes. Once critical thresholds have been passed, recovery of the system to its former state may take a long time, and indeed the former state may not be achieved.

### 8.3.1 Earth system thresholds

The definition of 'abrupt' change is necessarily subjective, since it depends on sample interval and the pattern of longer term variation within which the change is embedded. Nonetheless, rapid response to gradual forcing is a ubiquitous feature of past climatic and environmental change on all time and space scales, and must be expected to continue into the future. Many examples of threshold dynamics in paleoenvironmental reconstructions and models that are relevant to the future are presented in this book. Some selected examples include:

•    The upper and lower bounds within which atmospheric $CO_2$ concentration has varied over the past four glacial cycles appear to be thresholds within the global carbon cycle (Chapter 2). The ~200ppmv minimum during glacial times may represent reduced photosynthetic efficiency, below which the global biosphere is restricted in its ability to further lower atmospheric $CO_2$ levels (Chapter 5,

Section 5.3). The ~280 ppmv maximum during interglacials, though not fully understood, is probably controlled by oceanic processes. This limit has been drastically exceeded during the Anthropocene (Chapter1, figure 1.6) with as yet not fully understood consequences for the planet.

• The waxing and waning of massive continental glaciers during the Quaternary is reflected in a rich history of sea level variability (Chapter 3, Figure 3.2a). One simple but plausible model captures a great deal of this variability, simply by assuming a threshold response of the earth system to known high latitude insolation changes (Chapter 3, Figure 3.2b). On millennial timescales, the stadial/interstadial climatic transitions which punctuate glacial periods are modelled as being initiated when fresh water fluxes in the Northern North Atlantic cross a critical threshold amount required to shut down the global thermohaline circulation (Chapter 3, Section 3.3).

• Large regions of the modern ocean, despite ample nutrients, have very low marine productivity. Experiments have shown that with a supply of iron above some critical threshold (as is thought to have occurred in the past) productivity rises dramatically, providing a potentially large oceanic carbon sink. This 'biological pump' may have played a key role in past variations in atmospheric carbon dioxide concentrations (Chapter 4, Section 4.3)

• With some exceptions, the past record shows that many species are able to shift their ranges at a threshold rate of 200 to 2000 m a$^{-1}$ (Chapter 5, Section 5.4). If future climate change causes the potential range of species to shift faster than their threshold rate, this (combined with effects of a fractured, human-dominated landscape) will very likely lead to extinctions.

• Explosive volcanic eruptions that propel particles and gases high into the stratosphere may, if large enough, lower temperatures at the surface to the point that additional feedbacks (involving snow and sea-ice) may amplify the radiative effect of the aerosols, and cause the initial perturbation to persist (Chapter 6, section 6.11).

• Acidification of lakes in recent decades has been shown to be due primarily to industrial pollution and to lie outside of the range of natural variability (Chapter 7, Figure 7.3). The single most important factor determining vulnerability to acidification is the chemical buffering capacity of individual lake catchment systems (Chapter 7, Section 7,4). Chemical buffering serves to maintain stability of lake pH under natural variations in acid input. Given too large a pertubation however, the threshold buffering capacity may be overrun, resulting in rapid acidification and consequently

environmental degradation may occur.

## 8.3.2 Hysteresis

Hysteresis is a non-linear dynamical feature common to past climate and environmental processes, across the full range of temporal and spatial scales. In many systems, abrupt shifts associated with the crossing of critical threshold such as those described above, are either difficult or impossible to reverse. (Chapter 1). Figure 8.1 shows three selected examples of such hysteresis behavior. The upper panel shows an example from the past which is thought to have affected the entire earth system for thousands of years - shown here as thermohaline circulation shutdown by fresh water forcing in a numerical model. Even after fresh water forcing, such as that which triggered the Younger Dryas cold event, is reduced models suggest that it may take hundreds of years for the thermohaline circulation to return to its previous state. The second example, shown in the middle panel, is much smaller scale, and taken from the more recent past when natural and human effects interacted. Nutrient loading from fertilizer input in the 1970's caused catastrophic decline of charophyte vegetation in this Dutch lake and associated ecosystem change. In order to reverse this environmental degradation required that the nutrient loading of the lake be decreased far below the level that triggered the environmental catastrophe in the first place. In the lower panel, we show a schematic of the earth system as a whole. The horizontal axis shows the increases in atmospheric $CO_2$, combined with systemic human impacts on regional ecosystems, that are now driving the system over a critical threshold which separates the Holocene from the Anthropocene (Chapter 1). The degree of hysteresis in this transition is unknown, but certainly not zero.

Thus, any strategy to manage the earth system, or its components needs to account for an understanding of system behavior provided by past records. As discussed in Chapter 6, section 12, an understanding of such system behavior can not possibly be provided by predictive models alone. Rather, it must be gleaned from the past record. Policies which will benefit from a better understanding of past system dynamics and the inherent dangers of reaching critical thresholds in the climate system include those focused on:

• reducing greenhouse gases in an attempt to reduce associated global warming (Chapter
• understanding, monitoring and perhaps preventing changes in regional climate associated with variability of the thermohaline circulation (Chapter 3)

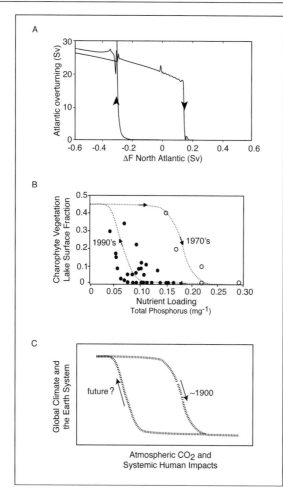

**Fig. 8.1.** Three examples of hysteresis in components of the earth system on vastly different spatial and temporal scales. Upper panel: Thermohaline circulation shutdown by fresh water forcing in a numerical model as a potential mechanism explaining the Younger Dryas or other abrupt climate changes (after Stocker and Wright 1991; Stocker et al. 2001). Middle panel: Lake vegetation response to fertilizer input changes in a Dutch lake over the past few decades (after Meier in Scheffer et al. 2001). Lower panel: A schematic of possible future hysteresis in earth system response to global anthropogenic climate change.

- sequestering fossil fuel carbon in either terrestrial or oceanic reservoirs (Chapter 4)
- developing natural reserves to maintain biodiversity and ensure species survival (Chapter 5)
- detecting and attributing global and regional change to natural and/or anthropogenic forcing (Chapter 6)
- managing lake catchments, coastal zones and landcover in order to protect them for future generations (Chapter 7)

## 8.4 Is the past irrelevant to the future?

It might be argued that future changes in the climate system resulting from the dramatic rise in anthropo-

genic greenhouse gases, will render evidence from the past irrelevant. This is not so. Times of rapid change in the past provide critical insights into how system responses and interactions can be expected to occur. For example, the rates at which various species migrate in response to climate change are primarily constrained by paleodata (Chapter 5, section 5.3). These parameters are critical for management strategies that seek to maintain biodiversity in the future.

Furthermore, there are periods in the past when conditions were warmer or drier than today (at least regionally) and these can be examined to assess how environments may change in the near future. For example, numerous high latitude records from the northern hemisphere point to the early Holocene as a time when summer temperatures were warmer than in the last millennium. It would be useful to know how permafrost responded to such conditions, what effect higher temperatures had on methane flux from northern wetlands, and what Arctic Ocean ice cover was like at that time. Of course, the driving force for such warmer conditions was not a high level of $CO_2$, so a precise analog should not be expected, but nevertheless, the environmental consequences of warmer summer temperatures (whatever their cause) are of interest as we prepare for a warmer future.

## 8.5 Human vulnerability to future climate change

Without water, civilization cannot exist. We are well reminded of this by the recently-documented collapse of Akkadian society in the Middle East during a prolonged arid episode 4,200 years ago, and the demise of the Maya in Central America some 1100 years ago when the region turned dry (Chapter 1, Figure 1.5; Chapter 7, Figure 7.11). Changes in the hydrological cycle in east central Africa during the last millennium (Chapter 1, Figure 1.5) have similarly driven societal upheavals (Chapter 6). Moist conditions indicated by century-long high lake stands in what is now Kenya supported three episodes of pronounced prosperity over the last 600 years, but the inhabitants of the region migrated away toward climatically more hospitable areas during a trio of prolonged dry intervals that also punctuated this time interval.

The last millennium has witnessed the severe impact of drought on civilizations elsewhere. Some 600,000 people died of starvation in northern India during the period 1790-1796 as a direct result of limited monsoon rainfall and low soil moisture. During such drought events, dust from the desiccated soils is carried by southerly winds onto the Tibetan Plateau north of the subcontinent. The airborne dust either settles onto snow-covered surfaces on the Plateau or is washed out of the atmosphere by precipitating snow. Progressive accumulation and compaction of seasonal dust-bearing snow layers gives rise to the Tibetan glaciers that mantle the high annulus of the Plateau. The dust accumulation history preserved in these ice deposits therefore yields an annual-resolution archive of the soil moisture deficit across northern India. The striking drought of 1790-96 is but one arid event recorded in this archive. Episodes of marked dust deposition have occurred intermittently over the past 600 years indicating that collapse of the monsoon is not an unknown occurrence in the region.

All these examples of unusual climatic conditions in the paleoclimatic record, have certain attributes in common: they were *abrupt* (and therefore unexpected); they were *unprecedented* in magnitude (in the experience of those societies affected), and they were *persistent* (lasting long enough that societies were unable to easily accommodate the change). Often, this led to people migrating from the affected region, as in the example noted earlier. But in the highly populated, internationally regulated world of today, such a response to climatic disruptions is unlikely to be an option and thus, if such climatic events were to recur again, the scene is set for societal disruptions on a massive scale. Furthermore, additional factors must be considered. The impact of droughts on agriculture can be mitigated by irrigation, providing that water reserves, primarily groundwater, are available. Agriculture in many parts of the world has flourished over the past century precisely because groundwater has been heavily exploited for agrarian purposes, thereby supporting an ever-increasing population. But groundwater is a critical buffer against episodic drought. Will such buffers be able to offset the next collapse of the monsoon in northern India, or future mega-droughts in the western and central U.S.? The answer almost certainly is no. The fossil groundwater that has been mined in Gujarat State is now almost exhausted; the water table over much of the region has dropped by over two hundred *meters* in much of the area. Similar problems are apparent in areas with more developed agricultural systems. In northwestern Texas, the High Plains Aquifer that has supported intensive agriculture in the region for more than a century has no more than a 30-year supply of water left at current rates of usage – the groundwater that accumulated over the course of the late Holocene has been nearly exhausted in the mere blink of a geological eye. The situation is as serious in many other parts of the world: southern Alberta, Canada, has diminishing stocks of groundwater; much of the sub-surface supply in middle eastern countries is being exhausted at rapid rates, leading to the desiccation of oases that had until recently supplied water for centuries. Another dimension to this issue exists in mountainous regions. The summer ice-melt that has sustained dry-season river flow for centuries in countries like Peru is now being seriously threatened by the recession of alpine glaciers (Chapter 6) as temperatures

Some of the general circulation models that are used to peer over the horizon suggest that as the concentrations of greenhouse gases in the atmosphere continue to increase, areas such as northwest India and the southwestern US will be drier in future. Because the fossil groundwater reserves that underlie these areas have been severely depleted, despite the relatively wet periods of much of the last century (Chapter 6), there is little prospect of future buffering of drought. Few options exist for substitution of groundwater with surface flow. As a result, acute impacts on societies in these regions can be anticipated, and need to be taken into account in climate-change policy decisions at local through international scales.

Where does this leave us? Not only are we living in unusual times in terms of global-scale anthropogenic climate change, but we have been unwittingly pre-conditioning the future by weakening our ability to adapt to *both* natural and human-induced change. Combining studies of the past with observations from the present and predictions of the future is yielding sobering messages, and they need to be heeded; we ignore them at our peril.

# Appendix A - The Past Global Changes (PAGES) Program

K. Alverson, I. Larocque, C. Kull
PAGES International Project Office, Bärenplatz 2, CH-3011 Bern, Switzerland

## A.1 The PAGES mission

PAGES is the International Geosphere Biosphere Program (IGBP) project charged with providing a quantitative understanding of the Earth's past climate and environment. One major obstacle standing in the way of producing reliable predictions of global climate change and its environmental impacts is a lack of data on time scales longer than the short instrumental record. Natural archives of past climate variability can provide relevant information over longer timescales. One major goal of PAGES is to provide this information to scientists, policy makers and to the general public.

## A.2 Research program

The PAGES research program is structured so as to bring researchers from a variety of disciplines and countries together to work on common themes. Thus, no explicit disciplinary or national structures exist within the program. Rather, the program elements are designed so as to reduce constraints imposed by geography and artificial disciplinary boundaries like those that commonly separate physical oceanography from continental paleoecology from archaeology. Although the five foci which make up the core of the PAGES program have been created to group types of research together, they are by no means restrictive, and collaboration between foci are strongly encouraged. The activities of each Focus are guided by a chair and a small steering group.

## Focus 1 - PANASH

The goal of the Paleoclimate and Environments of the Northern and Southern Hemispheres (PANASH) focus is to reconstruct paleoenvironments and paleoclimate along three Pole-Equator-Pole (PEP) terrestrial transects (Figure A.1) using a multiproxy data and modeling approach. One of the major roles of the PEP transects is to facilitate the development of north- south research partnerships and foster a unified sense purpose within the diverse international and interdisciplinary community addressing questions of past global change. The PEP transects have been extremely successful in achieving these goals as evidenced by the strong community interest in PEP activities (more than 300 people attended the PEP3 conference "Past Climate Variability Through Europe and Africa" in Aix-en-Provence, 27-31 August 2001), as well as the peer reviewed synthesis publications that have arisen from each transect (Markgraf 2001, Dodson and Guo in prep, Battarbee et al. in press).

In addition to supporting the PEP transects, PANASH as a whole stimulates exchange of information among marine, atmospheric, and terrestrial scientists, historical ecologists and environmental archaeologists globally. Its primary tasks are to:

- document the amplitude, phase and geographic extent of climate change in the two hemispheres,
- determine the history of potentially important forcing factors,
- identify the important feedbacks which amplify or reduce the influence the effect of these forcings,
- identify the mechanisms of climatic coupling between the two hemispheres.

The basis of all PANASH activities is understanding the modes of climatic and environmental variability by answering problem-oriented questions, through the use of global, multi-proxy based climate and environmental reconstruction. Further information about PANASH can be found here: www.pages-igbp.org/structure/focus1.html

## Focus 2 - The PAGES/CLIVAR Intersection

The PAGES/CLIVAR Intersection focus aims to improve the understanding of decadal to century scale climate variability, especially as relevant to improving predictability, through the use of high resolution paleoclimatic data. The activities within this focus are overseen by a joint working group shared between PAGES and the World Climate Research Program (WCRP) Climate Variability and Predictability Program (CLIVAR). The four principal areas of concentration, as outlined at the first international CLIVAR conference (Alverson et al. 1999), are:

- extending the instrumental climate record back in time with quantitative proxy data that can be accurately calibrated against instrumental records
- documenting and understanding rapid climate change
- documenting and understanding natural climate variability during the Holocene and other interglacial periods with background climatic states similar to those of today
- Testing the ability of climate models to capture known past climate variability.

Further information about the PAGES/CLIVAR intersection can be found at www.clivar.org/organization/pages/index.htm

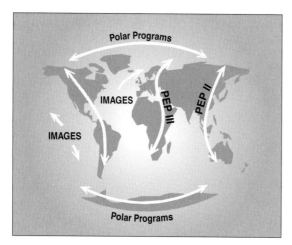

**Fig. A.1.** A schematic diagram showing some of the main Foci within PAGES. The Pole-Equator-Pole (PEP) transects within focus one, PANASH; focus three, IMAGES; and focus four, Polar Programs. The other two PAGES foci, the CLIVAR/PAGES intersection and Ecosystem Processes and Human Dimensions, do not have clear geographic boundaries.

## Focus 3 - IMAGES

The International Marine Past Global Changes Study (IMAGES) is the marine program shared between PAGES and the Scientific Committee for Ocean Research (SCOR). The principal aim of this program is to understand the mechanisms and consequences of past climate changes as long as ocean circulation, salinity, ventilation, carbon sequestration and flux, using oceanic sedimentary records. IMAGES supports a number of working groups, some of which are oriented around cruise planning and others around more general research questions. One major task of IMAGES is to organize international pooling of financial recourses and research expertise in order to enable cruises to be carried out

throughout the world oceans. All IMAGES activities are carried out with strong input and interaction from climate modelers, continental scientists and ice core researchers. Further information on the IMAGES program is available here: www.images-pages.org/

## Focus 4 - Polar Programs

The Polar Programs focus is bi-polar. Examples of research within the remit of this focus include the European and US N GRIP ice core programs, EPICA and the International Trans-Atlantic Scientific Expedition (ITASE) which seeks to map the spatial variability of Antarctic climate over the last millennium. This initiative is shared with the Scientific Committee on Antarctic Research (SCAR). In addition to ice core work, a wealth of other archives in polar regions are employed to provide a robust picture of high latitude environmental change. For example, the ESF-funded Quaternary Environment of the Eurasian North project (QUEEN) concentrates on mapping the extent of the last glaciation, the CircumArctic Paleo-Environments program (CAPE) facilitates integration of paleoenvironmental research on terrestrial and adjacent margins covering over the last few glacial cycles. Further information on the Polar Programs focus is available here: www.pages-igbp.org/structure/focus4.html

## Focus 5 - Past Ecosystem Processes and Human Environment Interactions

The Past Ecosystems and Human-Environment Interactions focus highlights PAGES concern with ecological responses to climate change and past human activities. The research within Focus 5 integrates past human-environment interactions at subcontinental scale with research and modeling based on present day ecosystems and watersheds. The focus is divided into three main activities: Human Impacts on Terrestrial Ecosystems (HITE), Land Use and Climate Impacts on Fluvial Systems during the Period of Agriculture (LUCIFS) and Human Impact on Lake Ecosystems (LIMPACS). These activities are case-study based and focus on ecosystems made vulnerable to global change through any combination of natural and human induced stresses. They also explore the basis for the durability of long-sustained ecosystems, questions of sensitivity, thresholds and non-linear responses. Further information about Focus 5 is available here: www.pages-igbp.org/structure/focus4.html

## A.3 Initiatives

The original list of tasks and activities that once underlay PAGES Foci (Oldfield 1998) has been almost entirely transformed. PAGES now supports initiatives driven by scientific questions. The PAGES Steering committee serves to critically ascertain if proposed initiatives should qualify for PAGES endorsement and support. Successful initiatives are expected to develop a clear research and workshop agenda over a 3-5 year period leading to a tangible goal. PAGES support for these initiatives is flexible but can include enhancing the profile of the initiative, advertising it to the international community and providing partial funding for workshops. One example of a successful initiative is the Environmental Processes of the Ice Age: Land, Oceans, Glaciers (EPILOG) program, which arose in 1999 as a multi-national working group of the PAGES marine program IMAGES and recently published an extensive special issue on ice sheets and sea level of the last glacial maximum (Clark and Mix 2002). The required qualifications to be considered as a PAGES initiative are:

- A question which seems likely, within a 3-5 year timeframe, to be tractable in the sense of leading to a peer reviewed product which advances the field.
- A clear reason why PAGES should be involved, for example to facilitate new international or interdisciplinary bridges and community building.

## A.4 Program Structure

PAGES activities are overseen by an international scientific steering committee (SSC) appointed by the Steering Committee of the IGBP. The sixteen members, who each serve for at most two consecutive three-year terms, are chosen to provide a balance of scientific expertise and national representation. This committee meets once a year to provide guidance for and oversight of the program as a whole. A subset of five committee members serves as an executive committee, which is in more regular contact with the International Project Office. As a general guideline the five member executive committee includes an American and a Swiss by virtue of the fact that these countries provide the bulk of PAGES core funding, and a member from a less developed country. By direction of the SSC, the staff of the small and efficient International Project Office (IPO) carries out the day to day running of the PAGES program as a whole. These activities include maintaining the PAGES website and database, organizing meetings and workshops, editing and writing PAGES publications, and serving as a liaison with other global change programs. The staff of the IPO nominally consists of three full time positions, executive director, scientific officer and office manager. In addition, the office regularly hosts both short and long term (sabbatical) visits from paleoscientists around the world.

## A.5 Links with other international programs

PAGES continues to be primarily concerned with understanding the past operation of the Earth system. As shown schematically in figure A.2, the PAGES remit includes the physical climate system, biogeochemical cycles, ecosystem processes and human dimensions. Thus, PAGES activities are not restricted to IGBP, but overlap substantially with IGBP's sister programs within the "earth system science partnership," the WCRP, International Human Dimensions Program on Global Environmental Change (IHDP) and Diversitas. As indicated at the base of the schematic in Figure 1, facilitating publicly accessible paleodata access, engaging with the climate modeling community and strengthening the role of developing countries in PAGES research continue to be three foundation blocks upon which the wider PAGES scientific program rests.

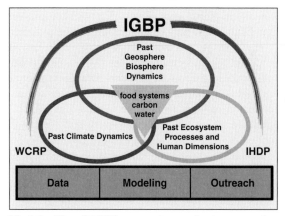

**Fig.A.2.** The PAGES research agenda is more than paleoclimate. It explicitly identifies the need to improve understanding of geosphere-biosphere changes at the global scale and to link human-environment interactions with studies of ecosystem processes. This schematic provides a simple visualization of the overarching themes that express:

- the major research areas with which the PAGES community is concerned
- the main areas of interaction with not only IGBP, but also the other principal organizations in the earth system science partnership, WCRP and IHDP and Diversitas
- the way these overlap and link into the three crosscutting themes within the earth system science partnership on sustainability issues of food systems, carbon and water.

the underpinning aspects of data, modeling and outreach that are essential to the success of the whole enterprise.

PAGES has built bridges with many other international scientific programs. Although only two of our Foci are officially shared, Focus 2 with WCRP-CLIVAR and Focus 3 with SCOR, all of them have substantial interactions with other programs. PAGES has launched joint initiatives with all of the other components of the IGBP, including both the core projects of IGBP phase 1 and the newly developing projects of IGBP phase 2, due to officially begin operation in 2003. A full listing of PAGES science partnerships is available on the PAGES website, some examples include the Global Network for Isotopes in Precipitation (GNIP), shared with the WCRP and the International Atomic Energy Agency (IAEA) and the International Mountain Research Initiative, co-sponsored by three other IGBP core projects, IHDP, GTOS and UNESCO. Through its intersection activities with other global change programs, PAGES provides the historical context for global change programs.

## A.6 Outreach and communications

One major task for PAGES is to provide easy access to paleoenvironmental information, to active researchers in paleosciences, researchers in other aspects of global change research, and the public. One of the most important communication platform is the website www.pages-igbp.org. This site is modified regularly and includes lists of new products, links to paleoenvironmental databases, science highlights, a calendar of upcoming events and information on how to become involved in PAGES activities. A set of overhead view graphs, based in part on figures from various PAGES publications, is available to download from the website. These overheads are regularly used by scientists for educational and public outreach lectures. The overheads are sent, on request, free of charge to scientists in developing countries who are unable to download and print them from the website. The PAGES website also supports an online database of paleoscientists, searchable by name and expertise.

Another important element in the PAGES communication strategy is the newsletter. PAGES NEWS is produced three times a year and sent free of charge to 3000 subscribing scientists in more than 70 countries. All are made available as pdf files on the PAGES website. Such wide distribution, coupled with a high degree of proactive submission by the research community has made the newsletter an important vehicle for the dissemination of research results, workshop reports and program news,

especially in countries with limited access to western journals. Each issue is developed around a specific theme, which might be a PAGES program or a particular paleoarchive. Ideas for themes for newsletter issues are always welcome.

In addition to its newsletter and website, PAGES strongly encourages publications in the peer reviewed literature as one outcome of all of its scientific activities. An exhaustive list of publications which have come about through research linked to PAGES is difficult to construct because of the very inclusive nature of PAGES organization. However, some examples of recent books and special journal issues which have arisen directly out of PAGES programs include (Dodson and Guo 1998, Alverson et al. 2000, Kroepelin and Petit-Maire 2000, Markgraf 2001, Mix et al. 2002, Dodson and Guo in prep, Battarbee et al. in press, Mix et al. in press).

## A.7 Data archives

Internationally accessible data archives are one of the primary foundations for paleoclimatic research that seeks to go beyond reporting results on a site by site basis. PAGES is committed to the IGBP principles of ensuring the preservation of all data needed for long-term, global change research and making them openly available as soon as they become widely useful. PAGES recognized the value of such data libraries at its outset and was an important early supporter of the World Data Center for paleoclimatology, Boulder (www.ngdc.noaa.gov/, appendix B). The WDC-Paleoclimatology is now a key source of paleoclimatic information that is made freely available to scientists across the globe. In addition to data repositories, PAGES, through the decisions of its data board, encourages the development of thematic and national relational paleoclimate databases such as PANGAEA initiative (www.pangaea.de). Future progress in understanding climate history will depend increasingly on the provision of well-documented data by such data centers. PAGES supports a data board with a primary responsibility for assuring compatibility and accessibility of available existing paleo-database. The PAGES data board is open to all interested participants and includes members from most major database centers and focus leaders.

## A.8 Capacity building - encouraging north-south research partnerships

PAGES has a strong interest in capacity building. The majority of paleo-environmental data are extracted in less developed countries. Recent high profile examples include ice cores from Kilimanjaro, tree rings and lake sediments from Siberia, loess records in China, tropical tree rings in SE Asia, and speleothems in Oman. Flying western scientists around the world to take these records home and analyze them, the current practice for the most part, is no more justifiable than bringing cultural artifacts back to the British Museum was in 1800. A better way to bring these records together into a synthesis view of past environmental change is to facilitate the careers of independent working scientists within the many countries in which these archives exist and to ensure that they are well tied into global efforts.

PAGES allocates approximately 20% of its budget directly to activities geared towards building the capacity of scientists residing in developing countries to carry out active participation in paleoenvironmental research. In addition, most PAGES workshop and summer school support is earmarked for participants from developing countries. In addition to finances, PAGES seeks to follow up one time support wherever possible by enhancing the number of young scientists from developing countries in our database, nominating outstanding individuals for various awards and entraining them directly in our major scientific initiatives. We occasionally host visits at the PAGES office usually when tied to academic visits at a department at the nearby University of Bern or the Swiss Climate summer school.

The PAGES Regional, Educational and Infra-structure Efforts (REDIE) project seeks to:

- Enlist scientists and technicians in developing countries in international paleoenvironmental research activities

- Promote the development of paleoscience research within developing countries.

Within the REDIE program, a number of approaches are used. Financial support is made available for the attendance of active young scientists at key conferences and summer schools. PAGES publications are made available free of charge to libraries and university laboratories in less-developed countries. Scientists from Asia, Africa and South America sit on the PAGES Science Committee and act as liaisons with their regional communities. PAGES scientific meetings are regularly organized in developing countries, with ample support and presentation time provided for the attendance of scientists from the region.

## A.9 How to get involved with PAGES activities

There are many ways to get involved with PAGES activities:

- **Join existing programs** by contacting the relevant focus leader directly or sending a description of your research project to the IPO, which can help link your project to complementary efforts around the world.

- **Propose a new initiative** as outlined above under 'initiatives' to the scientific steering committee by sending a brief written proposal to the IPO.

- **Propose a workshop** by sending a one-page description of the goals of the workshop, the expected products, likely speakers, the planned used of PAGES funds and likely additional sources of support.

- **Nominate someone for the SSC** by sending a brief cover letter and c.v. to the PAGES office (pages@pages.unibe.ch)

# Appendix B - The PAGES Data System

C. M. Eakin
NOAA/National Geophysical Data Center, World Data Center for Paleoclimatology, 325 Broadway E/GC, DSRC 1B139, Boulder, CO 80305-3328, U.S.A

M. Diepenbroek
Alfred Wegener Inst. for Polar & Marine Research, Am Alten Hafen 26, DE-27568 Bremerhaven, Germany

M. Hoepffner
CNES, MEDIAS- FRANCE, BPI 2102 18 avenue E. Belin, FR-31401 Toulouse-Cedex 4, France

## B.1 The PAGES data system and its components

When PAGES first began, centralized management of paleoclimatic data rested in the World Data Center (WDC) for Paleoclimatology (then called WDC-A for Paleoclimatology) and in disparate but burgeoning data management efforts. By the second PAGES Data Meeting in 1998 (Anderson and Webb, 1998), additional major data management efforts were taking shape. The PAGES Data Guide highlighted three main components with important roles in data management:

- Scientists, who generate, publish and contribute the paleoclimatic data
- Data cooperatives, project-level data management, thematic and regional archives, and
- The World Data Center for Paleoclimatology (and regional mirrors) that provide the long-term archive and access to the data.

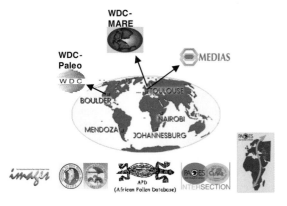

**Fig. B.1.** Many of the data centers and data activities that are part of the PAGES Data System.

Today, additional major data management efforts play essential roles in archiving and distributing paleoclimatic data (Figure B.1). The German PANGAEA group, host to the World Data Center for Marine Environmental Sciences serves as a major archive for paleoceanographic data and for major European paleoclimatic projects. The MEDIAS-France group provides archive and access to a variety of European paleoclimatic data efforts and plays a crucial role in developing data management and scientific infrastructure in Africa. Many other centers are developing data sharing for a wide range of paleoclimatic data. The diversity of the PAGES Data System has increased dramatically, and with it so has the challenge. All components of the PAGES Data System work to a common goal: to support the data management needs of the International Geosphere-Biosphere Programme (IGBP) core project on Past Global Changes (PAGES) and its parent body, the International Council for Science (ICSU).

## B.2 The World Data Center for Paleoclimatology

The World Data Center for Paleoclimatology in Boulder, Colorado (WDC-Paleoclimatology, http://www.ngdc.noaa.gov/paleo) is dedicated to providing the paleoclimatic data and information needed to understand and model interannual to centennial scale environmental variability. Focused on the principle of science-driven data management, WDC-Paleoclimatology relies upon the expertise of scientists at universities and institutions around the world to produce and share scientific data and information. Committed to the ICSU (International Council for Science) principle of full and open exchange of data and information for scientific and educational purposes, WDC-Paleoclimatology works to both enhance the compilation of paleoclimatic data within its holdings and to improve the accessibility and usability of these data. All data holdings are freely available via the Internet.

WDC-Paleoclimatology is the international component of the National Oceanic and Atmospheric Administration (NOAA) Paleoclimatology Program and is located at NOAA's National Geophysical Data Center in Boulder, Colorado. The NOAA Paleoclimatology Program and the WDC-Paleoclimatology has worked for over a decade to support the data management needs ICSU, IGBP-PAGES, and the U.S. Global Change Research Program (USGCRP). All three organizations have established similar policies promoting the free and open exchange of scientific data.

WDC-Paleoclimatology serves as the home to many large archives of paleoclimatic data, including the International Tree-Ring Databank, the Global Pollen Database, Global Database of Borehole Temperatures and Climate Reconstructions, the International Ice Core Data Cooperative and data for the PAGES project on Annual Records of Tropical Systems (ARTS). In addition it holds paleoclimatic data from most every discipline, including climate forcings, corals and sclerosponges, paleoceanography, paleolimnology, pollen and plant macrofossils, fauna, insects, model data, and reconstructions. It mirrors of data from the Paleoclimate Modelling Intercomparison Project and provides visualization tools for model data and a variety of paleoclimatic reconstructions. WDC-Paleoclimatology distributes free software and tools produced by many researchers. Major efforts are placed on tools for data browse and visualization (e.g., WebMapper) and providing information products such as the "Paleo Perspectives".

## B.3 The World Data Center for Marine Environmental Sciences / PANGAEA

The World Data Center for Marine Environmental Sciences (WDC-MARE, http://www.pangaea.de) is aimed at collecting, scrutinizing, and disseminating data related to global change in the fields of environmental oceanography, marine geology, paleoceanography, and marine biology. WDC-MARE uses the scientific information system PANGAEA (Network for Geosciences and Environmental Data) as its operating platform.

Essential services supplied by WDC-MARE / PANGAEA are project data management (e.g. for the PAGES project IMAGES, the International Marine Global Change Study), data publication, and the distribution of visualization and analysis software (freeware products). Organization of data management includes quality control and publication of data and the dissemination of metadata according to international standards. Data managers are responsible for acquisition and mainte-

nance of data. The data model used reflects the information processing steps in the earth science fields and can handle any related analytical data. A relational database management system (RDBMS) is used for information storage. Users access data from the database via web-based clients, including a simple search engine (PangaVista) and a comprehensive data-mining tool (ART). With its comprehensive graphical user interfaces and the built in functionality for import, export, and maintenance of information PANGAEA is a highly efficient system for scientific data management and data publication.

WDC-MARE / PANGAEA is operated as a permanent facility by the Centre for Marine Environmental Sciences at the Bremen University (MARUM) and the Alfred Wegener Institute for Polar and Marine Research (AWI) in Bremerhaven, Germany.

## B.4 MEDIAS-France

Located in Toulouse, MEDIAS-France (http://medias.obs-mip.fr:8000/) is a non-profit public corporation that works to develop cooperative research projects, set up permanent observation systems, build up data banks, develop models, and train and provide exchanges for students and researchers. MEDIAS-France has built databases in a wide range of scientific disciplines relating to climatic changes in the global environment, primarily paleoclimatology, hydrology, atmospheric chemistry in tropical region, meteorology (with rainfall forecasts covering North Africa) and oceanography (both dynamic and biological aspects). MEDIAS-France is also skilled in drawing up catalogues, developing integrated or specialized databases and educational products, and distributing its products. In addition MEDIAS-FRANCE provides training services to the scientific community, especially international institutional development such as: the SysTem for Analysis, Research and Training (START), ACMAD, the Sahara-Sahel Observatory, ENRICH, IGBP/DIS, etc. An important effort in developing START activities is directed at the Mediterranean and Africa, where Planning Committees have early been established. MEDIAS and START have evolved in close partnership, jointly sponsoring several activities. ENRICH promotes collaboration in Western Europe, encourages the endogenous research capabilities in developing countries, including in Africa and the Mediterranean Basin, and promotes support for relevant research initiatives in the countries of Central and

Eastern Europe and the New Independent States of the former Soviet Union (NIS).

MEDIAS-France develops and supports several major paleoclimatic database activities including the European and African Pollen Databases, Forest Modelling Assessment and Tree Rings (Format), and the European Diatom Database (EDDI, coordinated by Dr. S. Juggins, University of Newcastle). It distributes paleoclimatic data analysis software and hosts one of the mirror sites for WDC-Paleoclimatology.

### B.5 Mirror sites / World Data Center *partners*

While the major centers provide ready access to data and information, they can only work as well as the users' connection via the Internet. This is one of the reasons that WDC-Paleoclimatology has worked with partners in various countries to establish mirror sites (Figure B.2). These distributed archives hold complete copies of the web and ftp holdings of the WDC-Paleoclimatology, providing regional access points to the data. This means that data can be made more accessible, benefiting scientists in the region of the data's origin. Bringing the data close to home encourages scientists to contribute their data – increasing participation in PAGES-organized regional to global efforts to understand our climate system. Currently four mirror sites are operating, with more under consideration.

**Fig. B.2.** Map showing the locations of the World Data Center for Paleoclimatology, Boulder and its data mirrors around the world.

The idea of mirror sites has taken hold within the ICSU structure. The World Data Centers Panel has decided to begin new approaches to expand global data sharing. One of these, World Data Center *Partners*, entails the collaboration of a World Data Center with partner organizations in developing countries. The approach is intended to encourage data exchange without requiring the infrastructure commitments entailed in establishing new WDCs. While details of this program are still under development, the activities of the PAGES Data System and the WDC mirror sites established the model for this new ICSU program.

### B.6 Data cooperatives and project level data management

A key step between the scientists producing data and centralized data management are the data management efforts focused on disciplinary or regional programs. As each paleoclimatic data stream has individual characteristics and idiosyncrasies, data cooperatives have provided the expertise needed to develop proper protocols for handling data and metadata. These collaborations between expert university scientists, disciplinary data managers, and data management centers establish data management protocols, resolve taxonomic, methodological, dating, and data quality problems. These provide the procedures that scientists and data managers use for managing the data in the future. The role of data cooperatives and the activities of many of these were described in *The PAGES Data Guide* (Anderson and Webb 1998). Some of these include large international programs such as the International Ice Core Data Cooperative (for GRIP and GISP2 data), and data management groups for international programs such as the Paleoclimate Modelling Intercomparison Project (PMIP) and the International Marine Past Global Changes Study (IMAGES). Other data cooperatives pull together large but less structured groups of scientists and data such as the International Tree-Ring Data Bank and the Global Pollen Database. Once the data coop has developed appropriate data management procedures, the data are contributed to WDC holdings for permanent archive and access.

### B.7 Users of paleoclimatic data

While the primary users of paleoclimatic data are still members of the paleoclimatic research community, a much broader suite of users is now realizing the value of pre-instrumental records of climate. The most active of these have been the modern climate research and prediction communities. However, in recent years a growing number of policy makers, planners, and resource managers are using paleoclimatic data in planning and decision support. Additionally, paleoclimatic data now are accessed by a broad array of educators, students, the media, and the general public. While the research community mainly comes to the PAGES Data System to access data, these newer users typically seek information products that make past climate changes more understandable. In all, users in 147 countries have accessed data from the WDC-Paleoclimatology holdings (Figure B.3, as of December 2001).

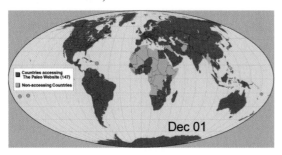

**Fig. B.3.** Countries that have downloaded paleoclimatic data from the World Data Center for Paleoclimatology, Boulder.

## B.8 Data access and information tools

Archived data are valuable only if they are accessible and used. For this reason, members of the PAGES Data System work to make sure that their data holdings are readily accessible. Text based search engines across the PAGES Data System provide the user with the ability to define a variety of search parameters to find desired data. These include searches by contributor, proxy, variables or expedition. Advanced text-based search tools such as PANGAEA's Advanced Retrieval Tool enables the user to retrieve and download data using user-defined configurations.

In addition to text-based information search and delivery tools, new geospatial tools have proven themselves as valuable ways to access the data. The WDC-Paleoclimatology's WebMapper (Figure B.4), a web-based browse, visualization and access tool and PANGAEA's PangaVista (Figure B.5) search engine are powerful new ways to search and access data from PAGES Data System archives.

**Fig. B.4.** The WDC-Paleoclimatology's WebMapper

For gridded data, online plotting tools have been applied to data from paleoclimatic models, and reconstructed temperature, drought and pressure

fields to provide users with visualizations of these large and complex datasets. Many data centers are applying new tools from the realm of Geographic Information Systems (GIS). GIS systems allow users to access a variety of different data types and superimpose them as "layers" to access and analyze disparate data.

So far each of the tools described above provides access to the holdings of only one of the PAGES data centers. New tools under development will improve access by allowing users to access data from multiple data centers, from a variety of proxies, and within selected time slices. By sharing metadata among centers, the PAGES Data System will provide a paleoclimatic data "portal" to make data from multiple centers more accessible.

While data access tools such as search engines, WebMapper and portals make data more accessible; many users come to PAGES seeking interpretations of the data. By providing information products (e.g., WDC-Paleoclimatology's *A Paleo Perspective on Global Warming*), PAGES data centers have provided access to paleoclimatic data in a format that explains the importance and meaning of many of these data.

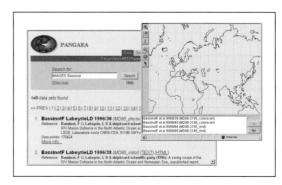

**Fig. B.5.** PANGAEA's PangaVista

## B.9 The PAGES Data Board and data sharing

The successful development of the various data efforts within the PAGES Data System has archived more data than was possible for any one center. However, this has led to some confusion on the part of the scientific community and other users. A recent editorial in Nature referred to the current situation as "something of a maze of publicly supported databases" (Anonymous, 2001). To address the need for greater coordination among paleoclimatic data efforts, the PAGES Data Board was formed. The PAGES data board is an open and equal partnership between international, national and thematic groups that archive paleoenvironmental data and provide tools to enhance their value. Its duties are to

- ensure common metadata formats
- develop a shared paleodata web portal
- promote data sharing, and
- recommend PAGES data policies

As a part of the IGBP, PAGES supports the free and open exchange of data as described by policies of IGBP and its parent organization ICSU (CODATA 2002). In particular, the PAGES Data Board has established policies that support the development and use of rules of good scientific practice (ESF, 2000), including

- making data and methods available for reproducibility of results,
- making data behind any published graphic or figure publicly available, and
- ethical use of data, including proper citation.

The data portal described above is being planned and implemented by the various PAGES data centers and its development and implementation is being facilitated through the PAGES Data Board. This includes the recent adoption of standards for archiving metadata through a standard profile and then exchanging them using extensible markup language (XML).

## B. 10 Future directions

New data access tools and methodologies continue to revolutionize the options for managing paleodata. Members of the PAGES Data System are working to advance development and application of databases, web portals and GIS technologies that will enable users to access, combine, and analyze data from a wide array of datasets across the Internet. This will allow data users to pinpoint tree-ring sites on digital elevation models, or to find lake-data from sites within drought-prone regions or time periods of interest. Implementation of standard metadata profiles will increase data sharing among data repositories, while XML and other methods to exchange metadata and data will facilitate data interchange and reuse. We plan to implement new data cooperatives, disciplinary and multidisciplinary data efforts to develop protocols for handling new proxies.

In a recent *Nature* correspondence, Alverson and Eakin (2001) pointed out that "PAGES, WDC-Paleo and the many scientists and institutions that support them have made great efforts to make data easily accessible and usable." As a community, we can all be proud that data sharing in paleoclimatology has come a long way over the last decade.

The world of data management is changing rapidly. As new developments and new ideas come forward, the PAGES Data System will continue to advance technology to manage and distribute paleoclimatic data and information. Through participation of the paleoclimatic research community, disciplinary data groups, and data management centers, the PAGES Data System will continue to make paleoclimatic data as accessible and usable as possible.

# References

Aaby B, Jacobsen J, Jacobsen OS (1979) Pb-210 dating and lead deposition in the ombrotrophic peat bog, Draved Mose, Denmark. In: *Danmarks Geologisk Ungersoglse Arbog*. pp. 5-43

Abyzov SS, Mitskevitch IN, Zhukova TY, Kuzhinovskii VA, Poglazova MN (1993) Microorganism numbers in the deep layers of the central Antarctic ice-sheet. In: Friedmann EI (ed). Antarctic Microbiology, Wiley and Sons, New York, pp. 133-134

Adam D (1998) Correlations of the Clear Lake, California, core CL-73-4 pollen sequence with other long climate records. In: Sims J (ed.) *Late Quaternary, Climate, Tectonism, and Sedimentation, in Clear Lake, Northern California Coast Ranges*. Geological Society of America Special Paper 214, pp. 81-89

Adams JM, Faure H (1998) A new estimate of changing carbon storage on land since the last glacial maximum, based on global land ecosystem reconstruction. *Global Planetary Change* 16-17:3-24

Adams JM, Post WM (1999) A preliminary estimate of changing calcrete carbon storage on land since the Last Glacial Maximum. *Global Planetary Change* 20:243-256

Adams JM, Faure H, Faure-Denard L, McGlade JM, Woodward FI (1990) Increases in terrestrial carbon storage from the Last Glacial Maximum to the present. *Nature* 348:711-714

Adelson JM, Helz GR (2001) Reconstructing the rise of recent coastal anoxia: Molybdenum in Chesapeake Bay sediments. *Geochimica & Cosmochimica Acta* 65:237-252

Adkins JF, Cheng H, Boyle EA, Druffel ERM, Edwards RL (1998) Deep-sea coral evidence for rapid change in ventilation of the deep North Atlantic 15,400 years ago. *Science* 280:725-730

Adkins JK, Schrag DP (2001) Pore fluid constraints on deep ocean temperature and salinity during the last glacial maximum. *Geophysical Research Letters* 28:771-774

Aeberhardt M, Blatter M, Stocker TF (2000) Variability on the century time scale and regime changes in a stochastically forced zonally averaged ocean-atmosphere model. *Geophysical Research Letters* 27:1303-1306

Alcoforado MJ, Nuñes MF, Garcia JC, Taborda JP (2000) Temperature and precipitation reconstructions in southern Portugal during the Late Maunder Minimum (1675 to 1715). *The Holocene* 10:333-340

Allen JRM, Watts WA, Huntley B (2000) Weichselian palynostratigraphy,palaeovegetation and palaeoenvironment: the record from Lago Grande di Monticchio, southern Italy. *Quaternary International* 73/74:91-110

Allen JRM, Brandt U, Brauer A, Hubberten H-W, Huntley B, Keller J, Kraml M, Mackensen A, Mingram J, Negendank JFW, Nowaczyk NR, Oberhänsli H, Watts WA, Wulf S, Zolitschka B (1999) Rapid environmental changes in southern Europe during the last glacial period. *Nature* 400:740-743

Alley RB (1998) Icing the North Atlantic. *Nature* 392:335-337

Alley RB (2000) The Younger Dryas cold interval as viewed from central Greenland. *Quaternary Science Reviews* 19:213-226

Alley RB, Ágústsdóttir AM (1999) Ice core evidence of late-Holocene reduction in North Atlantic ocean heat transport. In: Clark PU, Webb RS and Keigwin LD (eds.) *Mechanisms of Global Climate Change at Millennial Time Scales*. American Geophysical Union, Washington D.C pp. 301-312

Alley RB, Clark PU (1999) The deglaciation of the Northern Hemisphere: A global perspective. *Annual Review Earth Planetary Science* 27:149-182

Alley RB, Brook EJ, Anandakrishan S (2002) A northern lead in the orbital band: north-south phasing of Ice-Age events. *Quaternary Science Reviews* 21:431-441

Alley RB, Mayewski PA, Sowers T, Stuiver M, Taylor KC, Clark PU (1997) Holocene climatic instability: a prominent, widespread event 8200 yr ago. *Geology* 25:483-486

Alley RB, Meese DA, Shuman CA, Gow AJ, Taylor KC, Grootes PM, White JWC, Ram M, Waddington ED, Mayewski PA, Zielinski GA (1993) Abrupt increase in Greenland snow accumulation at the end of the Younger Dryas event. *Nature* 362:527-529

Allott TEH, Harriman R, Battarbee RW (1992) Reversibility of acidification at the Round Loch of Glenhead, Galloway, Scotland. *Environmental Pollution* 77:219-225

Alroy J (1999) Putting North America's end-Pleistocene megafaunal extinction in context. In: MacPhee (ed.) *Extinctions in Near Time*. Kluwer Academic/Plenum Publishers, New York

Altabet MA, François R (2001) Nitrogen isotope biogeochemistry of the antarctic polar frontal zone at 170 degrees W. *Deep-Sea Research* 48:4247-4273

Altabet MA, Francois R, Murray DW, Prell WL (1995) Climate-related variations in denitrification in the Arabian Sea from sediment $^{15}$N/$^{14}$N ratios. *Nature* 373:506-509

Altabet MA, François R, Murray DW, Prell WL (in press) Orbital period varations in $^{15}$N/$^{14}$N ratio in the Arabian Sea: an indicator of past changes in denitrification. *Nature* 373:506-509

Alverson K, Oldfield F (2000) PAGES - Past Global Changes and their Significance for the future: an Introduction. *Quaternary Science Reviews* 19:3-7

Alverson K, Kull C (in press) Understanding future climate change using paleorecords. In: Rodo X (ed.) *Current Uncertainties in the Climate System*. Springer Verlag

Alverson K, Oldfield F, Bradley R (2000) Past Global Changes and Their Significance for the Future. *Quaternary Science Reviews* 19:479 pp.

Alverson K, Bradley R, Pedersen T (2001) Environmental Variability and Climate Change. In: Eliott S (ed.) *IGBP Science volume 3*. IGBP Secretariat, Stockholm pp. 32

Alverson K, Duplessy JC, Jouzel J, Overpeck J (1999) The PAGES/CLIVAR Intersection. *World Climate Research Program Report Series* 108:42-47

Amman B, Oldfield F (2000) Biotic responses to rapid climatic changes around the Younger Dryas. *Palaeogeography Palaeoclimatology Palaeoecology* 159:175 pp.

An Z (2000) The history and variability of the East Asian paleomonsoon climate. *Quaternary Science Reviews* 19:171-187

An Z, Porter SC (1997) Millenial-scale oscillations during the last interglaciation in central China. *Geology* 25:603-606

An ZS, Kukula G, Porter SC, Xiao JL (1991) Magnetic susuceptibility evidence of monsoon variation on the Loess Plateau of central China during the last 130,000 years. *Quaternary Research* 36:29-36

Anderson DM, Webb RS (1998) *The PAGES Data Guide: Results from the second workshop on global paleoenvironmental data*. PAGES, Bern, Switzerland

Anderson NJ, Rippey B (1994) Monitoring lake recovery from point-source eutrophication: the use of diatom-inferred epilimnetic total phosphorus and sediment chemistry. *Freshwater Biology* 32:625-639

Anderson NJ, Renberg I, Segerstrom U (1995) Diatom production responses to the development of early agriculture in a boreal forest lake-catchment (Kassjön, N. Sweden). *Journal of Ecology* 83:802-822

Andren E, Shimmield G, Brand T (1999) Environmental changes of the last three centuries indicated by siliceous microfossil records from the Baltic Sea. *The Holocene* 9:25-38

Andren E, Andren T, Kunzendorf H (2000) Holocene history of the Baltic Sea as a background for assessing records of human impact in the sediments of the Gotland basin. *The Holocene* 10:687-702

Andrews JT, Jennings AE, Kerwin M, Kirby M, Manley W, Miller GH, Bond G, MacLean B (1995) A Heinrich-like event, H-0 (DC-0): Source(s) for detrital carbonate in the North Atlantic during the Younger Dryas chronozone. *Paleoceanography* 10:943-952

Anklin M, Barnola J-M, Schwander J, Stauffer B, Raynaud B (1995) Processes affecting the $CO_2$ concentrations measured in Greenland ice. *Tellus* 47 B:461-470

Anklin M, Schwander J, Stauffer B, Tschumi J, Fuchs A, Barnola JM, Raynaud D (1997) $CO_2$ record between 40 and 8 kyr B.P. from the Greenland Ice Core Project ice core. *Journal of Geophysical Research* 102:26,539-26,546

Anonymous (2001) Make the most of palaeodata. *Nature* 411:1

Appenzeller C, Stocker TF, Anklin M (1998) North Atlantic Oscillation dynamics recorded in Greenland ice cores. *Science* 282:446-449

Appleby PG, Oldfield F (1992) Application of $^{210}$Pb to sedimentation studies. In: Ivanovich M and Harmon R (eds.) *Uranium Series Disequilibrium: Applications to Earth, Marine and Environmental Studies*. Clarendon press, Oxford

Appleby PG, Richardson N, Nolan PJ (1991) $^{241}$Am dating of lake sediments. *Hydrobiologia* 214:35-42

Archer D (in press) Modeling $CO_2$ in the ocean: A review. In: Bouwman AF (ed.) *Scaling of Trace Gas Fluxes between Terrestrial and Aquatic Ecosystems and the Atmosphere.* Elsevier

Archer D, Maier-Reimer E (1994) Effect of deep-sea sedimentary calcite preservation on atmospheric $CO_2$ concentration. *Nature* 367:260-263

Arseneault D, Payette S (1997) Reconstruction of millennial forest dynamics from tree remains in a subarctic tree line peatland. *Ecology* 78:1873-1883

Ashworth AC (1997) The response of beetles to Quaternary climate changes. In: B. Huntley, Cramer W, Morgan AV, Prentice HC and Allen JRM (eds.) *Past and future rapid environmental changes: The spatial and evolutionary responses of terrestrial biota*. Springer-Verlag, Berlin pp. 119-127

Asioli A (1996) High resolution *Foraminifera* biostratigraphy in the Central Adriatic basin during the Last Deglaciation: a contribution to the PALICLAS project. In: Guilizzoni P and Oldfield F (eds.) *Palaeoenvironmental Analysis of Italian Crater Lake and Adriatic Sediments (PALICLAS)*. *Memorie del Istituto Italiano di Idrobiologia*. pp. 197 – 217

Asner GP, Townsend AR, Braswell BH (2000) Satellite observation of El Niño effects on Amazon forest phenology and productivity. *Geophysical Research Letters* 27:981-984

Atherden MA, Hall JA (1999) Human impact on vegetation in the White Mountains of Crete since AD 500. *The Holocene* 9:183-193

Austin M, Nicholls A, Margules C (1990) Measurement of the realized quantitative niche: environmental niches of five *Eucalyptus* species. *Ecological Monographs* 60:161-177

Avila A, Penuelas J (1999) Increasing frequency of Saharan rains over northeastern Spain and its ecological consequences. *The Science of Total Environment* 228:153-156

Bacon MP (1988) Tracers of chemical scavenging in the ocean: Boundary effects and large-scale chemical fractionation. *Philosohpical Transactions of the Royal Society, London* 320:187-200

Bahn P, Flenley J (1992) *Easter Island – Earth Island. A message from our past for the future of our planet* Thames and Hudson Ltd, London

Balbon E (2000) The last interglacial in the North Atlantic and Norwegian Sea :implications on ice sheet dynamic and glocal climate, UPS-Orsay

Balee W (1998) *Advances in Historical Ecology* Columbia University Press, Columbia

Balmford A, Moore JL, Brooks T, Burgess N, Hansen LA, Williams P, Rahbek C (2001) Conservation conflicts across Africa. *Science* 291:2616-2619

Barber DC, Dyke A, Hillaire-Marcel C, Jennings AE, Andrews JT, Kerwin MW, Bilodeau G, McNeely R, Southon J, Morehead MD, Gagnon JM (1999) Forcing of the cold event of 8200 years ago by catastrophic drainage of Laurentide lakes. *Nature* 400:344-348

Barber KE, Maddy D, Rose N, Stevenson AC, Stoneman R, Thompson R (2000) Replicated proxy-climate signals over the last 2000 yr from two distant UK peat bogs: new evidence for regional palaeoclimate teleconnections. *Quaternary Science Reviews* 19:481-487

Barber VA, Juday GP, Finney BP (2000) Reduced growth of Alaskan white spruce in the twentieth century from temperature-induced drought stress. *Nature* 405:668-673

Bard E, Hamelin B, Fairbanks RG (1990) U-Th ages obtained by mass spectrometry in corals from Barbados: sea level during the past 130,000 years. *Nature* 346:456-458

Bard E, Arnold M, Fairbanks RG (1993) $^{230}$Th-$^{234}$U and $^{14}$C ages obtained by mass spectrometry on corals. *Radiocarbon* 35:191-199

Bard E, Hamelin B, Fairbanks RG, Zindler A (1990) Calibration of the $^{14}$C timescale over the last 30,000 years using mass spectrometric U/Th ages from Barbados corals. *Nature* 345:405-410

Bard E, Raisbeck G, Yiou F, Jouzel J (2000) Solar irradiance during the last 1200 years based on cosmogenic nuclides. *Tellus* 52B

Bard E, Hamelin B, Fairbanks RG, Zindler A, Mathieu G, Arnold M (1990) U/Th and $^{14}$C ages of corals from Barbados and their use for calibrating the $^{14}$C time scale beyong 9000 years B.P. *Nuclear Instruments and Methods in Physics Research* B52:461-468

Bard E, Hamelin B, Arnold M, Montaggioni L, Cabioch G, Faure G, Rougerie F (1996) Deglacial sea-level record from Tahiti corals and the timing of global meltwater discharge. *Nature* 382:241-244

Bard E, Arnold M, Mangerud J, Paterne M, Labeyrie L, Duprat J, Mélières MA, Sonstegaard E, Duplessy JC (1994) The North Atlantic atmosphere-sea surface $^{14}$C gradient during the Younger Dryas climatic event. *Earth and Planetary Science Letters* 126:275-287

Barker G (1995) Land use and environmental degradation in Biferno Valley (Central Southern Italy) from prehistoric times to the Present Day. In: APCDA (ed.) *L'Homme et la Dégradation de l'Environnement. XVe Rencontres Internationales d'Archéologie d'Histoire d'Antibes*. Juan-les-Pins

Barlow LK, Sadler JP, Ogilvie AEJ, Buckland PC, Amorosi T, Ingimundarson T, Skidmore P, Dugmore AG, McGovern TH (1998) Interdisciplinary investigations of the end of the Norse western settlement in Greenland. *The Holocene* 7:489 - 500

Barnola J-M, Pimienta P, Raynaud D, Korotkevich YS (1991) $CO_2$-climate relationship as deduced from the Vostok ice core: A re-examination based on new measurements and on a re-evaluation of the air dating. *Tellus* 43:83-90

Barnola JM (1999) Status of the atmospheric $CO_2$ reconstruction from ice cores analyses. *Tellus Series B-Chemical and Physical Meteorology* 51 (2):151-155

Barnola JM, Raynaud D, Korotkevich YS, Lorius C (1987) Vostok ice core provides 160 000-year record of atmospheric $CO_2$. *Nature* 329:408-414

Barnosky C (1987) Response of vegetation to climatic changes of different duration in the late Neogene. *Trends in Ecology and Evolution* 2:247-250

Barriendos M (1997) Climatic variations in the Iberian Peninsula during the Late Maunder Minimum (AD 1675-1715): an analysis of data from rogation ceremonies. *The Holocene* 7:105-111

Bartlein P, Prentice I, Webb T III (1986) Climatic response surfaces from pollen data for some eastern North American taxa. *Journal of Biogeography* 13:35-57

Bartlein P, Whitlock C, Shafer S (1997) Climate in the Yellowstone National Park Region and its potential impact on vegetation. *Conservation Biology* 11:782-792

Bartlein PJ (1988) Late-Tertiary and Quaternary paleoenvironments. In: B. Huntley and WebbIII T (eds.) *Vegetation History*. Kluwer Academic Publishers, Dordrecht pp. 113-152

Bartlein PJ (1997) Past environmental changes: characteristic features of Quaternary climate variations. In: Huntley W, Cramer AV, Morgan HC, Prentice and Allen JRM (eds.) *Past and future rapid environmental changes: The spatial and evolutionary responses of terrestrial biota*. Springer, Berlin pp. 11-29

Bartley D, Morgan H (1990) The palynological record of the King's Pool, Stafford, England. *New Phytologist* 116:177 – 194

Basile I, Grousset FE, Revel M, Petit J-R, Biscaye BE, Barkov NI (1997) Patagonian origin of glacial dust deposited in East Antarctica (Vostok and Dome C) during glacial stages 2, 4 and 6. *Earth and Planetary Science Letters* 146:573-579

Bassinot F, Labeyrie LD, Vincent E, Quideleur X, Shackleton NJ, Lancelot Y (1994) The Astronomical Theory of climate and the age of the Brunhes-Matuyama magnetic reversal. *Earth and Planetary Science Letters* 126:91-108

Batjes NH (1996) Total carbon and nitrogen in the soils of the world. *European Journal of Soil Science* 47:151-163

Battarbee RW (1978) Observations on the recent history of Lough Neagh and its drainage basin. *Philosophical Transactions of the Royal Society* B 281:303-345

Battarbee RW (1990) The causes of lake acidification, with special reference to the role of acid deposition. *Philosophical Transactions of the Royal Society of London Series B* 327:339-347

Battarbee RW (1997) Freshwater quality, naturalness and palaeolimnology. In: Boon PJ and Howell DL (eds.) *Freshwater Quality: defining the Indefinable*. Scottish National Heritage, Edinburgh

Battarbee RW (1998) Lake management: the role of palaeolimnology. In: Harper DM (ed.) *The ecological basis for lake and reservoir management*. Wiley, London

Battarbee RW, Gasse F, Stickley C (in press) *Past Climate Variability through Europe and Africa*. Kluwer

Battarbee RW, Flower RJ, Stevenson AC, Rippey B (1990) Lake acidification in Galloway: a test of competing hypotheses. *Nature* 314:350 – 352

Battisti DS, Sarachik ES (1995) Understanding and predicting ENSO. *Reviews of Geophysics* Supplement:1367-1376

Battle M, Bender M, Sowers T, Tans PP, Butler JH, Elkins JW, Ellis JT, Conway T, Zhang N, Lang P, Clarke AD (1996) Atmospheric gas concentrations over the past century measured in air from firn at the South Pole. *Nature* 383:231-235

Bauch HA, Erlenkeuser H, Fahl K, Spielhagen RF, Weinelt M, Andruleit H, Henrich R (in press) Evidence for a steeper Eemian than Holocene sea surface temperature gradient between Arctic and sub-Arctic regions. *Palaeogeography, Palaeoclimatology, Palaeoecology*

Baumgartner S, Beer J, Masarik J, Wagner G, Meynadier L, Synal H-F (1998) Geomagnetic modulation of the $^{36}$Cl flux in the GRIP ice core, Greenland. *Science* 279:1330-1332

Beaufort L, Lancelot Y, Camberlin P, Cayre O, Vincent E, Bassinot F, Labeyrie L (1997) Insolation cycles as a major control of equatorial Indian Ocean primary production. *Science* 278:1451-1454

Beer J, Mende W, Stellmacher R, White OR (1996) Intercomparisons of proxies for past solar variability. In: Jones PD, Bradley RS and Jouzel J (eds.) *Climatic Variations and Forcing Mechanisms of the Last 2000 Years*. Springer Verlag, Berlin pp. 501-517

Beer J, Joos F, Lukasczyk C, Mende W, Rodriguez J, Siegenthaler U, Stellmacher R (1994) $^{10}$Be as an indicator of solar variability and climate. In: Nesmé-Ribes E (ed.) *The Solar Engine and its Influence on Terrestrial Atmosphere and Climate*. Springer-Verlag, Berlin pp. 221-233

Behling H (2000) A 2860-year high-resolution pollen and charcoal record from the Cordillera de Talamanca in Panama: a history of human and volcanic forest disturbance. *The Holocene* 10:387-395

Behrenfeld MJ, Bale AJ, Kolber ZS, Aitken J, Falkowski PG (1996) Confirmation of iron limitation of phytoplankton photosynthesis in the equatorial Pacific Ocean. *Nature* 383:508-511

Beltrami H, Mareschal J-C (1995) Resolution of ground temperature histories inverted from borehole temperature data. *Global and Planetary Change* 11:57-70

Beltrami H, Chapman DS, Archambault S, Bergeron Y (1995) Reconstruction of high-resolution ground temperature histories combining dendrochronological and geothermal data. *Earth and Planetary Science Letters* 136:437-445

Bender M, Sowers T, Dickson ML, Orchado J, Grootes P, Mayewski PA, Meese DA (1994) Climate connection between Greenland and Antarctica during the last 100,000 years. *Nature* 372:663-666

Benedict JB (1999) Effects of changing climate on game-animal and human use of the Colorado High Country (USA) since 1000 BC. *Arctic, Antarctic, and Alpine Research* 31:1-15

Beniston M (2000) *Environmental Change in Mountains and Uplands* Arnold, 172 pp.

Bennett KD (1990) Milankovitch cycles and their effects on species in ecological and evolutionary time. *Paleobiology* 16:11-21

Bennett KD (1997) *Evolution and ecology: The pace of life* Cambridge University Press, Cambridge, 241pp.

Bennion H, Juggins S, Anderson NJ (1996) Predicting epilimnetic phosphorus concentrations using an improved diatom-based transfer function and its application to lake eutrophication management. *Environmental Science & Technology* 30:2004-2007

Benson LV, Burdett JW, Kashgarian M, Lund SP, Phillips FM, Rye RO (1996) Climatic and hydrologic oscillations in the Owens Lake basin and adjacent Sierra Nevada, California. *Science* 274:746-749

Benson LV, Lund S, Burdett J, Kashgarian M, Rose T, Smoot J, Schwartz M (1998) Correlation of late-Pleistocene lake-level oscillations in Mono Lake, California, with North Atlantic climate events. *Quaternary Research* 49:1-10

Benson LV, Lund S, Paillet F, Kashgarian M, Smoot J, Mensing S, Dibb J (1999) A 2800-yr history of oscillations in surface-water supply to the Central Valley and to the Bay Area of Northern California, and to the Reno-Sparks area of Nevada. EOS

Berger A (1977) Long-term variations of the earth's orbital elements. *Celestial Mechanics* 15:53-74

Berger A, Loutre MF (1991) Insolation values for the climate of the last 10,000,000 years. *Quaternary Science Reviews* 10:297-317

Berger A, Loutre MF, Gallée H (1998) Sensitivity of the LLN climate model to the astronomical and $CO_2$ forcings over the last 200 ky. *Climate Dynamics* 14:615-629

Berger A, Tricot C, Gallée H, Fichefet T, Loutre MF (1994) The last two glacial-interglacial cycles simulated by the LLN model. In: Duplessy JC and Spyridakis MT (eds.) *Long-term climatic variations*. Springer-Verlag, Berlin

Berger AL (1978) Long-term variations of daily insolation and

Quaternary climate changes. *Journal of the Atmospheric Sciences* 35:2632-2637

Bergeron Y, Archambault S (1993) Decreasing frequency of forest fires in the southern boreal zone of Québec and its relation to global warming since the end of the "Little Ice Age". *The Holocene* 3:255-259

Berglund BE (1991) The cultural landscape during 6000 years in southern Sweden – the Ystad Project. *Ecological Bulletins* 41:495 pp

Bernabo JC, Webb T III (1977) Changing patterns in the Holocene pollen record from northeastern North America: a mapped summary. *Quaternary Research* 8:64-96

Bianchi GG, McCave IN (1999) Holocene periodicity in North Atlantic climate and deep ocean flow south of Iceland. *Nature* 397:515-517

Bigler M, Wagenbach D, Fischer H, Kipfstuhl J, Miller H, Sommer S, Stauffer B (in press) Sulphate record from a norheast Greenland ice core over the last 1200 years based on continuous flow analysis. *Annals of Glaciology* 35:

Biondi F, Gershunov A, Cayan DR (2001) North Pacific decadal climate variability since AD 1661. *Journal of Climate* 14:5-10

Biondi F, Lange CB, Hughes MK, Berger WH (1997) Interdecadal signals during the last millennium (A.D. 1117-1992) in the varve record of Santa Barbara Basin, California. *Geophysical Research Letters* 24:193-196

Biondi F, Isaacs C, Hughes MK, Cayan DR, Berger WH (2000) The near-1600 dry/wet knockout: linking terrestrial and near-shore ecosystems. In: *Proceedings of the Twenty-Fourth Annual Climate Diagnostics and Prediction Workshop.* US Department of Commerce, NOAA

Bird M, Cali J (1998) A million-year record of fire in sub-Saharan Africa. *Nature* 394:767-770

Bird MI, Lloyd J, Garquhar GD (1994) Terrestrial carbon storage at the LGM. *Nature* 371:566

Birks HJB (1989) Holocene isochrone maps and patterns of tree-spreading in the British Isles. *Journal of Biogeography* 16:503-540

Birks HJB (1996) Contributions of Quaternary palaeoecology to nature conservation. *Journal of Vegetation Science* 7:89-98

Birks HJB, Line JM (1992) The use of rarefaction analysis for estimating palynological richness from Quaternary pollen-analytical data. *The Holocene* 2: 1-10

Birks HJB, Berge F, Boyle JF, Cumming B (1990a) A palaeoecological test of the land use-hypothesis for recent lake acidification in South-West Norway using hill-top lakes. *Journal of Paleolimnology* 4:69–85

Birks HJB, Line JM, Juggins S, Stevenson AC, terBraak CJF (1990b) Diatoms and pH reconstruction. *Philosophical Transactions of the Royal Society of London Series B* 327:263-278

Biscaye PE, Grousset FE, Revel M, VanderGaast S, Zielinski GA, Vaars A, Kukla G (1997) Asian provenance of glacial dust (stage 2) in the Greenland Ice Sheet Project 2 ice core, Summit, Greenland. *Journal of Geophysical Research* 102:26765-26782

Bitz CM, Battisti DS (1999) Interannual to decadal variability in climate and the glacier mass balance in Washington, western Canada, and Alaska. *Journal of Climate* 12:3181-3196

Bjerknes J (1969) Atmospheric teleconnections from the equatorial Pacific. *Monthly Weather Review* 97:163-172

Björk S, Walker MJC, Cwynar LC, Knudsen KL, Lowe JJ, Wohlfarth B (1998) An event stratigraphy for the Last Termination in the North Atlantic region based on the Greenland ice core record: a proposal by the INTIMATE group. *Journal of Quaternary Science* 13:283-292

Black DE, Peterson LC, Overpeck JT, Kaplan A, Evans MN, Kashgarian M (1999) Eight centuries of North Atlantic Ocean atmosphere variability. *Science* 286:1709-1713

Blais JM (1998) Accumulation of persistent organochlorine compounds in mountains of western Canada. *Nature* 395:585-588

Blake DR, Rowland FS (1987) Continuing worldwide increase in tropospheric methane, 1978 to 1987. *Science* 239:1129-1131

Blamart D, Balbon E, Kissel C, Labeyrie L, Turpin L, Turon J-L, Robin E (1999) Geochemical (major elements) study of IMAGES Core MD 95-2009 in relation with deep water circulation variability in the southern Norwegian Sea during the last climatic cycle. *EOS* 80:12

Blunier T, Brook EJ (2001) Timing of millenial-scale climate change in Antarctica and Greenland during the Last Glacial period. *Science* 291:109-112

Blunier T, Chappellaz J, Schwander J, Stauffer B, Raynaud D (1995) Variations in atmospheric methane concentration during the Holocene epoch. *Nature* 374:46-49

Blunier T, Schwander J, Stauffer B, Stocker T, Dällenbach A, Indermühle A (1997) Timing of the Antarctic Cold Reversal and the atmospheric $CO_2$ increase with respect to the Younger Dryas event. *Geophysical Research Letters* 24:2683-2686

Blunier T, J. Chappellaz, J. Schwander, J.-M. Barnola, T. Desperts, B. Stauffer, Raynaud D (1993) Atmospheric methane, record from a Greenland ice core over the last 1000 years. *Journal of Geophysical Research* 20:2219-2222

Blunier T, Chappellaz J, Schwander J, Dällenbach A, Stauffer B, Stocker TF, Raynaud D, Jouzel J, Clausen HB, Hammer CU, Johnsen SJ (1998) Asynchrony of Antarctic and Greenland climate change during the last glacial period. *Nature* 394:739-743

Blunier T, Schwander J, Stauffer B, Stocker F, Dallenbach A, Indermühle A, Tschumi H, Chappellez J, Raynaud D, Barnola J-M, Jouzel J, Claussen HB, Hammer CU, Johnsen SJ (1997) Timing of temperature variations during the last deglaciation in Antarctica and the atmospheric $CO_2$ increase with respect to the Younger Dryas event. *Geophysical Research Letters* 24: 2683-2686

Bonan G, Pollard D, Thompson S (1992) Effects of boreal forest vegetation on global climate. *Nature* 359:716-718

Bond G, Lotti R (1995) Iceberg discharges into the North Atlantic on Millenial time scales during the last glaciation. *Science* 267:1005-1010

Bond G, Broecker W, Johnsen S, McManus J, Labeyrie L, Jouzel J, Bonani G (1993) Correlations between climate records from north Atlantic sediments and Greenland ice. *Nature* 365:143-147

Bond G, Showers W, Elliot M, Evans M, Lotti R, Hajdas I, Bonani G, Johnsen SJ (1999) The North Atlantic's 1-2 kyr climate rhythm: Relation to Heinrich events, Dansgaard/Oeschger cycles and the Little Ice Age. In: Clark PU, Webb RS and Keigwin LD (eds.) *Mechanisms of Global Climate Change at Millenial Time Scales.* AGU, Washington, DC pp. 35-58

Bond G, Showers W, Cheseby M, Lotti R, Almasi P, Menocal Pd, Priore P, Cullen H, Hajdas I, Bonani G (1997) A pervasive millennial-scale cycle in North Atlantic Holocene and glacial climates. *Science* 278:1257-1266

Bond G, Kromer B, Beer J, Muscheler R, Evans MN, Showers W, Hoffmann S, Lotti-Bond R, Hajdas I, Bonani G (2001) Persistent solar influence on North Atlantic climate during the Holocene. *Science* 294:2130-2136

Bond G, Heinrich H, Broecker W, Labeyrie L, McManus J, Andrews J, Huon S, Jantschik R, Clasen C, Simet C, Tedesco K, Klas M, Bonani G (1992) Evidence for massive discharges of icebergs into the glacial north Atlantic. *Nature* 360:245-249

Boninsegna JA (1992) South American dendroclimatological records. In: Bradley RS and Jones PD (eds.) *Climate Since A.D. 1500.* Routledge, London pp. 446-62

Boninsegna JA, Hughes MK (2001) Volcanic signals in temperature reconstructions based on tree-ring records for

North and South America. In: Markgraf V (ed.) *Interhemispheric Climate Linkages.* Academic Press, San Diego pp. 141-54

Borisenkov YP (1994) Climatic and other natural extremes in the European territory of Russia in the late Maunder Minimum (1675-1715) In: Frenzel B, Pfister C and Glaeser B (eds.) *Climatic trends and anomalies in Europe 1675-1715* Gustav Fischer Verlag, Stuttgart, Jena pp. 83-94

Botch MS, Kobak KI, Vinson TS, Kolchugina TP (1995) Carbon pools and accumulation in peatlands of the former Soviet Union. *Global Biogeochemical Cycles* 9:37-46

Botkin D (1972) Some ecological consequences of a computer model of forest growth. *Journal of Ecology* 60:849-872

Boulding K (1973) Foreword. In: Polak F (ed.) *The images of the future.* Elsevier, Amsterdam

Boutron CF, Görlach U, Candelone J-P, Bolshov MA, Delmas RJ (1991) Decrease in anthropogenic lead, cadmium and zinc in Greenland snows since the late 1960s. *Nature* 353:153-156

Bowman DMJS (1998) The impact of Aboriginal landscape burning on the Australian biota. *New Phytologist* 140:385-410

Boyd PW, Berges JA, Harrison PJ (1998) In vitro iron enrichment experiments at iron-rich and porr sites in the NE subarctic Pacific. *Journal of Experimental Marine Biology and Ecology* 227: 133-151

Boyd PW, Wong CS, Merrill J, Whitney F, Snow J, Harrisson PJ, Gower J (1998) Atmospheric iron supply and enhanced vertical carbon flux in the NE Subarctic Pacific: is there a connection? *Global Biogeochemical Cycles* 12: 429-441

Boyd P, Newton P (2000) Evidence of the potential influence of planktonic community structure on the interannual variability of material flux during spring in the NE Atlantic.

Boyle EA (1988) The role of vertical chemical fractionation in controlling late Quaternary atmospheric carbon dioxide. *Journal of Geophysical Research* 93:701-715

Boyle EA (1992) Cadmium and δ13C paleochemical ocean distributions during the stage 2 glacial maximum. *Annual Review of Earth and Planetary Sciences* 20: 245-287

Boyle EA (2000) Is ocean thermohaline circulation linked to abrupt stadial/interstadial transitions? *Quaternary Science Reviews* 19:255-272

Boyle EA, Keigwin LD (1982) Deep circulation of the North Atlantic over the last 200,000 years: Geochemical evidence. *Science* 218:784-787

Boyle EA, Keigwin LD (1985) Comparison of Atlantic and Pacific paleochemical records for the last 250,000 years: changes in deep ocean circulation and chemical inventories. *Earth Planetary Science Letters* 76:135-150

Boyle EA, Keigwin LD (1987) North Atlantic thermohaline circulation during the past 20,000 years linked to high latitude surface temperature. *Nature* 330:35-40

Boyle EA, Sclater FR, Edmond JM (1977) The distribution of dissolved copper in the Pacific. *Earth and Planetary Science Letters* 37:38-54

Braconnot P, Joussaume S, Marti O, de Noblet N (1999) Synergistic feedbacks from ocean and vegetation on the African monsoon response to mid-Holocene insolation. *Geophysical Research Letters* 26:2481-2484

Bradbury JP (in press) Limnological history of Lago de Patzcuaro, Michoacan, Mexico for the past 48,000 years: impacts of climate and man. *Palaeogeography Palaeoclimatology Palaeoecology*

Bradbury JP, Grosjean M, Stine S, Sylvestre F (2001) Full and Late Glacial lake records along the PEP1 transect: Their role in developing interhemispheric paleoclimate interactions. In: Markgraf V (ed.) *Interhemispheric Climate Linkages.* Academic Press, pp. 265-292

Bradley RS (1988) The explosive volcanic eruption signal in northern hemisphere continental temperature records.

*Climatic Change* 12:221-243

Bradley RS (1990) Holocene paleoclimatology of the Queen Elizabeth Islands, Canadian High Arctic. *Quaternary Science Reviews* 9:365-384

Bradley RS (1999) *Paleoclimatology: reconstructing climates of the Quaternary* Academic Press, San Diego, 610pp

Bradley RS (2000) Past global changes and their significance for the future. *Quaternary Science Reviews* 19:391-402

Bradley RS, Jones PD (1992) When was the "Little Ice Age"? In: Mikami T (ed.) *Proceedings of the International Symposium on the Little Ice Age Climate.* Dept. of Geography, Tokyo Metropolitan University, Tokyo pp. 1-4

Bradley RS, Jones PD (1992) Introduction. In: Bradley RS and Jones PD (eds.) *Climate Since A.D. 1500.* Routledge, London pp. 1-16

Bradley RS, Jones PD (1992) Records of explosive volcanic eruptions over the last 500 years. In: R.S. Bradley and Jones PD (eds.) *Climate Since A.D. 1500.* Routledge, London pp. 606-622

Bradley RS, Jones PD (1993) "Little Ice Age" summer temperature variations: their nature and relevance to recent global warming trends. *The Holocene* 3:367-376

Bradley RS, Hughes MK, Mann ME (2000) Comments on "Detection and Attribution of Recent climate change: a status report". *Bulletin of the American Meteorological Society* 81:2987-2990

Bradshaw R, Zackrisson O (1990) A two thousand year record of a northern Swedish boreal forest stand. *Journal of Vegetation Science* 1:519-528

Bradshaw R, Hannon G (1992) Climatic change, human influence and disturbance regimen in the control of vegetation dynamics within Fiby forest, Sweden. *Journal of Ecology* 80:625-632

Bradshaw R, Lindblagh M, Holmqvist BH, Cowling S (2000) Shift from deciduous to coniferous forests in Southern Scandinavia driven by climate change and land-use interactions. *PAGES Newsletter* 8:30-39

Brassell SC, Eglington G, Marlowe IT, Pflaumann U, Sarnthein M (1986) Molecular stratigraphy: a new tool for climatic assesment. *Nature* 320:129-133

Braswell BH, Schimel DS, Linder E, Moore B (1997) The response of global terrestrial ecosystems to interannual temperature variability. *Science* 278:870-872

Brázdil R, Kotyza O, Dobrovoln P (1999) *History of Weather and Climate in the Czech Lands, V. Period 1500-1599* Masaryk University, Brno

Brecher HH, Thompson LG (1993) Measurement of the retreat of Qori Kalis in the tropical Andes of Peru by terrestrial photogrammetry. *Photogrammetric Engineering and Remote Sensing* 59:1017-1022

Brenner M, Hodell DA, Curtis JH, M. F. Rosenmeier, Binford MW, Abbott MB (2001) Abrupt climate change and pre-Columbian cultural collapse. In: Markgraf V (ed.) *Interhemispheric Paleoclimate of the Americas.* Academic Press

Briffa KR (2000) Annual climate variability in the Holocene: interpreting the message of ancient trees. *Quaternary Science Reviews* 19:87-105

Briffa KR, Jones PD (1993) Global surface air temperature variations over the twentieth century: Part 2, implications for large-scale high-frequency palaeoclimatic studies. *The Holocene* 3:82-93

Briffa KR, Jones PD, Schweingruber FH, Osborn TJ (1998) Influence of volcanic eruptions on Northern Hemisphere summer temperature over the past 600 years. *Nature* 393:450-455

Briffa KR, Schweingruber FH, Jones PD, Osborn TJ, Shiyatov SG, Vaganov EA (1998) Reduced sensitivity of recent tree-growth to temperature at high northern latitudes. *Nature* 391:678-682

Briffa KR, Osborn TJ, Schweingruber FH, Harris IC, Jones PD, Shiyatov SG, Vaganov EA (2001) Low-frequency temperature variations from a northern tree ring density network. *Journal of Geophysical Research* 106D:2929-2941

Briffa KR, Jones PD, Bartholin TS, Eckstein D, Schweingruber FH, Karlén W, Zetterberg P, Eronen M (1992) Fennoscandian summers from A.D. 500: temperature changes on short and long timescales. *Climate Dynamics* 7:111-119

Briffa KR, Jones PD, Schweingruber FH, Shiyatov SG, Cook ER (1995) Unusual 20$^{th}$ century summer warmth in a 1000 year temperature record from Siberia. *Nature* 376: 156-159

Brimblecombe P (1987) *The big smoke* Methuen, London

Broccoli AJ (2000) Tropical cooling at the Last Glacial Maximum: An atmospheric-mixed layer ocean model simulation. *Journal of Climate* 13: 951-976

Broecker W, Klas M, Ragano-Beavan N, Mathieu G, Mix A (1988) Accelerator mass spectrometry radiocarbon measurements on marine carbonate samples from deep-sea cores and sediment traps. *Radiocarbon* 30:261-295

Broecker WS (1987) Unpleasant surprises in the greenhouse. *Nature* 328:123-126

Broecker WS (1998) Paleocean circulation during the last deglaciation: A bipolar seesaw? *Paleoceanography* 13:119-121

Broecker WS, Denton GH (1989) The role of ocean - atmosphere reorganisations in glacial cycles. *Geochim. Cosmochim. Acta* 53:2465-2501

Broecker WS, Donk JV (1970) Insolation changes, ice volumes, and the O-18 record in deep-sea cores. *Review of Geophysic and Space Physic* 8:169-198

Broecker WS, Henderson GM (1998) The sequence of events at termination II and their implications for the cause of glacial-interglacial $CO_2$ changes. *Paleoceanography* 13: 352-364

Broecker WS, Peng T-H (1982) *Tracers in the Sea*. Eldigio Press, Palisades, New York, 690 pp.

Broecker WS, Peng TH (1993) Evaluation of the $^{13}C$ constraint on the uptake of fossil fuel $CO_2$ by the ocean. *Global Biogeochemical cycles* 7:619-626

Broecker WS, Peteet DM, Rind D (1985) Does the ocean-atmosphere system have more than one mode of operation? *Nature* 315:21-26

Broecker WS, Sutherland S, Peng T-H (1999) A possible 20th-century slowdown of southern Ocean deep water formation. *Science* 286:1132-1135

Broecker WS, Bond GC, Klas M, Clark E, McManus J (1992) Origin of the Northern Atlantic Heinrich events. *Climate Dynamics* 6:265-273

Broecker WS, Thurber DL, Goddard J, Ku TL, Matthews RK, Mesolella KJ (1968) Milankovitch hypothesis supported by precise dating of coral reefs and deep sea sediments. *Science* 159:297-300

Broecker WS, Kennett JP, Flower BP, Teller JT, Trumbore S, Bonani G, Wolfli W (1989) Routing of meltwater from the Laurentide Ice Sheet during the Younger Dryas cold episode. *Nature* 341:318-321

Broecker WS, Andree M, Bonani G, Wolfi W, Oeschger H, Klas M, MIX A, CURRY W (1988) Preliminary estimates for the radiocarbon age of deep water in the glacial ocean. *Paleoceanography* 3: 659-669

Brook EJ, Harder S, Severinghaus J, Steig EJ, Sucher M (2000) On the origin and timing of rapid changes in atmospheric methane during the last glacial period. *Global Biogeochemical Cycles* 14:559-572

Broström A, Gaillard M-J, Ihse M, Odgaard B (1998) Pollen-landscape relationships in modern analogues of ancient cultural landscapes in southern Sweden – a first step towards quantification of vegetation openness in the past. *Vegetation History and Archaeobotany* 7:189-201

Broström A, Coe MT, Harrison SP, Gallimore R, Kutzbach JE, Foley J, Prentice IC, Behling P (1998) Land surface feedbacks and palaeomonsoons in northern Africa. *Geophysical Research Letters* 25:3615-3618

Brovkin V, Ganapolski A, Claussen M, Kubatski C, Petoukhov V (1999) Modelling climate responses to historical land cover change. *Global Ecology and Biogeography* 8:509 – 517

Brown PB, Hughes MK, Swetnam TW, Caprio AR (1992) Giant Sequoia ring-width chronologies from the central Sierra Nevada. *Tree-Ring Bulletin* 52:1-14

Brown SL, Bierman PR, Lini A, Southon J (2000) 10,000 yr record of extreme hydrologic events. *Geology* 28:335-338

Bugmann H, Pfister C (2000) Impacts of interannual climate variability on past and future forest composition. *Regional Environmental Change* 200:112-125

Bugmann HKM (1997) Gap models, forest dynamics and the response of vegetation to climate change. In: Huntley B, Cramer W, Morgan AV, Prentice HC and Allen JRM (eds.) *Past and Future Rapid Environmental Changes: The Spatial and Evolutionary Responses of Terrestrial Biota*. Springer Verlag NATO ASI Series, Berlin

Burn CR (1997) Cryostratigraphy, paleogeography and climate change during the early Holocene warm interval, western Arctic coast, Canada. *Canadian Journal of Earth Sciences* 34:912-925

Bush ABG (1999) Assessing the impact of mid-Holocene insolation on the atmosphere-ocean system. *Geophysical Research Letters* 26:99-102

Bush ABG, Philander SGH (1998) The role of Ocean-Atmosphere interactions in tropical cooling during the last glacial maximum. *Science* 279:1341-13474

Cacho I, Grimalt JO, Pelejero C, Canals M, Sierro FJ, Flores JA, Shackleton N (1999) Dansgaard-Oeschger and Heinrich event imprints in Alboran Sea paleotemperatures. *Paleoceanography* 14:698-705

Calvert S, Fontugne M (2001) On the late Pleistocene-Holocene sapropel record of climatic and oceanographic variability in the eastern Mediterranean. *Paleoceanography* 16:78-94

Camill P, Clark J (2000) Long-term perspectives on lagged ecosystem responses to climate change: permafrost in boreal peatlands and the grassland/woodland boundary. *Ecosystems* 3:534-544

Candelone J-P, Hong S, Pellone C, Boutron CF (1995) Post-industrial revolution changes in large-scale atmospheric pollution of the Northern Hemisphere for heavy metals as documented in central Greenland snow and ice. *Journal of Geophysical Research* 100:16605-16616

Cane M, Clement A, Gagan M, Ayliffe L, Tudhope S (2000) ENSO through the Holocene depicted in corals and a model simulation. *PAGES News* 5:3-7

Cane MA, Evans M (2000) Climate variability - Do the tropics rule? *Science* 290:1107-1108

Cane MA, Clement AC, Kaplan A, Kushnir Y, R. Murtugudde, Zebiak S (1997) Twentieth-century sea surface temperature trends. *Science* 275:957-960

Carcaillet C, Richard PJH (2000) Holocene changes in seasonal precipitation highlighted by fire incidence in eastern Canada. *Climate Dynamics* 16:549-559

Casseldine C, Hatton J (1993) The development of high moorland on Dartmoor: fire and the influence of Mesolithic activity on vegetation change. In: Chambers FM (ed.) *Climate Change and Human Impact on the Landscape*. Chapman and Hall, London pp. 119-131

Catubig NR, Archer DE, Francois R, deMenocal P, Howard W, Yu EF (1998) Global deep-sea burial rate of calcium carbonate during the last glacial maximum. *Paleoceanography* 13: 298-310

Cerling TE (1991) Carbon dioxide in the atmosphere: evidence from Cenozoic and Mesozoic Paleosols. *American Journal* of Science 291:377-400

Chadwick O, Derry L, Vitousek P, Huebert B, Hedin L (1999) Changing sources of nutrients during four million years of ecosystem development. Nature 397:491-497

Chang P, Ji L, Li H (1997) A decadal climate variation in the tropical Atlantic Ocean from thermodynamic air-sea interactions. Nature 385:516-518

Chao Y, Ghil M, McWilliams JC (2000) Pacific interdecadal variability in this century's sea surface temperatures. Geophysical Research Letters 27:2261-2264

Chapin FS, Zavaleta ES, Eviner VT, Naylor RL, Vitousek PM, Reynolds HL, Hooper DU, Lavorel S, Sala OE, Hobbie SE, Mack MC, Diaz S (2000) Consequences of changing biodiversity. Nature 405:234-242

Chappell J, Omura A, Esat T, McCulloch M, Pandolfi J, Ota Y, Pillans B (1996) Reconciliation of late Quaternary sea levels derived from coral terraces at Huon Peninsula with deep sea oxygen isotope records. Earth and Planetary Science Letters 141:227-236

Chappellaz J, Blunier T, Raynaud D, Barnola JM, Schwander J, Stauffer B (1993) Synchronous changes in atmospheric CH$_4$ and Greenland climate between 40 and 8 kyr BP. Nature 366:443-445

Chappellaz J, Blunier T, Kints S, Dällenbach A, Barnola J-M, Schwander J, Raynaud D, Stauffer B (1997) Changes in the atmospheric CH$_4$ gradient between Greenland and Antarctica during the Holocene. Journal of Geophysical Research 102:15987-15999

Charles CD, Fairbanks RG (1992) Evidence from Southern Ocean sediments for the effect of North Atlantic deep water flux on climate. Nature 355:416-419

Charles CD, Hunter DE, Fairbanks RG (1997) ENSO and the Monsoon in a coral record of Indian Ocean surface temperature. Science 277:925-928

Charles CD, Lynch-Stieglitz J, Ninnemann US, Fairbanks RG (1996) Climate connections between the hemisphere revealed by deep sea sediment core/ice core correlations. Earth and Planetary Science Letters 142:19-27

Charlson RJ, Lovelock JE, Andreae MO, Warren SG (1998) Oceanic phytoplankton, atmospheric sulphur, cloud albedo and climate. Nature 326:655-661

Charney J (1975) The dynamics of deserts and droughts. Quarterly Journal of the Royal Meteorological Society 101:109-202

Chase T, Pielke R, Kittel T, Nemani R, Running S (2000) Simulated impacts of historical land cover changes on global climate in northern winter. Climate Dynamics 16:93-105

Chase Z, Anderson RF, Fleisher MQ, Kubik P (2001) The influence of particle composition on scavenging of Th, Pa and Be in the ocean. EOS 82:F619

Chavez FP, Barber RT (1987) An estimate of new production in the equatorial Pacific. Deep-Sea Research 34: 1229-1243

Chavez FP, Strutton PG, Friederich GE, Feely RA, Feldman GC, Foley DG, McPhaden MJ (1999) Biological and chemical response of the equatorial Pacific ocean to the 1997-98 El Niño. Science 286:2126-2131

Chen FH, Wu RJ, Pompei D, Oldfield F (1995) Magnetic property and particle size variations in the late Pleistocene and Holocene parts of the Dadongling loess section near Xining. Quaternary Proceedings 4:27-40

Chen FH, Bloemendale J, Wang JM, Li JJ, Oldfield F (1997) High-resolution multi-proxy climate records from Chinese loess: evidence for rapid climatic changes over the last 75 kyr. Palaeogeography, Palaeoclimatology, Palaeoecology 130:323-335

Chepstow-Lusty AJ, Bennett KD, Fjeldsa J, Kendall A, Galiano W, Herrera AT (1998) Tracing 4,000 years of environmental history in the Cuzco area, Peru, from the pollen record. Mountain Research and Development 18:159-172

Christner BC, Mosley-Thompson E, Thompson LG, Zagorodnov

V (2000) Recovery and identification of viable bacteria immured in glacial ice. Icarus 144:479-485

Ciais P, Petit JR, Jouzel J, Lorius C, Barkov NI, Lipenkov V, Nicolaïev V (1992) Evidence for an early Holocene climatic optimum in the Antarctic deep ice-core record. Climate Dynamics 6:166-177

Clapperton CM, Sugden DE (1988) Holocene glacier fluctuations in South America and Antarctica. Quaternary Science Reviews 7:185-198

Clark CO, Cole JE, Webster PJ (2000) Relationship between Indian summer rainfall and Indian Ocean SST. Journal of Climate 13:2503-2519

Clark JS (1990) Fire and climate change during the last 750 years in northwestern Minnesota. Ecological Monographs 60:135 – 159

Clark JS, Royall P, Chumbley C (1996) The role of fire during climate change in an eastern deciduous forest at Devil's Bathtub, New York. Ecology 77:2148-2166

Clark JS, Stocks BJ, Richard PJH (1996) Climate implications of biomass burning since the 19th century in eastern North America. Global Change Biology 2:433-442

Clark JS, Lynch J, Stocks BJ, Goldammer JG (1998) Relationships between charcoal particles in air and sediments in west-central Siberia. The Holocene 8:19-31

Clark JS, Grimm E, Lynch J, Mueller P (2001) Effects of Holocene climate change on the C4 grassland/woodland boundary in the Northern Plains. Ecology 82:620-636

Clark JS, Fastie C, Hurtt G, Jackson ST, Johnson C, King GA, Lewis M, Lynch J, Pacala S, Prentice C, Schupp EW, WebbIII T, Wychoff P (1998) Reid's Paradox of Rapid Plant Migration: Dispersal theory and interpretation of palaeological records. BioScience 48:13-24

Clark PU, Mix AC (2002) Ice Sheets and Sea Level of the Last Glacial Maximum. Elsevier Science Ltd.

Clark PU, Alley RB, Pollard D (1999) Northern Hemisphere ice-sheet influences on global climate change. Science 286:1104-1111

Clarke GKC, Marshall SJ, Hillaire-Marcel C, Bilodeau G, Veiga-Pires C (1999) A glaciological perspective on Heinrich events. In: Clark PU, Webb RS and Keigwin LD (eds.) Mechanisms of Global Climate Change at Millenial Time Scales. American Geophysical Union, Washington DC pp. 243-262

Claussen M, Gayler V (1997) The greening of the Sahara during the mid-Holocene: results of an interactive atmosphere-biome model. Global Ecology and Biogeography Letters 6:369-377

Claussen M, Brovkin V, Ganopolski A, Kubatzki C, Petoukhov V (1998) Modelling global terrestrial vegetation-climate interaction. Philosohpical Transactions of the Royal Society, London 353:53-63

Claussen M, Kubatzki C, Brovkin V, Ganopolski A, Hoelzmann P, Pachur H-J (1999) Simulation of an abrupt change in Saharan vegetation in the mid-Holocene. Geophysical Research Letters 24:2037-2040

Clement A, Cane MA (1999) A role for the tropical Pacific coupled ocean-atmosphere system on Milankovich and millennial time scales. Part 1: A modeling study of tropical Pacific variability. In: Clark PU, Webb RS and Keigwin LD (eds.) Mechanisms of global climate change. American Geophysical Union, Washington DC pp. 363-371

Clement AC, Seager R, Cane MA (1999) Orbital controls on the El Niño/Southern Oscillation and tropical climate. Paleoceanography 14:441-456

Clement AC, Seager R, Cane MA (2000) Suppression of El Niño during the mid-Holocene by changes in the Earth's orbit. Paleoceanography 15:731-737

CLIMAP (1976) The surface of the ice-age earth. Science 191:1131-1137

CLIMAP (1981) Seasonal reconstructions of the Earth's surface at the last glacial maximum. Geological Society of America

Clow GD (1992) Temporal resolution of surface temperature histories inferred from borehole temperature measurements. *Palaeogeography, Palaeoclimatology, Palaeoecology* 98:81-86

Clow GD, Waddington ED (1999) *Abstract* IUGG, Birmingham

Clymo RS (1991) Peat growth. In: Shane LCK and Cushing EJ (eds.) *Quaternary Research*. Bellhaven, London pp. 76-112

Clymo RS, Oldfield F, Appleby PG, Pearson GW, Ratnesar P, Richardson N (1990) The record of atmospheric deposition on a rainwater-dependent peatland. *Philosophical Transactions of the Royal Society of London Series B* 327:331-338

Coale KH, Fitzwater SE, Gordon RM, Johnson KS, Barber RT (1996) Control of community growth and export production by upwelled iron in the equatorial Pacific Ocean. *Nature* 379:621-624

CODATA (2002) Scientific Data Policy Statements. (Originated from Report on ICSU Activities in Data and Information). In: *ICSU Ad Hoc Committee on Data Issues, CODATA, August, 1993*. International Council for Science Committee on Data for Science and Technology

Codispoti LA (1995) Is the ocean losing nitrate? *Nature* 376:724

Coe MT, Bonan G (1997) Feedbacks between climate and surface water in Northern Africa during the middle-Holocene. *Journal of Geophysical Research* 102:11087-11101

COHMAP (1988) Climatic changes of the last 18,000 years: observations and model simulations. *Science* 241:1043-1052

Cole DR, Monger HC (1994) Influence of atmospheric $CO_2$ on the decline of C4 plants during the last deglaciation. *Nature* 368:533-536

Cole J (2001) PALEOCLIMATE: Enhanced: A Slow Dance for El Niño. *Science* 291:1496-1497

Cole JE (2000) Coherent decadal variability in coral records from the tropical Indian and Pacific Oceans. 81:F38

Cole JE, Cook ER (1997) The coupling between ENSO and US drought: how stable is it? *EOS: Transactions* AGU, 78, F36

Cole JE, Cook ER (1998) The changing relationship between ENSO variability and moisture balance in the continental United States. *Geophysical Research Letters* 25:4529-4532

Cole JE, Fairbanks RG, Shen GT (1993) The spectrum of recent variability in the Southern Oscillation: results from a Tarawa Atoll coral. *Science* 262:1790-1793

Cole JE, Overpeck JT, Cook ER (submitted) Multiyear La Niñas and prolonged US drought. *Geophysical Research Letters*

Colinvaux PA, Oliveira PED, Bush MB (2000) Amazonian and Neotropical plant communities on glacial time-scales: The failure of the aridity and refuge hypotheses. *Quaternary Science Reviews* 19:141-169

Collatz G, Berry J, Clark J (1998) Effects of climate and atmospheric $CO_2$ partial pressure on the global distribution of C4 grasses: present, past and future. *Oecologia* 114:441-454

Collingham YC, Huntley B (2000) Impacts of habitat fragmentation and patch size upon migration rates. *Ecological Applications* 10:131-144

Collins M, Osborn TJ, Tett SFB, Briffa KR, Schweingruber FH (2000) A comparison of the variability of a climate model with palaeo-temperature estimates from a network of tree-ring densities. *Hadley Centre Technical Note* 16:41pp

Cook ER, Cole J (1991) On predicting the response of forests in eastern North-America to future climatic-change. *Climatic Change* 19:271-282

Cook ER, Meko DM, Stockton CW (1997) A new assessment of possible solar and lunar forcing of the bidecadal drought rhythm in the Western United States. *Journal of Climate* 10:1343-1356

Cook ER, D'Arrigo RD, Briffa KR (1998) The North Atlantic Oscillation and its expression in circum-Atlantic tree-ring chronologies from North America and Europe. *The Holocene* 8:9-17

Cook ER, Meko DM, Stahle DW, Cleaveland MK (1999) Drought reconstructions for the continental United States. *Journal of Climate* 12:1145-1162

Cooper SR (1995) Chesapeake Bay watershed historical land use: Impact on water quality and diatom communities. *Ecological Applications* 5:703-723

Cooper SR, Brush GS (1991) Long-term history of Chesapeake Bay anoxia. *Science* 254:992-996

Cooper SR, Brush GS (1993) A 2,500 year history of anoxia and eutrophication in Chesapeake Bay. *Estuaries* 16:617-626

Coplen TB, Winograd IJ, Landwehr JM, Riggs AC (1994) 500,000-year stable carbon isotopic record from Devils Hole, Nevada. *Science* 263:361-365

Cortijo E, Labeyrie L, Elliot M, Balbon E, Tisnerat N (2000) Rapid climatic variability of the North Atlantic Ocean and global climate: a focus of the IMAGES program. *Quaternary Science Reviews* 19:227-241

Cortijo E, Lehman S, Keigwin L, Chapman M, Paillard D, Labeyrie L (1999) Changes in meridional temperature and salinity gradients in the North Atlantic Ocean (30° to 72°N) during the Last Interglacial Period. *Paleoceanography* 14:23-33

Cortijo E, Labeyrie L, Vidal L, Vautravers M, Chapman M, Duplessy JC, Elliot M, Arnold M, Turon JL, Auffret G (1997) Sea surface temperature reconstructions during the Heinrich event 4 between 30 and 40 kyr in the North Atlantic Ocean (40-60°N). *E.P.S.L* 146:29-45

Coulthard TJ, Kirkby MJ, Macklin MG (2000) Modelling geomorphic response to environmental change in an upland catchment. *Hydrological Processes* 14:2031-2045

Couzin J (1999) Landscape changes make regional climate run hot and cold. *Science* 283:317 - 319

Cowling S (1999) Simulated effects of low atmospheric $CO_2$ on structure and composition of North American vegetation at the Last Glacial Maximum. *Global Ecology and Biogeography* 8:81-93

Cowling S, Sykes M (1999) Physiological significance of low atmospheric $CO_2$ for Plant-Climate Interactions. *Quaternary Research* 52:237-242

Cowling S, Maslin M, Sykes M (2001) Paleovegetation simulations of lowland Amazonia and implications for neotropical allopatry and speciation. *Quaternary Research* 55:140-149

Cragin JH, Herron MM, Langway CC, Jr. GK (1977) Interhemispheric comparison of changes in the composition of atmospheric precipitation during the late Cenozic era. In: Dunbar MJ (ed.) *Polar Oceans*. Arctic Institute of North America, Calgary pp. 617-631

Craig H, Chou CC (1982) Methane: the record in polar ice cores. *Geophysical Research Letters* 9:1221-1224

Craig H, Horibe Y, Sowers T (1988) Gravitational separation of gases and isotopes in polar ice caps. *Science* 242:1675-1678

Critchfield W (1984) Impact of the Pleistocene of the genetic structure of North American conifers. In: Lanner R (ed.) *Proceedings of the Eigth North American Forest Biology Workshop*. Utah State University, Logan pp. 70-119

Cronin TM (1985) Speciation and stasis in marine *Ostracoda*: climatic modulation of evolution. *Science* 227:60-63

Cronin TM, Dwyer GS, Baker PA, Rodriguez-Lazaro J, DeMartino DM (2000) Orbital and suborbital variability in North Atlantic bottom water temperature obtained from deep-sea ostracod Mg/Ca ratios. *Palaeogeography, Palaeoclimatology, Palaeoecology* 162:45-57

Crooks PRJ (1991) *The use of Chernobyl-derived radiocaesium for dating recent lake sediments* Unpublished PhD thesis, The University of Liverpool

Cross SL, Baker PA, Seltzer GO, Fritz SC, Dunbar RB (2000) Late Quaternary climate and hydrology of tropical South

America inferred from an isotopic and chemical model of Lake Titicaca, Bolivia and Peru. *Quaternary Research* 56:1-9

Crosta X, Pichon JJ, Burckle L (1998) Application of modern analog technique to marine Antarctic diatoms: reconstruction of maximum sea-ice extent at the Last Glacial Maximum. *Paleoceanography* 13:284-297

Crowley T (1990) Are there any satisfactory geologic analogs for a future greenhouse warming? *Journal of Climate* 3:1282-1292

Crowley TJ (1992) North Atlantic deep water cools the Southern Hemisphere. *Paleoceanography* 7:489-549

Crowley TJ (1995) Ice age terrestrial carbon changes revisited. *Global Biogeochemical Cycles* 8:366-389

Crowley TJ (2000) Causes of climate change over the past 1000 years. *Science* 289:270-277

Crowley TJ, Kim K-Y (1996) Comparison of proxy records of climate change and solar forcing. *Geophysical Research Letters* 23:359-362

Crowley TJ, Kim K-Y (1999) Modeling the temperature response to forced climate change over the last six centuries. *Geophysical Research Letters* 26:1901-1904

Crowley TJ, Lowery TS (2000) How warm was the Medieval Warm Period? *Ambio* 29:51-54

Crumley CL (1993) Analyzing historic ecotone shifts. *Ecological Applications* 3:3

Crumley CL (1994) *Historical Ecology: Cultural Knowledge and Changing Landscapes* School of American Research Press, Santa Fe

Crusius J, Calvert SE, Pedersen TF, Sage D (1996) Rhenium and molybdenum enrichments in sediments as indicators of oxic, suboxic and anoxic conditions of deposition. *Earth and Planetary Science Letters* 145:65-78

Crutzen PJ (2002) Geology of mankind. *Nature* 415:23

Crutzen PJ, Brühl C (1993) A model study of atmospheric temperatures and the concentrations of ozone, hydroxyl, and some other photochemically active gases during the glacial, the pre-industrial Holocene and the present. *Geophysical Research Letters* 20:1047-1050,

Cubasch U, Voss R, Hegerl GC, Waszkewitz J, Crowley TJ (1997) Simulation of the influence of solar radiation variations on the global climate with an ocean-atmosphere general circulation model. *Climate Dynamics* 13:757-767

Cuffey K, Marshall S (2000) Substantial contribution to sea-level rise during the last interglacial from the Breenland ice sheet. *Nature* 404:591-592

Cuffey KM, Clow GD, Alley RB, Stuiver M, Waddington ED, Saltus RW (1995) Large arctic temperature change at the Wisconsin-Holocene glacial transition. *Science* 270:455-458

Cullen HM, D'Arrigo RD, Cook ER, Mann ME (2001) Multiproxy reconstructions of the North Atlantic Oscillation. *Paleoceanography* 16:27-39

Cullen HM, DeMenocal PB, Hemming S, Hemming G, Brown FH, Guilderson T, Sirocko F (2000) Climate change and the collapse of the Akkadian empire: Evidence from the deep sea. *Geology* 28:379-382

Curran L, Caniago I, Paoli G, Astianti D, Kusneti M, Leighton M, Nirarita C, Haeruman H (1999) Impact of El Niño and logging on canopy recruitment in Borneo. *Science* 286: 22,184-22,188

Curry RG, McCartney MS, Joyce TM (1998) Oceanic transport of subpolar climate signals to mid-depth subtropical waters. *Nature* 391:575-577

Curry W, Oppo DW (1997) Synchronous, high frequency oscillations in tropical sea surface temperature and North Atlantic deep-water production during the last glacial cycle. *Paleoceanography* 12:1-14

Curry WB, Lohmann GP (1983) Reduced advection into Atlantic deep eastern basins during last glacial maximum. *Nature* 306:577-580

Curry WB, Duplessy J-C, Labeyrie LD, Shackleton NJ (1988) Changes in the distribution of $\delta^{13}C$ of deep water $SCO_2$ between the last glaciation and the Holocene. *Paleoceanography* 3:317-341

D'Arrigo R (1998) The Southeast Asian dendro workshop, 1998. *PAGES News* 6:14-15

D'Arrigo R, Villalba R, Wiles G (submitted) Tree-ring estimates of Pacific decadal climate variability. *Climate Dynamics*

D'Arrigo R, Jacoby G, Free M, Robock A (1999) Northern Hemisphere temperature variability for the past three centuries: tree-ring and model estimates. *Climatic Change* 42:663-675

D'Arrigo RD, Jacoby GC (1993) Secular trends in high northern latitude temperature reconstructions based on tree-rings. *Climatic Change* 25:163-177

D'Arrigo RD, Cook ER, Jacoby GC, Briffa KR (1993) NAO and sea-surface temperature signatures in tree-ring records from the North-Atlantic sector. *Quaternary Science Reviews* 12:431-440

Dahle SO, Nesje A (1996) A new approach to calculating Holocene winter precipitation by combining glacier equilibrium line altitudes and pine-tree limits: a case study from Hardangerjøkulen, central southern Norway. *The Holocene* 6:381-398

Dahl-Jensen D, V. Morgan, Elcheikh A (1999) Monte Carlo inverse modelling of the Law Dome temperature profile. *Annals of Glaciology* 29:45-150

Dahl-Jensen D, Mosegaard K, Gunderstrup N, Clow GO, Johnsen SJ, Hansen AW, Balling N (1998) Past temperature directly from the Greenland Ice sheet. *Science* 252:268-271

Dalfes N, G. Kukla, Weiss H (1997) *Third Millennium Climate Change and Old World Social Collapse* Springer, Berlin

Dällenbach A, Blunier T, Flückiger J, Stauffer B, Chappellaz J, Raynaud D (2000) Changes in the atmospheric $CH_4$ gradient between Greenland and Antarctica during the Last Glacial and the transition to the Holocene. *Geophysical Research Letters*

Dansgaard W (1964) Stable isotopes in precipitation. *Tellus* 16:436-468

Dansgaard W, Oeschger H (1989) Past environmental long-term records from the Arctic. In: H. Oeschger and Langway CC (eds.) *The Environmental Record in Glaciers and Ice Sheets*. John Wiley and Sons Limited, Chichester, UK, pp. 287-318

Dansgaard W, Johnsen SJ, Clausen HB, Gunderstrup N (1973) Stable isotope glaciology. *Meddelelser om Grönland* 197:321

Dansgaard W, Johnsen SJ, Clausen HB, Dahl-Jensen D, Gundestrup N, Hammer CU (1984) North Atlantic climatic oscillations revealed by deep Greenland ice cores. In: Hansen JE and Takahashi T (eds.) *Climate Processes and Climate Sensitivity*. American Geophysical Union, Washington, D.C pp. 288-298

Dansgaard W, Johnsen SJ, Clausen HB, Dahl-Jensen D, Gundestrup NS, Hammer CU, Hvidberg CS, Steffensen JP, Sveinbjornsdottir AE, Jouzel J, Bond G (1993) Evidence for general instability of past climate from a 250-kyr ice-core record. *Nature* 364:218-220

Davis A, Jenkinson L, Lawton J, Shorrocks B, Wood S (1998) Making mistakes when predicting shifts in species range in response to global warming. *Nature* 391:783-786

Davis M, Botkin D (1985) Sensitivity of cool-temperate forests and their fossil pollen record to rapid temperature change. *Quaternary Research* 23:327-340

Davis M, Shaw R (2001) Range shifts and adaptive responses to Quaternary climate change. *Science* 292:673-679

Davis MB (1976) Erosion rates and land use history in Southern Michigan. *Environmental Conservation* 3:139-148

Davis MB (1976) Pleistocene biogeography of temperate deciduous forests. *Geoscience and Man* 13:13-26

Davis MB (1989) Insights from paleoecology on global change.

*Ecological Society of America Bulletin* 70:220-228

Davis MB (1990) Biology and paleobiology of global climate change: Introduction. *Trends in Ecology and Evolution* 5:269-270

Davis OK (1998) Palynological evidence for vegetation cycles in a 1.5 million year pollen record from the Great Salt Lake, Utah, USA. *Palaeogeography Palaeoclimatology Palaeoecology* 138:175-185

Davis OK, Moutoux TE (1998) Tertiary and Quaternary vegetation history of the Great Salt Lake, Utah, USA. *Journal of Paleolimnology* 19:417-427

De Angelis M, Legrand M (1995) Preliminary investigations of post depositional effects on HCl, $HNO_3$ and organic acids in polar firn layers. In: Delmas RJ (ed.) *Ice CoreStudies of Global Biogeochemical Cycles*. Springer Verlag, pp. 369-390

De Angelis M, Barkov NI, Petrov VN (1987) Na aerosol concentrations over the last climatic cycle (160 kyr) from an Antarctic ice core. *Nature* 325:318-321

De Beaulieu J-L, Reille M (1984) A long upper Pleistocene pollen record from Les Echets, near Lyon, France. *Boreas* 13:111-132

De Beaulieu J-L, Reille M (1992) The last climatic cycle at La Grande Pile (Vosges, France). A new pollen profile. *Quaternary Science Reviews* 11:431-438

De La Mare WK (1997) Abrupt mid-twentieth century decline in Antarctic sea-ice extent from whaling records. *Nature* 57-60

De La Rocha CL, Brzezinski MA, DeNiro MJ, Shemesh A (1998) Silicon-isotope composition of diatoms as an indicator of past oceanic change. *Nature* 395:680-683

De Menocal P (1995) Plio-Pleistocene African climate. *Science* 270:53-59

De Menocal PB (2001) Cultural responses to climate change during the Late Holocene. *Science* 292:667-673

De Menocal PB, Rind D (1993) Sensitivity of Asian and African climate to variations in seasonal insolation, glacial ice cover, sea-surface temperature, and Asian orography. *Journal of Geophysical Research-Atmospheres* 98:7265-7287

De Menocal PB, Ortiz J, Guilderson T, Sarnthein M (2000) Coherent high- and low-latitude climate variability during the Holocene warm period. *Science* 288:2198-2202

De Menocal PB, Oritz J, Guilderson T, Adkins J, Sarnthein M, Baker L, Yarusinsky M (2000) Abrupt onset and termination of the African Humid Period: rapid climate responses to gradual insolation forcing. *Quaternary Science Reviews* 19:347-361

De Noblet N, Prentice IC, Joussaume S, Texier D, Botta A, Haxeltine A (1996) Possible role of atmospheric biosphere interactions in triggering the last glaciation. *Geophysical Research Letters* 23:3191-3194

De Vernal A, Hillaire-Marcel C (2000) Sea-ice cover, sea-surface salinity and halo-thermocline structure of the northwest North Atlantic: modern versus full glacial conditions. *Quaternary Science Reviews* 19:65-85

De Vernal A, Hillaire-Marcel C, Bilodeau G (1996) Reduced meltwater outflow from the Laurentide ice margin during the Younger Dryas. *Nature* 381:774-777

De Vries T, Ortlieb L, Diaz A, Wells L, Hillaire-Marcel C (1997) Determining the early history of El Niño. *Science* 276:965-966

Dean JS, Euler RC, Gumerman GJ, Plog F, Hevly RH, Kartstrom TNV (1985) Human behaviour, demography, and paleoenvironment on the Colorado Plateaus. *American Antiquity* 50:537-554

Dean JS, Gumerman GJ, Epstein JM, Axtell RL, Swedlund AC, Parker MT, McCaroll S (1999) Understanding Anasazi culture change through age-based modeling. In: Kohler T and Gumerman G (eds.) *Dynamics in Human and Primate Societies*. Oxford University Press, pp. 179-205

Dearing JA (1994) Reconstructing the history of soil erosion. In:

Roberts N (ed.) *The Changing Global Environment*. Blackwells, pp. 242-261

Dearing JA, Zolitschka B (1999) System dynamics and environmental change: an exploratory study of Holocene lake sediments at Holzmaar, Germany. *The Holocene* 9:531-540

Dearing JA, Jones RT (in press) Coupling temporal and spatial dimensions of global sediment flux through lake and marine sediment records. *Global and Planetary Change*

Dearing JA, Alström K, Bergman A, Regnell J, Sandgren P (1990) Recent and long-term records of soil erosion from southern Sweden. In: Boardman J, Foster IDL and Dearing JA (eds.) *Soil Erosion on Agricultural Land*. John Wiley & Sons Ltd, London pp. 173-191

Deevey ES (1967) Coaxing history to conduct experiments. *BioScience* 19:40 – 43

Deevey ES, Rice DS, Vaughan HH, Brenner M, Flannery MS (1979) Mayan urbanism: Impact on a tropical karst environment. *Science* 206:298 – 306

Dehairs F, Fagel N, Antia AN, Peinert R, Elskens M, Goeyens L (2000) Export production in the Bay of Biscay as estimated from barium - barite in settling material: a comparison with new production. *Deep-Sea Research* 47:583-601

Delaney ML (1998) Phosphorus accumulation in the marine sediments and the oceanic phosphorus cycle. *Global Biogeochemical Cycles* 12:563-572

Delcourt HR, Delcourt PA, Webb T (1983) Dynamic plant ecology: The spectrum of vegetation change in space and time. *Quaternary Science Reviews* 1:153-175

Delcourt PA, Delcourt HR (1987) *Long-term forest dynamics of the temperate zone: a case study of late-Quaternary forests in eastern North America* Springer-Verlag, New York, 439 pp.

Deleage JP, Hemery D (1990) From ecological history to world ecology. In: Brimblecombe P and Pfister C (eds.) *The Silent Countdown. Essays in European Environmental History.* Springer-Verlag, Berlin/ Heidelberg/ New York pp. 21–36

Delmas RJ (1993) A natural artefact in Greenland ice-core $CO_2$ measurements. *Tellus* 45B:391-396

Delmonte B, Petit J-R, Maggi V (in press) Glacial to Holocene implications of the new 27,000 year dust record from the EPICA Dome C (East Antarctica) ice core. *Climate Dynamics*

Delmonte B, Petit J-R, Maggi V (in press) LGM-Holocene changes and Holocene millennial-scale oscillations of dust particles in the EPICA-Dome C ice core (East Antarctica). *Annals of Glaciology*

Delworth TL, Mann ME (2000) Observed and simulated multidecadal variability in the northern hemisphere. *Climate Dynamics* 16:661-676

Dentener FJ, Carmichael GR, Zhang Y, Lelieveld J, Crutzen PJ (1996) Role of mineral aerosol as a reactive surface in the global troposphere. *Journal of Geophysical Research* 101:22869-22889

Deser C (2000) On the teleconnectivity of the "Arctic Oscillation". *Geophysical Research Letters* 27:779-782

Dettinger MD, Cayan DR, Diaz HF, Meko DM (1998) North-south precipitation patterns in Western North America on interannual-to-decadal time scales. *Journal of Climate* 11:3095-3111

Deutsch C GN, Key RM, Sarmiento JL, Ganachaud A (2001) Denitrification and $N_2$ fixation in the Pacific Ocean. *Global Biogeochemical Cycles* 15:483-506

Diaz HF, Graham NE (1996) Recent changes in tropical freezing heights and the role of sea surface temperature. *Nature* 383:152-155

Dibb JE, Rasmussen RA, Mayewski PA, Northern HG (1993) Northern Hemisphere concentrations of methane and nitrous oxide since 1800: results from the Mt Logan and 20D ice cores. *Chemosphere* 27:2413-2423

Dickinson RE, Kennedy P (1992) Impacts on regional climate of Amazon deforestation. *Geophysical Research Letters* 19:1947-1950

Dickson R, Lazier J, Meincke J, Rhines P, Swift J (1996) Long-term coordinated changes in the convective activity of the North Atlantic. *Progress In Oceanography* 38:241-295

Ding ZL, Rutter N, Yu ZW, Guo ZT, Zhu RX (1995) Ice-volume forcing of East Asian winter monsoon variations in the past 800,000 years. *Quaternary Research* 44:149-159

Dixit AS, Dixit SS, Sushil S, Smol JP (1996) Setting goals for an acid and metal-contaminated lake: A paleolimnological study of Daisy Lake (Sudbury, Canada). *Journal of Lake and Reservoir Management* 12:323-330

Dixit SS, Dixit AS, Smol JP (1989) Lake acidification recovery can be monitored using chrysophycean microfossils. *Canadian Journal of Fisheries and Aquatic Sciences* 46:1309-1312

Dodson J, Guo ZT (in prep) Past environmental variability in Austral-Asia. *Quaternary International*

Dodson JR, Guo ZT (1998) Past Global Change. *Global and Planetary Change* 18:202pp

Doherty R, Kutzbach J, Foley J, Pollard D (2000) Fully coupled climate/dynamical vegetation model simulations over Northern Africa during the mid-Holocene. *Climate Dynamics* 16:561-573

Dokken TM, Jansen E (1999) Rapid changes in the mechanism of ocean convection during the last glacial period. *Nature* 401:458-46

Doney SC, Wallace DWR, Ducklow HW (2000) The North Atlantic Carbon Cycle: New Perspectives from JGOFS and WOCE. In: Hanson RB, Ducklow HW and Field JG (eds.) *The Changing Ocean Carbon Cyelce*. Cambridge University Press, pp. 375-391

Dorale J, Edwards R, Ito E, Gonzalez L (1998) Climate and vegetation history of the midcontinent from 75 to 25 ka: a speleothem record from Crevice Cave, Missouri, USA. *Science* 282:1871-1874

Douglas I (1967) Man, vegetation and the sediment yields of rivers. *Nature* 215:25-28

Droxler AW (2000) Marine Isotope stage 11 (MIS 11): new insights for a warm future. *Global and Planetary Change* 24:1-5

Duck RW, McManus J (1990) Relationships between catchment characteristics, land use and sediment yield in the midland valley of Scotland. In: Boardman J, Foster IDL and Dearing JA (eds.) *Soil Erosion on Agricultural Land*. John Wiley & sons Ltd, London

Dugdale RC, Wilkerson FP (1998) Silicate regulation of new production in the eastern equatorial Pacific. *Nature* 391:270-273

Dumont HJ, Cocquyt C, Fontugne M, Arnold M, Reyss J-L, Bloemendal J, Oldfield F, Steenbergen CLM, Korthals HJ, Zeeb B (1998) The end of Moai quarrying and its effect on lake Rano Raraku, Easter Island. *Journal of Paleolimnology* 20:409-422

Dunbar RB, Cole JE (1999) Annual Records of Tropical Systems (ARTS). *A PAGES/CLIVAR Initiative: Recommendations for Research.*

Dunbar RB, Wellington GM, Colgan MW, Glynn PW (1994) Eastern Pacific sea surface temperature since 1600 A.D.: The $\delta^{18}O$ record of climate variability in Galapagos corals. *Paleoceanography* 9:291-316

Dunwiddie PW, LaMarche VC (1980) A climatically responsive tree-ring record from *Widdringtonia cedabergensis*, Cape Province, South Africa. *Nature* 286:796-797

Duplessy JC, Ivanova E, Murdmaa I (2001) Holocene paleoceanography of the Northern Barents Sea and variations of the northward heat transport by the Atlantic Ocean. *Boreas* 30:2-13

Duplessy JC, Bard E, Labeyrie L, Duprat J, Moyes J (1993) Oxygen isotope records and salinity changes in the Northeastern Atlantic Ocean during the last 18,000 years. *Paleoceanography* 8:341-350

Duplessy JC, Shackleton NJ, Fairbanks RG, Labeyrie LD, Oppo D, Kallel N (1988) Deepwater source variations during the last climatic cycle and their impact on the global deepwater circulation. *Paleoceanography* 3:343-360

Dyke AS, Morris TF (1990) Postglacial history of the bowhead whale and of driftwood penetration: implications for paleoclimate, central Canadian Arctic. *Geological Survey of Canada* Paper 89-2:17pp

Dyke AS, Savelle JM (2001) Holocene history of the Bering Sea bowhead whale (*Balaena mysticetus*) in its Beaufort Sea summer grounds off southwestern Victoria Island, western Canadian Arctic. *Quaternary Research* 55:371-379

Dymond J, Lyle M (1985) Flux comparisons between sediments and sediment traps in the eastern tropical Pacific: Implications for atmospheric $CO_2$ variations during the Pleistocene. *Limnology and Oceanography* 30:699-712

Dyurgerov MB, Meier MF (2000) Twentieth century climate change: evidence from small glaciers. *Proceedings of the National Academy of Sciences* 97:1406-1411

Ebbesmeyer CC, Cayan DR, McLain DR, F.H. Nichol, Peterson DH, Redmond KT (1991) 1976 step in the Pacific: Forty environmental changes between 1968-1975 and 1977-1984. In: Betancourt JL and Sharp VL (eds.) *1976 step in the Pacific: Forty environmental changes between 1968-1975 and 1977-1984*. California Department of Water Resources, Pacific Grove, CA

Eden DN, Page MJ (1998) Palaeoclimatic implications of a storm erosion record from late Holocene lake sediments, North Island, New Zealand. *Palaeogeography, Palaeoclimatology, Palaeoecology* 139:37-58

Edgington DN, Robbins JA (1976) Records of lead deposition in lake Michigan sediments since 1800. *Environmental Science and Technology* 10:266-274

Edmunds WM, Fellman E, BabaGoni I (1999) Environmental change, lakes and groundwater in the Sahel of Northern Nigeria. *Journal of the Geological Society London* 156:345-355

Edwards KJ, MacDonald GM (1991) Holocene palynology: part 2, human influence and vegetation change. *Progress in Physical Geography* 15:364-391

Egan D, Howell E (2001) *Handbook of Historical Ecology* Island Press, Washington DC

Elderfield H, Rickaby REM (2000) Oceanic Cd/P ratio and nutrient utilization in the glacial Southern Ocean. *Nature* 405:305-310

Elderfield H, Ganssen G (2000) Past temperature and d $^{18}O$ of surface ocean waters inferred from foraminiferal Mg/Ca ratios. *Nature* 405:442-445

Eldredge N (1999) Cretaceous meteor showers, the human ecological "niche," and the sixth extinction. In: MacPhee (ed.) *Extinctions in Near Time*. Kluwer Academic/Plenum Publishers, New York

Elias SA (1994) *Quaternary insects and their environments* Smithsonian Institution Press, Washington D.C, 284 pp.

Elliot M, Labeyrie L, Bond G, Cortijo E, Turon JL, Tisnerat N, Duplessy JC (1998) Millenial scale iceberg discharges in the Irminger Basin during the last glacial period: relationship with the Heinrich events and environmental settings. *Paleoceanography* 13:433-446

Elliott MB, Flenley JR, Sutton DG (1998) A late Holocene pollen record of deforestation and environmental change from the Lake Tauanui catchment, Northland, New Zealand. *Journal of Paleolimnology* 19:23-32

Enzel Y, Cayan DR, Anderson RY, Wells SG (1989) Atmospheric circulation during Holocene lake stands in the Mojave Desert: evidence of regional climate change. *Nature* 341:44-47

Eriksson MG, Sandgren P (1999) Mineral magnetic analyses of sediment cores recording recent soil erosion history in central Tanzania, *Palaeogeography Palaeoclimatology Palaelecology* 152: 365-384

ESF (2000) Good scientific practice in research and scholarship. In: *European Science Foundation Policy Briefing, December 2000*. ISRN ESF-SPB-00-10-FR+ENG, pp. in particular, see section 27, page 6

Esser G, Lautenschlager M (1994) Estimating the change of carbon in the terrestrial biosphere from 18,000 BP to present using a carbon cycle model. *Environmental Pollution* 83:45-53

Etheridge DM, Pearman GI (1988) Atmospheric trace-gas variations as revealed by air trapped in an ice-core from Law Dome, Antarctica. *Annals of Glaciology* 10: 28-33

Etheridge DM, G.I. Pearman, Fraser PJ (1992) Changes in tropospheric methane between 1841 and 1978 from a high accumulation-rate Antarctic ice core. *Tellus* 44 B:282-294

Etheridge DM, Steele LP, Francey RJ, Langenfields RL (1998) Atmospheric methane between 1000 A.D. and present: Evidence of anthropogenic emissions and climatic variability. *Journal of Geophysical Research* 103:15979-15993

Etheridge DM, Steele LP, Langenfields RL, Francey RJ, Barnola J-M, Morgan VI (1996) Natural and anthropogenic changes in atmospheric $CO_2$ over the last 1000 years from air in Antarctic ice and firn. *Journal of Geophysical Research* 101:4115-4128

Eugster W, Rouse W, Pielke R, McFadden J, Baldocchi D, Kittel T, Chapin F, Liston G, Vidale P, Vaganov E, Chambers S (2000) Land-atmosphere energy exchange in Arctic tundra and boreal forest: available data and feedbacks to climate. *Global Change Biology* 6:84-115

Evans MN, Fairbanks RG, Rubenstone JL (1999) The thermal oceanographic signal of ENSO reconstructed from a Kiritimati Island coral. *Journal of Geophysical Research* 104:409-13

Evans MN, Cane MA, Schrag DP, Kaplan A, Linsley BK, Villala R, Wellington GM (2001) Support for tropically-driven Pacific decadal variability based on paleoproxy evidence. *Geophysical Research Letters* 28: 3,689-3,692

Fairbanks RG (1989) A 17,000 year glacio-eustatic sea level record : influence of glacial melting rates on the Younger Dryas event and deep ocean circulation. *Nature* 342:637-642

Fairhead J, Leach M (1998) *Reframing Deforestation. Global analysis and local realities: studies in West Africa* Routledge, London

Falkowski P, Scholes RJ, Boyle E, Canadell J, Canfield D, Elser J, Gruber N, Hibbard K, Högberg P, Linder S, Mackenzie FT, III BM, Pedersen T, Rosenthal Y, Seitzinger S, Smetacek V, Steffen W (2000) The global carbon cycle: a test of our knowledge of earth as a system. *Science* 290:291-296

Falkowski PG (1997) Evolution of the nitrogen cycle and its influence on the biological pump in the ocean. *Nature* 387:272-275

Fang X (1995) The origin and provenance of Malan loess along the eastern margin of Qinghai-Xizang (Tibetan) Plateau and its adjacent area. *Science in China (Series B)* 38:876-887

Fang XM, Ono Y, Fukusawa H, Pan BT, Li J, Guan DH, Oi K, Tsukamoto S, Torii M, Mishima T (1999) Asian summer monsoon instability during the past 60,000 years: magnetic susceptibility and pedogenic evidence from the western Chinese Loess Plateau. *Earth and Planetary Science Letters* 168: 219-232

Fanning PC (1999) Recent landscape history in arid western new South Wales, Australia: a model for regional change. *Geomorphology* 5:401-428

Farley KA, Patterson DB (1995) A 100-kyr periodicity in the flux of extraterrestrial $^3$He to the sea floor. *Nature* 378:600-603

Farrera I, Harrisson S, Prentice I, Ramstein C, Guiot J, Bartlein P, Bonnefille R, Bush M, Cramer W, vonGrafenstein U, Holmgren K, Hooghiemstra H, Hope G, Jolly D, Lauritzen S-E, Ono Y, Pinot S, Stute M, Yu G (1999) Tropical climates at the Last Glacial Maximum: a new synthesis of terrestrial palaeoclimate data. I. Vegetation, lake-levels and geochemistry. *Climate Dynamics* 15:823-856

FAUNMAP Working Group (1996) Spatial response of mammals to late Quaternary environmental fluctuations. *Science* 272:1601-1606

Feely RA, Wanninkhof R, Takahashi T, Tans P (1999) Influence of El Niño on the equatorial Pacific contribution to atmospheric $CO_2$ accumulation. *Nature* 398:597-601

Finney BP, Gregory-Eaves I, Sweetman J, Douglas MSV, Smol JP (2000) Impacts of climatic change and fishing on Pacific salmon abundance over the past 300 years. *Science* 290:795-799

Fisher D, Koerner RM, Kuivinen K, Clausen HB, Johnsen SJ, Steffensen J-P, Gunderstrup N, Hammer CU (1996) Inter-comparisons of ice core $\delta^{18}O$ and precipitation records from sites in Canada and Greenland over the last 3500 years and over the last few centuries in detail using EOF techniques. In: P.D. Jones, R.S. Bradley and Jouzel J (eds.) Springer-Verlag, Berlin pp. 297-328

Fisher DA, Koerner RM (1994) Signal and noise in four ice-core records from the Agasiz Ice Cap, Ellsemere Island Canada: details of the past millenium for stable isotopes and solid conductivity. *The Holocene* 4:113-120

Fisher DA, Koerner RM (2002) Holocene ice core climate history: a multi-variable approach. In: Mackay AW, Battarbee RW, Birks HJB and Oldfield F (eds.) *Global Change in the Holocene: approaches to reconstructing fine-resolution climate change*. Arnold, London

Fisher DA, Koerner RM, Reeh N (1995) Holocene climatic records from the Agassiz Ice Cap, Ellesmere Island, N.W.T., Canada. *The Holocene* 5:19-24

Fitzpatrick FA, Knox JC, Whitman HE (1999) Effects of historical land-cover changes on flooding and sedimentation, North Fish Creek, Wisconsin USGS Water-Resources Investigations pp. 3999-4083

Flenley J (in press) The history of human presence in islands of south-east Asia and the South Pacific

Flückiger J, Dällenbach A, Blunier T, Stauffer B, Stocker TF, Raynaud D, Barnola JM (1999) Variations in atmospheric $N_2O$ concentration during abrupt climatic changes. *Science* 285:227-230

Flückiger J, Monnin E, Stauffer B, Schwander J, Stocker TF, Chappellaz J, Raynaud D, Barnola JM (2001) High resolution Holocene $N_2O$ ice core record and its relationship with $CH_4$ and $CO_2$. *Global Biogeochemical Cycles*

Foley J (1994) The sensitivity of the terrestrial biosphere to climatic change: A simulation of the middle Holocene. *Global Biogeochemical Cycles* 8:505-525

Foley J, Kutzbach J, Coe M, Levis S (1994) Feedbacks between climate and boreal forests during the Holocene epoch. *Nature* 371:52-54

Foley JA, Prentice IC, Ramankutty N, Levis S, Pollard D, Stich S, Haxeltine A (1996) An integrated biosphere model of land surface processes, terrestrial carbon balance and vegetation dynamics. *Global Biogeochemical cycles* 10:603-628

Fontes JC, Gasse F, Andrews JN (1993) Climatic conditions of Holocene groundwater recharge in the Sahel zone of Africa. In: *Isotope techniques in the study of past and current environmental changes in the Hydrosphere and the Atmosphere*. International Atomic Agency, Vienna

Fontugne M, Usselmann P, Lavallée D, Julien M, Hatté C (1999) El Niño variability in the coastal desert of Southern Peru during the Mid-Holocene. *Quaternary Research* 52:171-179

Fontugne M, Arnold M, Labeyrie L, Calvert SE, Paterne M, Duplessy JC (1994) Palaeoenvironment, sapropel chronology and Nile river discharge during the last 20,000 years as indicated by deep sea sediment records in the Eastern Mediterranean. In: Ofer Bar Y and Kra R (eds.) *Late Quaternary Chronology and Paleoclimates of Eastern Mediterranean.* pp. 75-88

Forest C, Stone P, Sokolov A, Allen M, Webster M (2002) Quantifying uncertainties in climate system properties with the use of recent climate observations. *Science* 295:113-117

Foster GC, Dearing JA, Jones RT, Crook DC, Siddle DS, Appleby PG, Thompson R, Nicholson J, Loizeaux J-L (in press) Meteorological and land use controls on geomorphic and fluvial processes in the pre-Alpine environment: an integrated lake-catchment study at the Petit Lac d'Annecy. *Hydrological Processes*

Foster IDL (1995) Lake and reservoir bottom sediments as a source of soil erosion and sediment transport data in the UK. In: Foster IDL, Gurnell AM and Webb BW (eds.) *Sediment and Water Quality in River Catchments.* Wiley, London

Foster IDL, Dearing JA, Grew R (1988) Lake-catchments: an evaluation of their contribution to studies of sediment yield and delivery processes. *Sediment Budgets, Proceeding of the Porto Alegre Symposium I.A.H.S. Publn* 174:413-424

Foster IDL, Mighall TM, Wotton C, Owens PN, Walling DE (2000) Evidence for Medieval soil erosion in the South Hams region of Devon, UK. *The Holocene* 10:261-271

François L, Kaplan J, Otto D, Roelandt C, Harrison SP, Prentice IC, Warnant P, Ramstein G (2000) Comparison of vegetation distributions and terrestrial carbon budgets reconstructed for the last glacial maximum with several biosphere models PMIP, 2000: Paleoclimate Modelling Intercomparison Project (PMIP), Proceedings of the Third PMIP workshop Canada, 4-8 October 1999

François LM, Delire C, Warnant P, Munhoven G. (1998) Modelling the glacial-interglacial changes in the continental biosphere. *Global Planetary Change* 16-17:37-52

François LM, Goddéris Y, Warnant P, Ramstein G, Noblet Nd, Lorenz S (1999) Carbon stocks and isotopic budgets of the terrestrial biosphere at mid-Holocene and last glacial maximum times. *Chemical Geology* 159:163-189

François R (1990) Marine sedimentary humic substances: structure, genesis, and properties. *Reviews in Aquatic Sciences* 3:41-80

François R, Bacon MP (1991) Variations in terrigenous input into the deep equatorial Atlantic during the past 24,000 years. *Science* 251:1473-1476

François R, Altabet M, Yu E-F, Sigman DM, Bacon MP, Frankl M, Bohrmann G, Bareille G, Labeyrie LD (1997) Contribution of southern ocean surface-water stratification to low atmospheric $CO_2$ concentrations during the last glacial period. *Nature* 389: 929-935

François R, Honjo S, Manganini SJ, Ravizza GE (1995) Biogenic barium fluxes to the deep sea: implications for paleoproductivity reconstruction. *Global Biogeochemical Cycles* 9: 289-303

Frank M, Gersonde R, Mangini A (1999) Sediment redistribution, $^{230}$Thex-normalization and implications for the reconstruction of particle flux and export paleoproductivity. In: Fischer G and Wefer G (eds.) *Proxies in Paleoceanography.* University of Bremen, Bremen pp. 409-426

Franzén LG (1994) Are wetlands the key to the ice-age cycle enigma? *Ambio* 23:300-308

Free M, Robock A (1999) Global warming in the context of the Little Ice Age. *Journal of Geophysical Research-Atmospheres* 104:19057-19070

Friedlingstein P, Delire C, Müller J-F, Gérard J-C (1992) The climate-induced variation of the continental biosphere : A model simulation of the Last Glacial Maximum. *Geophysical Research Letters* 19:897-900

Fritts HC (1991) *Reconstructing Large-Scale Climatic Patterns From Tree-Ring Data* The University of Arizona Press, Tucson

Fritts HC, Lofgren GR, Gordon GA (1980) Past climate reconstructed from tree-rings. *Journal of Interdisciplinary History* 10:773-793

Fritts HC, T.J. Blasing, B.P. Hayden, Kutzbach JE (1971) Multivariate techniques for specifying tree-growth and climate relationships and for reconstructing anomalies in paleoclimate. *Journal of Applied Meteorology* 10:845-864

Fritz SC, Ito E, Yu Z, Laird K, Engström DR (2000) Hydrologic variation in the Northern Great Plains during the last two millennia. *Quaternary Research* 53:175-84

Froelich PN, Blanc V, Mortlock RA, Chillrud SN, Dunstan A, Udomkit A, Peng T-H (1992) River fluxes of dissolved silica to the ocean were higher during glacials: Ge/Si in diatoms, rivers, and oceans. *Paleoceanography* 7:739-767

Fuhrer K, Neftel A, Anklin M, Maggi V (1993) Continuous measurements of hydrogen peroxide, formaldehyde, calcium and ammonium concentrations along the new GRIP ice core from Summit, Central Greenland. *Atmospheric Environment* 27A:1873-1880

Fuller JL, Foster DR, McLachlan JS, Drake N (1998) Impact of human activity on regional forest composition and dynamics in central New England. *Ecosystems* 1:76-95

Fung I, John J, Lerner J, Matthews E, Prather M, Steele LP, Fraser PJ (1991) Three-dimensional model synthesis of the global methane cycle. *Journal of Geophysical Research* 96:13033-13065

Gaffen DJ, Santer BD, Boyle JS, Christy JR, Graham NE, Ross RJ (2000) Multidecadal changes in the vertical temperature structure of the tropical troposphere. *Science* 287:1242-1245

Gagan MK, Ayliffe LK, Beck JW, Cole JE, Druffel ERM, Dunbar RB, Schrag DP (2000) New views of tropical paleoclimates from corals. *Quaternary Science Reviews* 19:45-64

Gagan MK, Ayliffe LK, Hopley D, Cali JA, Mortimer GE, Chappell J, McCulloch MT, Head MJ (1998) Temperature and surface-ocean water balance of the mid-Holocene tropical western Pacific. *Science* 279:1014-1018

Gallée H, Ypersele JPv, Fichefet T, Tricot C, Berger A (1991) Simulation of the last glacial cycle by a coupled, sectorially averaged climate-ice sheet model, 1. The climate model. *Journal of Geophysical Research* 96:13139-13161

Gallée H, Ypersele JPv, Fichefet T, Marsiat I, Tricot C, Berger A (1992) Simulation of the last glacial cycle by a coupled, sectorially averaged climate-ice sheet model, 2. Response to insolation and $CO_2$ variation. *Journal of Geophysical Research* 97:15, 713-15,740

Galloway JN, Likens GE (1979) Atmospheric enhancement of metal deposition in Adirondack lake sediment. *Limnology and Oceanography* 24:427-433

Gallup CD, Edwards RL, Johnson RG (1994) The timing of high sea levels over the past 200,000 years. *Science* 263:796-800

Ganachaud A, Wunsch C (2000) Improved estimates of global ocean circulation, heat transport and mixing from hydrographic data. *Nature* 408:453-457

Ganapolski A, Kubatski C, Claussen M, Brovkin V, Petouhkov V (1998) The influence of vegetation-atmosphere-ocean interactions on climate during the mid-Holocene. *Science* 280:1916 – 1919

Ganeshram RS, Pedersen TF, Calvert SE, Murray JW (1995) Large changes in oceanic nutrient inventories from glacial to interglacial periods. *Nature* 376:755-758

Ganeshram RS, Calvert SE, Pedersen TF, Cowie. GA (2000) Factors controlling the burial of organic carbon in laminated and bioturbated sediments off NW Mexico: Implications for hydrocarbon preservation. *Geochimica et Cosmochimica Acta*

Ganeshram RS, Pedersen TF, Calvert SE, Francois R (2002) Reduced nitrogen fixation in the glacial ocean inferred from changes in marine nitrogen and phosphorus inventories. *Nature* 415:156-159

Ganeshram RS, Pedersen TF, Calvert SE, McNeill GW, Fontugne MR (2002) Glacial-interglacial variability in denitrification in the World's Oceans: Causes and consequences. *Paleoceanography*

Ganopolski A, Rahmstorf S (2001) Rapid changes of glacial climate simulated in a coupled climate model. *Nature* 409:153-158

Ganopolski A, Rahmstorf S, Petoukhov V, Claussen M (1998) Simulation of modern and glacial climates with a coupled global model of intermediate complexity. *Nature* 391:351

Garreaud RD, Battisti DS (1999) Interannual (ENSO) and interdecadal (ENSO-like) variability in the Southern Hemisphere tropospheric circulation. *Journal of Climate* 12:2113-2123

Gasse F (2000) Hydrological changes in the African tropics since the Last Glacial Maximum. *Quaternary Science Reviews* 19:189-211

Gasse F, Van Campo E (1994) Abrupt post-glacial climate events in West Asia and North Africa monsoon domains. *Earth and Planetary Science Letters* 126:435-456

Gasse F, Fontes JC, Plaziat JC, Carbonel P, Kaczmarska I, De Deckker P, Soulie-Marsche I, Callot Y, Dupeuple PA (1987) Biological remains, geochemistry and stable isotopes for the reconstruction of environmental and hydrological changes in the Holocene lakes from North Sahara. *Palaeogeography, Paleoclimatology, Paleoecology* 60:1-46

Gates WL, Henderson-Sellers A, Boer G, Folland CK, Kitoh A, McAvaney B, Semazzi F, Smith NE, Weaver AJ, Zeng Q-C (1995) Climate models - Evaluation in climate change 1995: The science of climate change. In: J.T. Houghton LGMF, B.A. Callander, N. Harris, A. Kattenberg (eds) Cambridge University Press, Cambridge, U.K.

Gear AJ, Huntley B (1991) Rapid changes in the range limits of Scots Pine 4000 years ago. *Science* 251:544-547

Gedye SJ, Jones RT, Tinner W, Ammann A, Oldfield F (2000) The use of mineral magnetism in the reconstruction of fire history: a case study from Lago di Origlio, Swiss Alps. *Palaeogeography, Palaeoclimatology, Palaeoecology* 164:101 – 110

Geladov Z, Smith DJ (2001) Interdecadal climate variability and regime-scale shifts in Pacific North America. *Geophysical Research Letters* 28:1515-1518

Gershunov A, Barnett TP (1998) Interdecadal modulation of ENSO teleconnections. *Bulletin of the American Meteorological Society* 79:2715-2725

Glueck M, Stockton CW (in press) Reconstruction of the North Atlantic Oscillation, 1429-1983. *International Journal of Climatology*

Godwin H (1944) Age and origin of the Breckland Heaths. *Nature* 154:6-8

Goodland R (1995) The concept of environmental sustainability. *Annual Review of Ecology and Systematics* 26:1-24

Goolsby DA (2000) Mississippi basin nitrogen flux believed to cause gulf hypoxia. *EOS* 29:321-327

Gordon AL (1986) Interocean exchange of thermocline water. *Journal of Geophysical Research* 91:5037-5046

Gorham E (1958) The influence and importance of daily weather conditions in the supply of chloride sulphate and other ions to freshwaters from atmospheric precipitation. *Philosohpical Transactions of the Royal Society, London* B, 241:147-178

Gorham E (1975) Acid precipitation and its influence upon aquatic ecosystems – an overview. In: Dockinger LS and Selinga TA (eds.) *First International Symposium on Acid Precipitation and the Forest Ecosystem*. Forest Service, USDA, Columbus, Ohio

Gorham E (1990) Biotic impoverishment in Northern Peatlands.

In: Woodwell GM (ed.) *The Earth in Transition: Patterns and Processes of Biotic Impoverishment*. CUP, New York

Gorham E (1991) Northern peatlands; role in carbon cycle and probable responses to climate change. *Ecological Applications* 1:182-195

Goslar T, Arnold M, Bard E, Kuc T, Pazdur M, Ralska-Jasiewiczowa M, Rozanski K, Tisnerat N, Walanus A, Wicik B, Wieckowski K (1995) High concentration of atmospheric $^{14}C$ during the Younger Dryas cold episode. *Nature* 377:414-417

Govindasamy B, Duffy P, Calderia K (2001) Land use changes and northern hemisphere cooling. *Geophysical Research Letters* 28:291-294

Grabherr G, Pauli H, Gottfried M (1994) Climate effects on mountain plants. *Nature* 369:448

Graham NE, Barnett TP, Wilde R, Ponater M, Schubert S (1994) On the roles of tropical and midlatitude SSTs in forcing interannual to interdecadal variability in the winter Northern Hemisphere circulation. *Journal of Climate* 7:1416-1441

Graham RW (1992) Late Pleistocene faunal changes as a guide to understanding effects of greenhouse warming on the mammalian fauna of North America. In: Peters RL and Lovejoy TE (eds.) *Global Warming and Biological Diversity*. Yale University Press, New Haven pp. 76-87

Graham RW (1997) The spatial response of mammals to Quaternary climate changes. In: Huntley B, Cramer W, Morgan AV, Prentice HC and Allen JRM (eds.) *Past and future rapid environmental changes: The spatial and evolutionary responses of terrestrial biota*. Springer-Verlag, Berlin pp. 153-162

Graham RW, Lundelius EL (1984) Coevolutionary disequilibrium and Pleistocene extinctions. In: Martin PS and Klein RG (eds.) *Quaternary Extinctions: A prehistoric revolution*. University of Arizona Press, Tucson pp. 223-249

Graham RW, Mead JI (1987) Environmental fluctuations and evolution of mammal faunas during the last deglaciation in North America. In: Ruddiman WF and H.E. Wright J (eds.) *North America and adjacent oceans during the last deglaciation*. Geological Society of America, Boulder, Colorado pp. 371-402

Graumlich LJ (1993) A 1000-year record of temperature and precipitation in the Sierra Nevada. *Quaternary Research* 39:249-255

Graumlich LJ (1993) Response of tree growth to climatic variation in the mixed conifer and deciduous forests of the Upper Great-Lakes Region. *Canadian Journal of Forest Research* 23:133-143

Graybill DA, Shiyatov SG (1992) Dendroclimatic evidence from the northern Soviet Union. In: Bradley RS and Jones PD (eds.) *Climate Since A.D. 1500*. Routledge, London pp. 393-414

Graybill DA, Funkhouser G (1999) Dendroclimatic reconstructions during the past millennium in the Southern Sierra Nevada and Owens Valley, California. In: Lavenberg R (ed.) *Southern California Climate: Trends and Extremes of the Past 2000 Years*. Natural History Museum of Los Angeles County, Los Angeles, CA pp. 239-269

Greenland Summit (1997) The Greenland Summit Ice Cores CD-ROM National Snow and Ice Data Center at University of Colorado at Boulder and World Data Center-A for Paleoclimatology at National Geophysical Data Center

Grigg L, Whitlock C (1998) Late-Glacial vegetation and climate change in Western Oregon. *Quaternary Research* 49:287-298

Grigg LD, Whitlock C, Dean WE (2001) Evidence for millennial-scale climate change during marine isotope stages 2 and 3 at Little Lake, western Oregon, USA. *Quaternary Research* 56:10-22

Grime JP (1978) *Plant strategies and vegetation processes* John Wiley & Sons, New York, 222 pp.

Grimm EC (2001) Warm wet Heinrich events in Florida. In: ARCHES NC (ed.) *Heinrich Events Miniconference*. Lamont-Doherty Earth Observatory, Palisades, New York pp. 28

Grimm EC, Jacobson GL, Watts JWA, Hansen BCS, Maasch KA (1993) A 50,000-year record of climate oscillations from Florida and its temporal correlation with the Heinrich Events. *Science* 261:198-200

Grissino-Mayer HD (1996) A 2129-year annual reconstruction of precipitation for Northwestern New Mexico, USA. In: D.M. Meko, Swetnam TW and Dean JS (eds.) *Tree-Rings, Environment and Humanity*. University of Arizona Press, Tucson pp. 191-204

Grootes PM, Stuiver M (1997) Oxygen 18/16 variability in Greenland snow and ice with 133 to 105-year resolution. *Journal of Geophysical Research* 102:26 455-26 470

Grootes PM, Stuiver M, White JWC, Johnsen S, Jouzel J (1993) Comparison of oxygen isotopes records from the GISP 2 and GRIP Greenland ice cores. *Nature* 466:552-554

Grosjean M, Geyh MA, Messerli B, Schotterer U (1995) Late-glacial and early Holocene lake sediments, ground-water formation and climate in the Atacama Altiplano 22-24°S. *Journal of Paleolimnology* 14:241-252

Grousset F, Pujol C, Labeyrie L, Auffret G, Boelaert A (2000) Were the North Atlantic Heinrich events triggered by the behavior of the European ice sheets? *Geology* 28:123-126

Grousset F, Labeyrie L, Sinko J, Cremer M, Bond G, Duprat J, Cortijo E, Huon S (1993) Patterns of ice rafted detritus in the Glacial North Atlantic (40-55°N). *Paleoceanography* 8:175-192

Grove JM (1988) *The Little Ice Age* Methuen, London, 498pp

Grove JM (2001) The initiation of the "Little Ice Age" in regions round the North Atlantic. *Climatic Change* 48:53-82

Grove JM (2001) The onset of the Little Ice Age. In: Jones PD, Ogilvie AEJ, Davies TD and Briffa KR (eds.) *History and Climate: Memories of the Future?* Kluwer Academic/Plenum, pp. 153-185

Grove JM, Switsur R (1994) Glacial geological evidence for the Medieval Warm Period. *Climatic Change* 26:143-169

Gruber N, Sarmiento JL (1997) Global patterns of marine nitrogen fixation and denitrification. *Global Bigeochemical Cycles* 11:235-266

Gu D, Philander SGH (1997) Interdecadal climate fluctuations that depend on exchanges between the tropics and extratropics. *Science* 275:805-807

Guilderson TP, Schrag DP (1998) Abrupt shifts in subsurface temperatures in the tropical Pacific associated with changes in El Niño. *Science* 281:240-243

Guilderson TP, Fairbanks RG, Rubenstone JL (1994) Tropical temperature variations since 20,000 years ago: modulating interhemispheric climate change. *Science* 263:663-665

Guiot J, deBeaulieu JL, Cheddadi R, David F, Ponel P, Reille M (1993) Climate in Western Europe during the last Glacial/Interglacial cycle derived from pollen and insect remains. *Palaeogeography, Paleaeoclimatology, Palaeoecology* 103:73-93

Guo ZT, Liu TS, Guiot J, Wu N, Lu H, Han J, Liu J, Gu Z (1996) High frequency pulses of East Asian monsoon climate in the two glaciations: link with the North Atlantic. *Climate Dynamics* 12:701-709

Guo ZT, Liu TS, Fedroff N, Wei LY, Ding ZL, Wu NQ, Lu HY, Jiang WY, An ZS (1998) Climate extremes in loess of China coupled with the strength of deep-water formation in the North Atlantic. *Global and Planetary Change* 18:113-128

Gwiazda RH, Hemmings SR, Broecker WS (1996) Tracking the sources of icebergs with lead isotopes: the provenance of ice-rafted debris in Heinrich layer 2. *Paleoceanography* 11:77-93

Haan D, Raynaud D (1998) Ice core record of CO variations during the last two millennia: atmospheric implications and

chemical interactions within the Greenland ice. *Tellus Series B-Chemical and Physical Meteorology* 50:253-262

Haberle S (1994) Anthropogenic indicators in pollen diagrams: problems and prospects for late Quaternary palynology in New Guinea. In: Hather J (ed.) *Tropical Archaeobotany: Applications and New Developments*. Routledge, London pp. 172-201

Haberle SG (1998) Late Quaternary vegetation change in the Tari Basin, Papua New Guinea. *Palaeogeography Palaeoclimatology Palaeoecology* 137:1-24

Hagemeijer EJM, Blair MJ (1997) *The EBCC Atlas of European Breeding Birds: Their distribution and abundance* T. & A.D. Poyser, London, 903 pp.

Hanebuth T, Stattegger K, Grootes PM (2000) Rapid flooding of the Sunda shelf: A late-Glacial Sea-Level record. *Science* 288:1033-1035

Hansen B, Turrell WR, Østerhus S (2001) Decreasing overflow from the Nordic seas into the Atlantic Ocean through the Faroe Bank channel since 1950. *Nature* 411: 927 - 930

Hansen BCS, Engström DR (1996) Vegetation history of Pleasant Island, southeastern Alaska, since 13,000 yr BP. *Quaternary Research* 46:161-175

Hansen J, Lacis A, Rind D, Russell G, Stone P, Fung I, Ruedy R, Lerner J (1984) Climate sensitivity: analysis of feedback mechanisms. In: Hansen J and Takahashi T (eds.) *Climate Processes and Climate Sensitivity*. American Geophysical Union, Washington, DC pp. 130-163

Hare S, Mantua N (2000) Empirical evidence for North Pacific regime shifts in 1977 and 1989. *Progress in Oceanography* 47:103-145

Harris RN, Chapman DS (2001) Mid-latitude (30°-60°N) climatic warming inferred by combining borehole temperatures with surface air temperatures. *Geophysical Research Letters* 28:747-750

Hassan FA (1994) Population ecology and civilization in ancient Egypt. In: Crumley CL (ed.) *Historical Ecology: Cultural Knowledge and Changing Landscapes*. School of American Research Press, Santa Fe

Hastenrath S, Greischar L (1993) Circulation mechanisms related to Northeast Brazil rainfall anomalies. *Journal of Geophysical Research* 98: 5093-5102

Hatté C, Antoine P, Fontugne M, Lang A, Rousseau D-D, Zöller L (2001) $\delta^{13}$C of loess organic matter as a potential proxy for paleoprecipitation reconstruction. *Quaternary Research* 55:33-38

Hatté C, Fontugne M, Rousseau D, Antoine P, Zöller L, Tisnérat-Laborde N, Bentaleb I (1998) $\delta^{13}$C variations of loess organic matter as a record of the vegetation response to climatic changes during the Weichselian. *Geology* 26:583-586

Haug GH, Hughen KA, Sigman DM, Peterson LC, Röhl U (2001) Southward migration of the Intertropical Convergence Zone through the Holocene. *Science* 293:1304-1308

Haxeltine A, Prentice IC (1996) BIOME3: An equilibrium terrestrial biosphere model based on ecophysiological constraints, resource availability and competition among plant functional types. *Global Biogeochemical Cycles* 10:693-710

Hayden B (1998) Ecosystem feedbacks on climate at the landscape scale. *Philosohpical Transactions of the Royal Society, London* 353:5-18

Hays JD, Imbrie J, Shackleton NJ (1976) Variations in the Earth's orbit: pacemakers of the ice ages. *Science* 194:1121-1132

Hedges JI, Keil RG (1995) Sedimentary organic matter preservation: An assessment and speculative synthesis. *Marine Chemistry* 49:81-115

Heim C, Nowaczyk NR, Negendank JFW, Leroy SAG, BenAvraham Z (1997) Near East desertification: Evidence from the Dead Sea. *Naturwissenschaften* 84:398-401

Heinrich H (1988) Origin and consequences of cyclic ice rafting in the Northeast Atlantic Ocean during the past 130,000 years. *Quaternary Research* 29:142-152

Heinze C, Maier-Reimer E (1991) Glacial $pCO_2$ reduction by the world ocean: experiments with the Hamburg carbon cycle model. *Paleoceanography* 6:395-430

Heiss GA (1994) Coral reefs in the Red Sea: Growth, production and stable isotopes GEOMAR pp. 1-141

Henderson G, Slowey N (2000) Evidence from U-Th dating against Northern Hemisphere forcing of the penultimate deglaciation. *Nature* 404:61-66

Henderson GM, Martel DJ, O'Nions RK, Shackleton NJ (1994) Evolution of seawater $^{87}Sr/^{86}Sr$ over the last 400ka: The absence of glacial/interglacial cycles. *Earth and Planetary Science Letters* 128:643-651

Hendy EJ, Gagan MK, Alibert CA, McCulloch MT, Lough JM, Isdale PJ (2002) Abrupt decrease in tropical Pacific sea surface salinity at end of Little Ice Age. *Science* 295:1511-1514

Hendy IL, Kennett JP (1999) Latest Quaternary North Pacific surface-water responses imply atmosphere driven climate instability. *Geology* 27:291-294

Hewitt C, Mitchell J (1998) A fully coupled GCM simulation of the climate of the mid-Holocene. *Geophysical Research Letters* 25:361-364

Heyerdahl EK, Card V (2000) Implications of paleorecords for ecosystem management. *Trends in Ecology and Evolution* 15:48-50

Hicock SR, Lian OB, Mathewes RW (1999) 'Bond cycles' recorded in terrestrial Pleistocene sediments of southwestern British Columbia, Canada. *Journal of Quaternary Science* 14:443-449

Higgitt SE, Oldfield F, Appleby PG (1991) The record of land use change and soil erosion in the late Holocene sediments of the Petit Lac d'Annecy, eastern France. *The Holocene* 1:14-28

Hill JK, Thomas CD, Huntley B (1999) Climate and habitat availability determine 20th century changes in a butterfly's range margin. *Proceedings of the Royal Society of London* 266:1197-1206

Hill JK, Thomas CD, Huntley B (in press) Modelling present and potential future ranges of European butterflies using climate response surfaces. In: Boggs C, Watt W and Ehrlich P (eds.) *Ecology and evolution taking flight: Butterflies as model study systems.* University of Chicago Press, Chicago

Hites RA (1981) Sources and fates of polycyclic aromatic hydrocarbons. *American Chemical Society Symposium Series* 167:187-196

Hodell DA, Curtis JH, Brenner M (1995) Possible role of climate in the collapse of classic Maya civilization. *Nature* 375:391-394

Hodell DA, Brenner M, Curtis JH, Guilderson T (2001) Solar forcing of drought frequency in the Maya lowlands. *Science* 292:1367-1370

Hoerling MP, Hurrell JW, Xu TY (2001) Tropical origins for recent North Atlantic climate change. *Science* 292:90-92

Holmgren K, Tyson PD, Moberg A, Svanered O (2001) A preliminary 3000-year regional temperature reconstruction for South Africa. *South African Journal of Science* 97:1-3

Holmgren K, Karlén W, Lauritzen SE, Lee-Thorp JA, Partridge TC, Piketh S, Repinski P, Stevenson C, Svanered O, Tyson PD (1999) A 3000-year high resolution stalagmite-based record of paleoclimate for northeastern South Africa. *The Holocene* 9:295-309

Hong S, Candelone J-P, Patterson CC, Boutron CF (1994) Greenland ice evidence of hemispheric lead pollution two millennia ago by Greek and Roman civilizations. *Science* 265: 1841-1843

Hooke JM (1977) The distribution and nature of changes in river channel patterns: the example of Devon. In: Gregory KJ (ed.) *River Channel Changes.* Wiley, Chichester

Hostetler SW, Mix AC (1999) Reassessment of ice-age cooling of the tropical ocean and atmosphere. *Nature* 399:673-676

Houghton JT, Filho LGM, Callander BA, Harris N, A. Kattenberg (1995) *Climate Change 1995: The Science of Climate Change* Cambridge University Press, Cambridge, U.K

Houghton JT, Ding Y, Griggs DG, Noguer M, Linden PJvd, Dai X, Maskell K, Johnson CA (2001) *Climate Change 2001: The Scientific Basis. Contribution of Working Group I to the Third Assessment Report of the IPCC, 2001.* Cambridge University Press

Houghton RW, Tourre YM (1992) Characteristics of low-frequency sea surface temperature fluctuations in the tropical Atlantic. *Journal of Climate* 5:756-771

Hoyt DV, Schatten KH (1993) A discussion of plausible solar irradiance variations, 1700-1992. *Journal of Geophysical Research-Space Physics* 98A

Hu FS, Brubaker LB, Anderson PM (1995) Postglacial vegetation and climate-change in the Northern Bristol Bay-Region, Southwestern Alaska. *Quaternary Research* 43:382-392

Huang G, Yim W (2001) An 8000 year record of typhoons in the northern South China Sea. *PAGES News* 9:7-8

Huang S, Pollack HN (1997) Late Quaternary temperature changes seen in worldwide continental heat flow measurements. *Geophysical Research Letters* 24:1947-1950

Huang S, Shen P-Y, Pollack HN (1996) Deriving century-long trends of surface temperature change from borehole temperatures. *Geophysical Research Letters* 23:257-260

Huang S, Pollack HN, Shen P-Y (2000) Temperature trends over the past five centuries reconstructed from borehole temperatures. *Nature* 403:756-758

Huang Y, Street-Perrott FA, Metcalfe SE, Brenner M, Moreland M, Freeman KH (2001) Climate change as the dominant control on glacial-interglacial variations in C-3 and C-4 plant abundance. *Science* 293:1647-1651

Hughen KA, Overpeck JT, Peterson LC, Trumbore S (1996) Rapid climate changes in the tropical Atlantic region during the last deglaciation. *Nature* 380:51-54

Hughen KA, Southon JR, Lehman SJ, Overpeck JT (2000) Synchronous radiocarbon and climate shifts during the last deglaciation. *Science* 290:1951-1954

Hughen KA, Overpeck JT, Lehman SJ, Kashgarian M, Southon J, Peterson LC, Alley R, Sigman DM (1998) Deglacial changes in ocean circulation from an extended radiocarbon calibration. *Nature* 391:65-68

Hughes MK, Brown PM (1992) Drought frequency in central California since 101 B.C. recorded in giant Sequoia tree rings. *Climate Dynamics* 6:161-167

Hughes MK, Diaz HF (1994) Was there a "Medieval Warm Period" and if so, where and when? *Climatic Change* 26:109-142

Hughes MK, Graumlich LJ (1996) Multimillennial dendoclimatic records from Western North America. In: P.D. Jones, Bradley RS and Jouzel J (eds.) *Climatic Variations and Forcing Mechanisms of the Last 2000 Years.* Springer Verlag, Berlin pp. 109-124

Hughes MK, Funkhouser G (1998) Extremes of moisture availability reconstructed from tree rings for recent millennia in the Great Basin of western North America. In: M. Innes and Beniston JL (eds.) *The Impacts of Climate Variability on Forests.* Springer, Berlin pp. 99-107

Hughes MK, Vaganov EA, Shiyatov S, Touchan R, Funkhouser G (1999) Twentieth-century summer warmth in northern Yakutia in a 600- year context. *Holocene* 9:629-634

Hughes PDM, Mauquoy D, Barber KE, Langdon PG (in press) Mire development pathways and palaeoclimatic records from a full Holocene peat archive at Walton Moss, Cumbria,

England. *The Holocene*

Huijzer B, Vandenberghe J (1998) Climate reconstruction of the Weichselian Pleniglacial in northwestern and central Europe. *Journal of Quaternary Science* 13:391-417

Hulme M, Osborn TJ, Jones J, Briffa KR, Jones PD (1999) Climate observations and GCM validation. Unpublished report to UK Dept. of Environment, Transport and the Regions 66 pp.

Huntley B (1988) Glacial and Holocene vegetation history: Europe. In: B. Huntley and T. Webb I (eds.) *Vegetation History*. Kluwer Academic Publishers, Dordrecht pp. 341-383

Huntley B (1991) How plants respond to climate change: migration rates, individualism and the consequences for plant communities. *Annals of Botany* 67:15-22

Huntley B (1995) Plant species' response to climate change: implications for the conservation of European birds. *Ibis* 137:127-138

Huntley B (1999a) The dynamic response of plants to environmental change and the resulting risks of extinction. In: Mace GM, Balmford A and Ginsberg JR (eds.) *Conservation in a changing world*. Cambridge University Press, Cambridge pp. 69-85

Huntley B (1999b) Species distribution and environmental change: considerations from the site to the landscape scale. In: Maltby E, Holdgate M, Acreman M and Weir A (eds.) *Ecosystem Management: Questions for Science and Society*. Royal Holloway Institute for Environmental Research, Virginia Water, UK pp. 115-129

Huntley B, Birks HJB (1983) *An atlas of past and present pollen maps for Europe: 0-13000 B.P.* Cambridge University Press, Cambridge, 667pp.

Huntley B, Webb T III (1988) Vegetation History. In: Lieth H (ed.) *Handbook of Vegetation Science*. Kluwer Academic Publishers, Dordrecht pp. 803

Huntley B, Berry PM, Cramer WP, McDonald AP (1995) Modelling present and potential future ranges of some European higher plants using climate response surfaces. *Journal of Biogeography* 22:967-1001

Huon S, Jantschik R (1993) Detrital silicates in Northeast Atlantic deep-sea sediments during the Late Quaternary: Major elements, REE and Rb-Sr isotopic data. *Ecologiae Geologica Helvetica* 86:195-218

Hurrell JW (1995) Decadal trends in the North Atlantic Oscillation: regional temperatures and precipitation. *Science* 269:676-679

Hutchins DA, Bruland KW (1998) Iron-limited diatom growth and Si:N uptake ratios in a coastal upwelling regime. *Nature* 393:561-564

Imbrie J, Kipp NG (1971) A new micropaleontological method for paleoclimatology: application to a late Pleistocene Caribbean core. In: Turekian KK (ed.) *The Late Cenozoic Glacial Ages*. Yale University Press, New Haven, CT pp. 71-181

Imbrie J, Imbrie JZ (1980) Modelling the climatic response to orbital variations. *Science* 207:943-953

Imbrie J, McIntyre A, Mix A (1989) Oceanic response to orbital forcing in the late Quaternary: observational and experimental strategies. In: Berger A, Schneider SH and Duplessy JC (eds.) *Climate and Geosciences*. Kluwer Academic, Dordrecht pp. 121-164

Imbrie J, Hays JD, Martinson DG, McIntyre A, Mix AC, Morley JJ, Pisias NG, Prell WL, Shackleton NJ (1984) The orbital theory of Pleistocene climate: support from a revised chronology of the marine $\delta^{18}$O record. In: Berger AL, Imbrie J, Hays J, Kukla G and Saltzman B (eds.) *Milankovitch and Climate, Part 1*. D. Riedel, Hingham, MA pp. 269-305

Imbrie J, Boyle E, Clemens S, Duffy A, Howard W, Kukla G, Kutzbach J, Martinson D, McIntyre A, Mix A, Molfino B, Morley J, Peterson L, Pisias N, Prell W, Raymo M,

Shackleton N, Toggweiler J (1992) On the structure and origin of major glaciation cycles, 1: linear responses to Milankovitch forcing. *Paleoceanography* 7:701-738

Imbrie J, Berger A, Boyle E, Clemens S, Duffy A, Howard W, Kukla G, Kutzbach J, Martinson D, McIntyre A, Mix A, Molfino B, Morley J, Peterson L, Pisias N, Prell W, Raymo M, Shackleton N, Toggweiler J (1993) On the structure and origin of major glaciation cycles, 2: The 100,000 years cycle. *Paleoceanography* 8:699-735

Immirzi CP, Maltby E (1992) *The global status of peatlands and their role in carbon cycling*. Friends of the Earth Trust Ltd., London

Indermühle A, Monnin E, Stauffer B, Stocker TF, Wahlen M (2000) Atmospheric $CO_2$ concentration from 60 to 20 kyr BP from the Taylor Dome ice core, Antarctica. *Geophysical Research Letters* 27:735-738

Indermühle A, Stocker TF, Fischer H, Smith HJ, Joos F, Wahlen M, Deck B, Mastroianni D, Tschumi J, Blunier T, Meyer R, Stauffer B (1999) High-resolution Holocene $CO_2$-record from the Taylor Dome ice core (Antarctica). *Nature* 398:121-126

Indermühle A, Stocker TF, Joos F, Fischer H, Smith HJ, Wahlen M, Deck B, Mastroianni D, Tschumi J, Blunier T, Meyer R, Stauffer B (1999) Holocene carbon-cycle dynamics based on $CO_2$ trapped in ice at Taylor Dome, Antarctica. *Nature* 398:121-126

IPCC (2001) *Climate Change 2001: The Scientific Basis. Contribution of Working Group I to the Third Assessment Report of the Intergovernmental Panel on Climate Change* Cambridge University Press, Cambridge, 881 pp.

Isaksen K, Holmlund P, Sollid JL, Harris C (2001) Three deep Alpine-permafrost boreholes in Svalbard and Scandinavia. *Permafrost and Periglacial Processes* 12:13-25

Iversen J (1941) Landnam i Danmark's stenalder (Land occupation in Denmark's Stone Age). *Danmarks Geologiske Undersöglse* Series II:1-68

Jackson S, Weng C (1999) Late Quaternary extinction of a tree species in eastern North America. In: *National Academy of Sciences USA*. pp. 13847-13852

Jackson S, Overpeck J, Webb T III, Keattch S, Anderson K (1997) Mapped plant macrofossil and pollen records of Late Quaternary vegetation changes in eastern North America. *Quaternary Science Reviews* 16:1-70

Jackson ST, Overpeck JT (2000) Responses of plant populations and communities to environmental changes of the late Quaternary. *Paleobiology* 26:194-220

Jackson S, Webb R, Anderson K, Overpeck J, WebbT III, Williams J, Hansen B (2000) Vegetation and environment in eastern North America during the last glacial maximum. *Quaternary Science Reviews* 19: 489-508

Jacobsen T, Adams R (1958) Salt and silt in ancient Mesopotamian agriculture. *Science* 128:1251 – 1258

Jacobson G, Webb T III, Grimm E (1987) Patterns and rates of vegetation change during the deglaciation of eastern North America. In: Ruddiman W and Wright HJ (eds.) *North American and adjacent oceans during the last deglaciation*. Geological Society of America, Boulder, CO pp. 277-288

Jacoby GC, D'Arrigo R (1989) Reconstructed northern hemisphere annual temperature since 1671 based on high-latitude tree-ring data from North-America. *Climatic Change* 14:39-59

Jacoby GC, D'Arrigo RD (1995) Tree-ring width and density evidence of climatic and potential forest change in Alaska. *Global Biogeochemical Cycles* 9:227-234

Jalas J, Suominen J (1976) *Atlas Florae Europaeae* Societas Biologica Fennica Vanamo, Helsinki, 122 pp.

Jalut G, Amat AE, Bonnet L, Gauquelin T, Fontugne M (2000) Holocene climatic changes in the Western Mediterranean, from south-east France to south-east Spain. *Palaeogeography, Palaeoclimatology, Palaeecology*

160:255-290

Jenkins A, Whitehead PG, Cosby BJ, Birks HJB (1990) Modeling long-term acidification: a comparison with diatom reconstructions and the implications for reversibility. *Philosophical Transactions of the Royal Society* B 327:209-214

Jenkins A, Renshaw M, Helliwell R, Sefton C, Ferrier R, Swingewood P (1997) *Modelling surface water acidification in the UK: application of the MAGIC model to the Acid Waters Monitoring Network*

Jin H, Li S, Cheng G, Wang S, Li X (2000) Permafrost and climatic change in China. *Global and Planetary Change* 26:387-404

Johns TC, Gregory JM, Stott PA, Mitchell JFB (2001) Correlations between patterns of 19th and 20th century surface temperature change and HadCM2 climate model ensembles. *Geophysical Research Letters* 28:1007-1010

Johnsen SJ, Dansgaard W, Clausen HB, C.C. Langway J (1972) Oxygen isotope profiles through the Antarctic and Greenland ice sheets. *Nature* 235:429-434

Johnsen SJ, Dahl-Jensen D, Dansgaard W, Gundestrup N (1995) Greenland paleotemperatures derived from GRIP bore hole temperature and ice core isotope profiles. *Tellus* 47B:624-629

Johnsen SJ, Dahl-Jensen D, Gundestrup N, Steffensen JP, Clausen HB, Miller H, Masson-Delmotte V, Sveinbjornsdottir AE, 2001 JW (2001) Oxygen isotope and paleotemperature records from six Greenland ice-core stations: Camp Century, Dye-3, GRIP, GISP2, Renland and NorthGRIP. *Journal of Quaternary Science* 16:299-307

Johnsen SJ, Clausen HB, Dansgaard W, Fuhrer K, Gundestrup N, Hammer CU, Iversen P, Jouzel J, Stauffer B, Steffensen JP (1992) Irregular glacial interstadials recorded in a new Greenland ice core. *Nature* 359:311-313

Jolly D, Haxeltine A (1997) Effect of low glacial atmospheric $CO_2$ on tropical African montane vegetation. *Science* 786-788

Jones PD, Hulme M (1997) The changing temperature of 'Central England'. In: Hulme M and Barrow E (eds.) *Climates of the British Isles: Present, Past and Future*. Routledge, London pp. 173-196

Jones PD, Briffa KR (1998) Global surface air temperature variations during the twentieth century: Part 1, spatial, temporal and seasonal details. *The Holocene* 2:174-188

Jones PD, Osborn TJ, Briffa KR (2001) The evolution of climate over the last millennium. *Science* 292:662-667

Jones PD, Briffa KR, Barnett TP, Tett SFB (1998) High-resolution palaeoclimatic records for the last millennium: interpretation, integration and comparison with General Circulation Model control-run temperatures. *The Holocene* 8:455-471

Jones PD, Briffa KR, Osborn TJ, Bergstrom H, Moberg A (2002) Relationships between circulation strength and the variability of growing season and cold season climate in northern and central Europe. *The Holocene*, in press

Jones PD, New M, Parker DE, Martin S, Rigor IG (1999) Surface air temperature and its changes over the past 150 years. *Reviews of Geophysics* 37:173-199

Joos F, Bruno M, Fink R, Siegenthaler U, Stocker TF, LeQuéré C, Sarmiento JL (1996) An efficient and accurate representation of complex oceanic and biospheric models of anthropogenic carbon uptake. *Tellus* 48B:397-417

Joussaume S, Taylor KE (2000) The Paleoclimate Modelling Intercomparison Project (PMIP). *Third PMIP Workshop (WMO-TD 1007)* 9-24

Joussaume S, Taylor KE, Braconnot P, Mitchell JFB, Kutzbach JE, Harrison SP, Prentice IC, Broccoli AJ, Abe-Ouchi A, Bartlein PJ, Bonfils C, Dong B, Guiot J, Herterich K, Hewitt CD, Jolly D, Kim JW, Kislov A, Kitoh A, Loutre MF, Masson V, McAvaney B, McFarlane N, Noblet Nd, Peltier

WR, Peterschmitt JY, Pollard D, Rind D, Royer JF, Schlesinger ME, Syktus J, Thompson S, Valdes P, Vettoretti G, Webb RS, Wyputta U (1999) Monsoon changes for 6,000 years ago: Results of 18 simulations from the Paleoclimate Modeling Intercomparison Project (PMIP). *Geophysical Research Letters* 26:859-862

Jouzel J, Hoffmann G, Koster RD, Masson V (2000) Water isotopes in precipitation: data/model comparison for present-day and past climates. *Quaternary Science Reviews* 19:363-379

Kabat K, Claussen M, Dirmeyer PA, Gash JC, de Guenni LB, Meybeck M, Pielke RAS, Vörösmarty CJ, Hutjes RWA, Lütkemeier S (2001) *Vegetation, water, humans and the climate: a new perspective on an interactive system: A synthesis of the IGBP core project biospheric aspects of the hydrological cycle*. Springer-Verlag, Berlin

Kageyama M, Valdes PJ, Ramstein G, Hewitt C, Wyputta U (1999) Northern hemisphere storm-tracks in present day and last glacial maximum climate simulations: a comparison of the European PMIP models. *Journal of Climate* 12:742-760

Kageyama M, Peyron O, Pinot S, Tarasov P, Guiot J, Joussaume S, Ramstein G (2001) The Last Glacial Maximum climate over Europe and western Siberia: a PMIP comparison between models and data. *Climate Dynamics* 17:23-43

Kallel N, Labeyrie LD, Juillet-Leclerc A, Duplessy JC (1988) A deep hydrological front between intermediate and deep-water masses in the glacial Indian Ocean. *Nature* 33:651-655

Kallel N, Paterne M, Duplessy JC, Vergnaud-Grazzini C, Pujol C, Labeyrie L, Arnold M, Fontugne M, Pierre C (1997) Enhanced rainfall in the mediterranean region during the last sapropel event. *Oceanologica Acta* 20:697-712

Kammen DM, Marino BD (1993) On the origin and magnitude of pre-industrial anthropogenic $CO_2$ and $CH_4$ emissions. *Chemosphere* 26:69-86

Kanfoush SL, Hodell DA, Charles CD, Guilderson TP, Mortyn PG, Ninnemann US (2000) Millenial-scale instability of the Antarctic Ice Sheet during the last glaciation. *Science* 288:1815-1818

Kaplan JO (2000) Wetlands at the Last Glacial Maximum: Distribution and methane emissions. *Geophysical Research Letters* 29:1-4

Karl D, Letelier R, Tupas L, Dore J, Christian J, Hebel D (1997) The role of nitrogen fixation in biogeochemical cycling in the subtropical North Pacific Ocean. *Nature* 388:533-538

Karlsen AW, Cronin TM, Ishman SE, Willard DA, Holmes CW, Marot M, Kerhin R (in press) Historical trends in Chesapeake Bay dissolved oxygen based on benthic *Foraminifera* from sediment cores. *Estuaries*

Kasting JF, Walker JCG (1992) The geochemical carbon cycle and the uptake of fossil fuel $CO_2$. In: Levi BG, Hafemeister D, Scribner R (eds). *AIP Conference Proceedings* 247. Global Warming: Physics and Facts.

Kealhofer L, Penny D (1998) A combined pollen and phytolith record for fourteen thousand years of vegetation change in northeastern Thailand. *Review of Palaeobotany and Palynology* 103:83-93

Keefer DK, DeFrance SD, Moseley ME, Satterlee DR, Day-Lewis A (1998) Early maritime economy and El Niño events at Quebrada Tacahuay, Peru. *Science* 281:1833-1835

Keeling CD, Whorf TP (2000) Atmospheric $CO_2$ records from sites in the SIO air sampling network. In: *Trends: A Compendium of Data on Global Change*. Carbon Dioxide Information Analysis Center, Oak Ridge National Laboratory, U.S. Department of Energy, Oak Ridge, Tennessee

Keeling CD, Whorf TP, Wahlen M, Vanderplicht J (1995) Interannual extremes in the rate of rise of atmospheric carbon-dioxide since 1980. *Nature* 375:666-670

Keeling CD, Bacastow RB, Carter AF, Piper SC, Whorf TP,

Heimann M, Roeloffzen H (1989) A three dimensional model of atmospheric $CO_2$ transport based on observed winds: 1. Analysis of observational data. In: Peterson DH (ed.) *Aspects of climate variability in the Pacific and Western Americas*. AGU, Washington pp. 165-236

Keigwin LD (1996) The Little Ice Age and Medieval Warm Period in the Sargasso Sea. *Science* 274:1504-1508

Keigwin LD, Pickert RS (1999) Slope water current over the Laurentian Fan on interannual to millennial time scales. *Science* 286:520-523

Keigwin LD, Boyle EA (2000) Detecting Holocene changes in thermohaline circulation. *Proceedings of the National Academy of Sciences* 97:1343-1346

Kelly PM, Wigley TML (1992) Solar-cycle length, greenhouse forcing and global climate. *Nature* 360:328-330

Kennett JP, Shackleton NJ (1975) Laurentide ice sheet meltwater recorded in Gulf of Mexico deep-sea cores. *Science* 188:147-150

Kennett JP, Cannariato KG, Hendy IL, Behl RJ (2000) Carbon isotopic evidence for methane hydrate instability during Quaternary interstadials. *Science* 288:128-133

Kerr RA (1997) A new driver for the Atlantic's moods and Europe's weather? *Science* 275:754-5

Kerwin M, Overpeck J, Webb R, DeVernal A, Rind D, Healy R (1999) The role of oceanic forcing in mid-Holocene northern hemisphere climatic change. *Paleoceanography* 14:200-210

Khalil MAK, Rasmussen RA (1982) Secular trends of atmospheric methane ($CH_4$). *Chemosphere* 11:877-883

Khodri M, Leclainche Y, Ramstein G, Braconnot P, Marti O, Cortijo E (2001) Simulating the amplification of orbital forcing by ocean feedbacks in the last glaciation. *Nature* 410:570-574

Kiladis G, Diaz HF (1989) Global climatic anomalies associated with extremes in the Southern Oscillation. *Journal of Climate* 2:1069-1090

Kissel C, Laj C, Mazaud A, Dokken T (1998) Magnetic anisotropy and environmental changes in two sedimentary cores from the Norwegian Sea and the North Atlantic. *Earth and Planetary Science Letters* 164:617-626

Kissel C, Laj C, Labeyrie L, Dokken T, Voelker A, Blamart D (1999) Rapid climatic variations during marine isotopic stage 3: magnetic analysis of sediments from Nordic Seas and North Atlantic. *EPSL* 171:489-502

Kitagawa H, van der Plicht J (1998) Atmospheric radiocarbon calibration to 45,000 yr B.P.: Late Glacial fluctuations and cosmogenic isotope production. *Nature* 279:1187-1190

Kleeman R, Power SB (2000) Modulation of ENSO variability on decadal and longer time scales. In: Diaz HF and Markgraf V (eds.) *El Niño and the Southern Oscillation: Multiscale variability and global and regional impacts.* Cambridge University Press, Cambridge, U.K pp. 413-441

Kleidon A, Fraedrich K, Heimann M (2000) A green planet versus a desert world: estimating the maximum effect of vegetation on the land surface climate. *Climatic Change* 44:471-493

Klein AG, Seltzer GO, Isacks BL (1999) Modern and Last Glacial maximum snowlines in the central Andes of Peru, Bolivia and Northern Chile. *Quaternary Science Reviews* 18:65-84

Klein-Goldewijk K (in press) Estimating global land use change over the past 300 years: the HYDE database. *Global Biogeochemical Cycles*

Knox JC (2000) Sensitivity of modern and Holocene floods to climate change. *Quaternary Science Reviews* 19:439-457

Kober B, Wessels M, Bollhöfer A, Mangini A (1999) Pb isotopes in sediments of Lake Constance, Central Europe constrain the heavy metal pathways and the pollution history of the catchment, the lake and the regional atmosphere. *Geochimica et Cosmochimica Acta* 63:1293-1303

Koç N, Jansen E, Haflidason H (1993) Paleoceanographic reconstruction of surface ocean conditions in the Greenland, Iceland and Norwegian Seas through the last 14 ka based on diatoms. *Quaternary Science Reviews* 12:115-140

Koç N, Jansen E, Hald M, Labeyrie L (1996) Late glacial-Holocene sea surface temperatures and gradients between the north Atlantic and the Norwegian Sea: Implications for the Nordic heat pump. In: J.T. Andrews, W.E.N. Austin, H. Bergsten and Jennings AE (eds.) *The Late Glacial paleoceanography of the North Atlantic margins*. Geological Society Special Publication, pp. 177-185

Koerner RM, Fisher DA (1990) A record of Holocene summer climate from a Canadian High Arctic ice core. *Nature* 343:630-631

Kohfeld KE, Harrison SP (2000) How well can we simulate past climates? Evaluating the models using global palaeoenvironmental datasets. *Quaternary Science Reviews* 19:321-346

Kohfeld KE, Fairbanks RG, Smith SL, Walsh ID (1996) *Neogloboquadrina pachyderma* (sinistral coiling) as paleoceanographic tracers in polar oceans: Evidence from northeast water polynya plankton tows, sediment traps, and surface sediments. *Paleoceanography* 11:679-699

Kohler TA (1992) Prehistoric human impact on the environment in upland North American Southwest. *Population and Environment: A Journal of Interdisciplinary Studies* 13:255–268

Kolber ZS, Barber RT, Coale KH, Fitzwater SE, Greene RM, Johnson KS, Lindley S, Falkowski PG (1994) Iron limitation of phytoplankton photosynthesis in the equatorial Pacific Ocean. *Nature* 371:145-149

Kreitz SF, Herbert TD, Schuffert JD (2000) Alkenone paleothermometry and orbital-scale changes in sea-surface temperatue at site 1020, Northern California margin. *Proceedings of the Ocean Drilling Program. Scientific results* 167:153-161

Kring D (2000) Impact events and their effect on the origin, evolution, and distribution of life. *GSA Today* 10:1-6

Kroepelin S, Petit-Maire N (2000) Paleomonsoon variations and terrestrial environmental change during the Late Quaternary. *Special Issue of Global and Planetary Change* 26:316pp

Kroeze C, Mosier A, Bowman L (1999) Closing the global $N_2O$ budget: A retrospective analysis 1500-1994. *Global Biogeochemical Cycles* 13:1-8

Kuhnert H, Patzold J, Hatcher B, Wyrwoll K-H, Eisenhauer A, Collins LB, Zhu ZR, Wefer G (1999) A 200-year coral stable oxygen isotope record from a high-latitude reef off Western Australia. *Coral Reefs* 18:1-12

Kukla G, Heller F, Liu XM, Xu TC, Liu TS, An ZS (1988) Pleistocene climates in China dated by magnetic susceptibility. *Geology* 16:811-814

Kullman L (1989) Tree-limit history during the Holocene in the Scandes Mountains, Sweden inferred from sub-fossil wood. *Reviews of Paleobotany and Palynology* 58:163-171

Kumar A, Hoerling MP (1997) Interpretation and implications of the observed inter-El Niño variability. *Journal of Climate* 10:83-91

Kumar KK, Rajagopalan B, Cane MA (1999) On the weakening relationship between the Indian monsoon and ENSO. *Science* 284:2156-2159

Kumar KK, Kleeman R, Cane MA, Rajagopalan B (1999) Epochal changes in Indian monsoon-ENSO precursors. *Geophysical Research Letters* 26:75-78

Kumar N, Anderson RF, Mortlock RA, Froelich PN, Kubik P, Dittrich-Hannen B, Suter M (1995) Increased biological productivity and export production in the glacial Southern Ocean. *Nature* 378:675-680

Kutzbach J, Bartlein P, Foley J (1996a) Potential role of vegetation feedback in the climate sensitivity of high-latitude regions: A case study at 6000 years B.P. In: *Global Biogeochemical Cycles*. The American Geophysical Union, pp. 727-736

Kutzbach J, Bonan G, Foley J, Harrison S (1996b) Vegetation and soil feedbacks on the response of the African monsoon to orbital forcing in the early to middle Holocene. *Nature* 384:623-626

Kutzbach JE (1981) Monsoon climate of the early Holocene - climate experiment with the earths orbital parameters for 9,000 years ago. *Science* 214:59-61

Kutzbach JE, Otto-Bliesner BL (1982) The sensitivity of the African-Asian monsoonal climate to orbital parameter changes for 9,000 years B.P. *Journal of the Atmospheric Sciences* 39:1177-1188

Kutzbach JE, Street-Perrott FA (1985) Milankovitch forcing of fluctuations in the level of tropical lakes from 18 to zero kyr BP. *Nature* 317:130-134

Kutzbach JE, Liu Z (1997) Response of the African monsoon to orbital forcing and ocean feedbacks in the middle Holocene. *Science* 278:440-443

Labeyrie L (2000) Glacial climate instability. *Science* 290:1905-1907

Labeyrie L, Leclaire H, Waelbroeck C, Cortijo E, Duplessy JC, Vidal L, Elliot M, Lecoat B, Auffret G (1999) Temporal variability of the surface and deep waters of the North West Atlantic Ocean at orbital and millenial scales. In: P. Clark, Webb RS and Keigwin LD (eds.) *Mechanisms of Global Climate Change at millenial Time scales*. AGU, Washington pp. 77-98

Labeyrie L, Vidal L, Cortijo E, Paterne M, Arnold M, Duplessy JC, Vautravers M, Labracherie M, Duprat J, Turon JL, Grousset F, Weering Tv (1995) Surface and deep hydrography of the Northern Atlantic Ocean during the last 150 kyr. *Philosophical Transcripts Royal Society London* 348:255-264

Labeyrie L, Labracherie M, Gorfti N, Pichon JJ, Duprat J, Vautravers M, Arnold M, Duplessy JC, Paterne M, Michel E, Caralp J, Turon JL (1996) Hydrographic changes of the Southern Ocean (south-east Indian sector) over the last 230 ka. *Paleoceanography* 11:57-76

Labeyrie LD, Duplessy JC, Blanc PL (1987) Variations in mode of formation and temperature of oceanic deep waters over the past 125,000 years. *Nature* 327:477-482

Labeyrie LD, Duplessy JC, Duprat J, Juillet-Leclerc AJ, Moyes J, Michel E, Kallel N, Shackleton NJ (1992) Changes in the vertical structure of the north Atlantic Ocean between glacial and modern times. *Quaternary Science Reviews* 11:401-413

Labracherie M, Labeyrie LD, Duprat J, Bard E, Arnold M, Pichon JJ, Duplessy JC (1989) The last deglaciation in the Southern Ocean. *Paleoceanography* 4:629-638

Laird KR, Fritz SC, Maasch KA, Cumming BF (1996) Greater drought intensity and frequency before AD 1200 in the Northern Great Plains, USA. *Nature* 384:552-554

Laj C, Mazaud A, Duplessy JC (1996) Geomagnetic intensity and $^{14}C$ abundance in the atmosphere and ocean during the past 50 kyr. *Geophysical Research Letters* 23:2045-2048

Laj C, Kissel C, Mazaud A, Channell JET, Beer J (2000) North Atlantic paleointensity stack since 75 ka (NAPIS-75) and the duration of the Laschamp event. *Phil. Transcripts Royal Society London* 358:1009-1025

La Marche VC, Fritts HC (1971) Anomaly patterns of climate over the Western United States, 1700-1930, derived from principal component analysis of tree-ring data. *Monthly Weather Review* 99:138-142

La Marche VCJ (1973) Holocene climatic variations inferred from treeline fluctuations in the White Mountains. *Quaternary Research* 3:632-660

La Marche VCJ (1974) Paleoclimatic inferences from long tree-ring records. *Science* 183:1043-1048

La Marche VCJ, Pittock AB (1982) Preliminary temperature reconstructions for Tasmania. In: Hughes MK, Kelly PM, Pilcher JR and LaMarche VC (eds.) *Climate from Tree Rings*. Cambridge University Press, Cambridge pp. 177-185

Lamb H, Damblon F, Maxted RW (1991) Human impact on the vegetation of the Middle Atlas, Morocco, during the last 5000 years. *Journal of Biogeography* 18:519-532

Lamb HF, Gasse F, Benkaddour A, Hamouti NE, Kaars Svd, Perkins WT, Pearce NJ, Roberts CN (1995) Relation between century-scale Holocene arid intervals in tropical and temperate zones. *Nature* 373:134-137

Lamb HH (1963) What can we learn about the trend of our climate? *Weather* 18:194-216

Lamb HH (1965) The early Medieval warm epoch and its sequel. *Palaeogeography, Palaeoclimatology, Palaeoecology* 1:13-37

Lambeck K, Chappell J (2001) Sea level change through the last glacial cycle. *Science* 292:679-686

Lang C, Leuenberger M, Schwander J, Johnsen S (1999) 16°C rapid temperature variation in central Greenland 70,000 years ago. *Science* 286:934-937

Lara A, Villalba R (1993) A 3620-year temperature record from *Fitzroya cupressoides* tree rings in southern South America. *Science* 260:1104-1106

Laskar J (1990) The chaotic motion of the solar system: A numerical estimate of the chaotic zones. *Icarus* 88:266-291

Latif M (2001) Tropical Pacific/Atlantic ocean interactions on multidecadal time scales. *Geophysical Research Letters* 28:539-542

Latif M, Barnett TP (1996) Decadal climate variability over the North Pacific and North America: Dynamics and predictability. *Journal of Climate* 9:2407-2423

Lauritzen S-E (1996) Calibration of speleothem stable isotopes against historical records: a Holocene temperature curve for north Norway? In: *Climatic Change: the Karst Record*. Karst Waters Institute Special Publication 2, Charles Town, West Virginia pp. 78-80

Lawton JH (1999) What to conserve – species or ecosystems? In: Maltby E, Holdgate M, Acreman M and Weit A (eds.) *Ecosystem management: Questions for science and society*. Royal Holloway Institute for Environmental Research, Royal Holloway, University of London, Egham, UK pp. 55-62

Lea DW, Spero HJ (1994) Assessing the reliability of paleochemical tracers: Barium uptake in the shells of planktonic *Foraminifera*. *Paleoceanography* 9:445-452

Lea DW, Pak DK, Spero HJ (2000) Climate impact of late Quaternary equatorial Pacific sea surface temperature variations. *Science* 289:1719-1724

Lean J, A. Skumanich, White O (1992) Estimating the sun's radiative output during the Maunder Minimum. *Geophysical Research Letters* 19:1591-1594

Lean J, Beer J, Bradley RS (1995) Reconstruction of solar irradiance since 1610: implications for climate change. *Geophysical Research Letters* 22:3195-3198

Ledru MP, Salgado-Labouriau ML, Lorscheitter ML (1998) Vegetation dynamics in southern and central Brazil during the last 10,000 yr BP. *Review of Palaeobotany and Palynology* 99: 131-142

Ledwell JR, Watson AJ, Law CS (1993) Evidence for slow mixing across the pycnocline from an open-ocean tracer-release experiment. *Nature* 364:701-703

Lee JA (1998) Unintentional experiments with terrestrial ecosystems: ecological effects of sulphur and nitrogen pollutants. *Journal of Ecology* 86:1-12

Leemans R, Cramer W (1991) *The IIASA database for mean monthly values of temperature, precipitation and cloudiness of a global terrestrial grid* International Institute for Applied Systems Analysis (IIASA, Laxenburg, Austria, 62

Leemans R, Halpin P (1992) Global change and biodiversity. In: Groombridge B (ed.) *Biodiversity 1992: Status of the Earth's Living Resources*. Chapman and Hall, London pp. 254-255

Legrand M, Feniet-Saigne C (1991) Strong El Niño revealed by methanesulphonic acid in South Polar snow layers? *Geophysical Research Letters* 18:187-190

Legrand M, Lorius C, Barkov NI, Petrov VN (1988) Atmospheric chemistry changes over the last climatic cycle (160,000 yr) from Antarctic ice. *Atmospheric Environment* 22:317-331

Legrand M, Feniet-Saigne C, Saltzman ES, Germain C (1992) Spatial and temporal variations of methanesulfonic acid and non sea salt sulfate in Antarctic ice. *Journal of Atmospheric Chemistry* 14:245-260

Legrand M, Feniet-Saigne C, Saltzman ES, Germain C, Barkov NI, Petrov VN (1991) Ice-core record of oceanic emissions of dimethylsulphide during the last climatic cycle. *Nature* 350:144-146

LeGrand P, Wunsch C (1995) Constraints from paleotracer data on the North Atlantic circulation during the last glacial maximum. *Paleoceanography* 10:1011-1045

LeGrand P, Alverson K (2001) Variations in atmospheric $CO_2$ during glacial cycles from an inverse ocean modeling perspective. *Paleoceanography* 16:604-616

Lehtonen H, Huttunen P (1997) History of forest fires in eastern Finland from the fifteenth century AD – the possible effects of slash-and-burn cultivation. *The Holocene* 7:223-228

Leuenberger M, Lang C, Schwander J (1999) $\delta^{15}N$ measurements as a calibration tool for the paleothermometer and gas-ice age differences. A case study for the 8200 B. P. event on GRIP ice. *Journal of Geophysical Research* 104:22163-22169

Lemoine F (1998) Changements de l'hydrologie de surface de l'ocean austral en relation avec les variations de la circulation thermohaline au cours des deux derniers cycles climatiques. UPS-Orsay

Leuschner DC, Sirocko F (2000) The low-latitude monsoon climate during Dansgaard-Oeschger cycles and Heinrich Events. *Quaternary Science Reviews* 19:243-254

Levine JS (1991) *Global biomass burning: atmospheric, climatic, and biospheric implications* The MIT Press, Cambridge, MA

Levine JS (1996a) Biomass burning and global change (vol I). In: *Remote sensing, modeling and inventory development, and biomass burning in Africa.* The MIT Press, Cambridge, MA

Levine JS (1996b) Biomass burning and global change (vol II). In: *Biomass burning in South America, Southeast Asia, and temperate and boreal ecosystems, and the oil fires of Kuwait.* The MIT Press, Cambridge, MA

Levis S, Foley J (1999) $CO_2$, climate, and vegetation feedbacks at the Last Glacial Maximum. *Journal of Geophysical Research* 104:31,191-31,198

Levitus S, Antonov JI, Boyer TP, Stephens C (2000) Warming of the World Ocean. *Science* 287:2225-2229

Levitus S, Antonov JI, Wang J, Delworth TL, Dixon KW, Broccoli AJ (2001) Anthropogenic warming of Earth's climate system. *Science,* 292:267-270

Lewis TJ (1998) The effect of deforestation on ground surface temperature. *Global and Planetary Change* 18:1-13

Lewis TJ, Wang K (1998) Geothermal evidence for deforestation induced warming: implications for the climatic impact of land development. *Geophysical Research Letters* 25:535-538

Li H, Bischoff JL, Ku TL, Lund SP, Stott L (in press) Climate variability in east central California during the past 1000 years. *Quaternary Research*

Li J-J, Fang M-X, Wang J, Zhong W, Cao J, Pan B, Ma H, Zhu J, Zhou S, Chen F, Wang J, Ma Y (1995) *Uplift of the Qinghai-Xizang (Tibet) plateau and global change* Lanzhou University Press, 207 pp

Likens GE (1972) Eutrophication in aquatic ecosystems. In: Likens GE (ed.) *Nutrients and Eutrophication.* American Society of Limnology and Oceanography, pp. 3-14

Likens GE, Driscoll CT, Buso DC (1996) Long-term effects of acid rain: Response and recovery of a forest ecosystem. *Science* 272:244 – 246

Lindblagh M, Bradshaw R, Holmqvist BH (2000) Pattern and process in south Swedish forests during the last 3000 years, sensed at stand and regional scales. *Journal of Ecology* 88:113-128

Linsley BK, Wellington GM, Schrag DP (2000) Decadal sea surface temperature variability in the subtropical South Pacific from 1726 to 1997 AD. *Science* 290:1145-1148

Linsley BK, Ren L, Dunbar RB, Howe SS (2000) El Niño Southern Oscillation (ENSO) and decadal-scale climate variability at 10°N in the eastern Pacific from 1893 to 1994: A coral-based reconstruction from Clipperton Atoll. *Paleoceanography* 15:322-335

Liss PS, Hatton AD, Malin G, Nightingale PD, Turner SM (1997) Marine sulphur emissions. *Philosohpical Transactions of the Royal Society, London* 352:159-168

Lister AM (1993) 'Gradualistic' evolution: Its interpretation in Quaternary large mammal species. *Quaternary International* 19:77-84

Lister AM, Sher AV (1995) Ice cores and mammoth extinction. *Nature* 378:23-24

Litt T, Brauer A, Goslar T, Merkt J, Balaga K, Muller H, Ralska-Jasiewiczowa M, Stebich M, Negendank JFW (2001) Correlation and synchronisation of Lateglacial continental sequences in northern central Europe based on annually laminated lacustrine sediments. *Quaternary Science Reviews* 20:1233-1249

Little MG, Schneider RR, Kroon D, Price B, Summerhayes CP, Segl M (1997) Trade wind forcing of upwelling, seasonality, and Heinrich events as a response to sub-Milankovitch climate variability. *Paleoceanography* 12:568-576

Liu KB, Fearn ML (1993) Lake-sediment record of Late Holocene hurricane activities from coastal Alabama. *Geology* 2:793-796

Liu KB, Fearn ML (2000) Reconstruction of prehistoric landfall frequencies of catastrophic hurricanes in northwestern Florida from lake sediment records. *Quaternary Research* 54:238-245

Liu KB, Shen CM, Louie KS (2001) A 1,000-year history of typhoon landfalls in Guangdong, southern China, reconstructed from Chinese historical documentary records. *Annals of the Association of American Geographers* 91:453-464

Liu Z, Kutzbach JE, Wu L (2000) Modeling climate shift of El Niño variability in the Holocene. *Geophysical Research Letters* 27:2265-2268

Lloyd AH, Graumlich LJ (1997) Holocene dynamics of treeline forests in the Sierra Nevada. *Ecology* 78:1199-1210.

Loaiciga HA, Valdes JB, Vogel R, Garvey J, Schwarz H (1996) Global warming and the hydrological cycle. *Journal of Hydrology* 174:83-127

Lohmann GP (1995) A model for variation in the chemistry of planktonic *Foraminifera* due to secondary calcification and selective dissolution. *Paleoceanography* 10:445-457

Long C, Whitlock C, Bartlein P, Millspaugh S (1998) A 9000-year fire history from the Oregon Coast Range, based on a high-resolution charcoal study. *Canadian Journal for Research* 28:774-787

Longhurst A, Sathyendranath S, Platt T, Claverhill C (1995) An estimate of global primary production in the ocean from satellite radiometer data. *Journal of Plankton Research* 17:1245-1271

Loreau M, Naeem S, Inchausti P, Bengtsson J, Grime JP, Hector A, Hooper DU, Huston MA, Raffaelli D, Schmid B, Tilman D, Wardle DA (2001) Ecology - Biodiversity and ecosystem functioning: Current knowledge and future challenges. *Science* 294:804-808

Lotter AF (1998) The recent eutrophication of Baldeggersee

(Switzerland) as assessed by fossil diatom assemblages. *The Holocene* 8:395-405

Lotter AF (1999) Late-glacial and Holocene vegetation history and dynamics as shown by pollen and plant macrofossil analyses in annually laminated sediments from Soppensee, central Switzerland. *Vegetation History and Archaebotany* 8:165-184

Loubere P, 1999 MM-A (1999) A multiproxy reconstruction of biological productivity and oceanography in the eastern equatorial Pacific for the past 30,000 years. *Marine Micropaleontology* 37:173-198

Lough JM, Fritts HC (1985) The Southern Oscillation and tree-rings: 1660-1961. *Journal of Climate and Applied Meteorology* 24:952-965

Lough JM, Fritts HC (1987) An assessment of the possible effects of volcanic eruptions on North American climate using tree-ring data, 1602 to 1900 A.D. *Climatic Change* 219-239

Lu H, VanHuissteden K, An Z, Nugteren G, Vandenberghe J (1999) East Asia winter monsoon variations on a millennial time-scale before the last glacial-interglacial cycle. *Jounal of Quaternary Science* 14:101-110

Luckman B, Kavanagh T (2000) Impact of Climate Fluctuations on Mountain Environments in the Canadian Rockies. *Ambio* 29:

Luckman BH (1994) Evidence for climatic conditions between ca. 900-1300 A.D. in the southern Canadian Rockies. *Climatic Change* 26:171-182

Luckman BH (1996) Reconciling the glacial and dendrochronological records for the last millennium in the Canadian Rockies. In: Jones PD, Bradley RS and Jouzel J (eds.) *Climatic variations and forcing mechanisms of the last 2000 years.* Springer-Verlag, Berlin pp. 85-108

Ludwig W, Amiotte-Suchet P, Probst JL (1999) Enhanced chemical weathering of rocks during the last glacial maximum: a sink for atmospheric $CO_2$? *Chemical Geology* 159: 147-161

Luterbacher J, Schmutz C, Gyalistras D, Xoplaki E, Wanner H (1999) Reconstruction of monthly NAO and EU indices back to AD 1675. *Geophysical Research Letters* 26:2745

Luterbacher J, Schmutz D, Gyalistras D, Jones PD, Davies TD, Wanner H, Xoplaki E (2000) Reconstruction of highly resolved NAO and EU indices back to AD 1500. *Geophysical Research Abstracts* 2:OA 34

Lynch-Stieglitz J, Curry WB, Slowey N (1999) Weaker Gulf Stream in the Florida straits during the last glacial maximum. *Nature* 402:644-648

MacAyeal DR (1993) Binge/purge oscillations of the Laurentide Ice sheet as a cause of the North Atlantic's Heinrich events. *Paleoceanography* 8:775-784

MacDonald GM, Gervais BR, Snyder JA, Tarasov GA, Borisova OK (2000) Radiocarbon dated *Pinus sylvestris* L. wood from beyond treeline on the Kola Peninsula, Russia. *The Holocene* 10:143-148

MacDonald GM, Velichko AA, Kremenetski CV, Borisova OK, Goleva AA, Andreev AA, Cwynar LC, Riding RT, Forman SL, Edwards TWD, Aravena R, Hammarlund D, Szeicz JM, Gattaulin VN (2000) Holocene treeline history and climate change across northern Eurasia. *Quaternary Research* 53:302-311

Machida T, Nakazawa T, Fujii Y, Aoki S, Watanabe O (1995) Increase in the atmospheric nitrous oxide concentration during the last 250 years. *Geophysical Research Letters* 22:2921-2924

Mack RN (1981) Invasion of *Bromus tectorum* L. into western North America: An ecological chronicle. *Agro-Ecosystems* 7:145-165

Macklin MG (1999) Holocene river environments in prehistoric Britain: human interaction and impact. *Quaternary Proceedings* 7:521-530

Macklin MG, Bonsall C, Davies FM, Robinson MR (2000) Human-environment interactions during the Holocene: new data and interpretations from the Oban area, Argyll, Scotland. *The Holocene* 10:109-121

Magnuson JJ, Robertson DM, Benson BJ, Wynne RH, Livingstone DM, Arai T, Assel RA, Barry RG, Card V, Kuusisto E, Granin NG, Prowse TD, Stewart KM, Vuglinski VS (2000) Historical trends in lake and river ice cover in the northern hemisphere. *Science* 289:1743-1746

Magny M (1993) Solar influences on Holocene climatic changes. *Quaternary Research* 40:1-9

Mahowald N, Kohfeld K, Mansson M (1999) Dust sources and deposition during the last glacial maximum and current climate: A comparison of model results with paleodata from ice cores and marine sediments. *Journal of Geophysical Research* 104:15,895-15,916

Maley J, Brenac P (1998) Vegetation dynamics, palaeoenvironments and climatic changes in the forests of western Cameroon during the last 28,000 years BP. *Review of Palaeobotany and Palynology* 99:157-187

Maloney BK (1980) Pollen analytical evidence for early forest clearance in North Sumatra. *Nature* 287:324-326

Maloney BK (1981) A pollen diagram from Tao Sipingyan, a lake site in the Batak Highlands of North Sumatra, Indonesia. *Modern Quaternary Research in South East Asia* 6:57-76

Manabe S, Stouffer RJ (1993) Century-scale effects of increased atmospheric $CO_2$ on the ocean-atmosphere system. *Nature* 364:215-218

Manabe S, Stouffer RJ (1997) Coupled ocean-atmosphere model response to fresh water input: Comparison to Younger Dryas event. *Paleoceanography* 12:321-336

Mangan J, Overpeck J, Webb R, Wessman C, Goetz A (in press) Response of Nebraska Sand Hills natural vegetation to drought, fire, grazing, and plant functional type shifts as simulated by the CENTURY model. *Climatic Change*

Mangerud J, Andersen ST, Berglund BE, Donner J (1974) Quaternary stratigraphy of Norden, a proposal for terminology and classification. *Boreas* 3:109-127

Manley G (1974) Central England temperatures: monthly means 1659 to 1973. *Quarterly Journal of the Royal Meteorological Society* 100:389-405

Mann ME (in press) Large-scale climate variability and connections with the Middle East in past centuries. *Climatic Change*

Mann ME, Park J (1996) Joint spatio-temporal modes of surface temperature and sea level pressure variability in the Northern Hemisphere during the last century. *Journal of Climate* 9:2137-2162

Mann ME, Park J, Bradley RS (1995) Global inter-decadal and century-scale climate oscillations during the past five centuries. *Nature* 378:266-270

Mann ME, Bradley RS, Hughes MK (1998) Global-scale temperature patterns and climate forcing over the past six centuries. *Nature* 392:779-787

Mann ME, Bradley RS, Hughes MK (1999) Northern hemisphere temperatures during the past millennium: inferences, uncertainties, and limitations. *Geophysical Research Letters* 26:759-762

Mann ME, Bradley RS, Hughes MK (2000) Long-term variability in the El Niño Southern Oscillation and associated teleconnections. In: H.F. Diaz and Markgraf V (eds.) *El Niño and the Southern Oscillation. Multiscale Variability and Global and Regional Impacts.* Cambridge Univ. Press, Cambridge, U.K pp. 357-412

Mann ME, Rutherford S, Bradley RS, Hughes MK, Keimig FT (submitted) Reconciling borehole and other proxy-based estimates of temperature trends in past centuries.

Mann ME, Gille E, Bradley RS, Hughes MK, Overpeck JT, Webb RS, Keimig FT (2000) Annual temperature patterns in past centuries: an interactive presentation. *Earth Interactions* 44:1-29

Manthé S (1998) Variabilité de la circulation thermohaline glaciaire et interglaciaire en Atlantique Nord, tracée par les foraminifères planctoniques et la microfaune benthique. Bordeaux I.

Mantua NJ, Hare SR, Zhang Y, Wallace JM, Francis RC (1997) A Pacific interdecadal climate oscillation with impacts on salmon production. *Bulletin of the American Meteorological Society* 78:1069-79

Marchal O, Stocker TF, Joos F (1998) Impact of oceanic reorganisations on the ocean carbon cycle and atmospheric carbon dioxide content. *Paleoceanography* 13:225-244

Marchal O, Stocker TF, Joos F, Indermühle A, Blunier T, Tschumi J (1999) Modelling the concentration of atmospheric $CO_2$ during the Younger Dryas climate event. *Climate Dynamics* 15:341-354

Marchant R, Taylor D, Hamilton AC (1997) Late glacial and Holocene history of Bwindi-Impenetrable Forest, southwest Uganda. *Quaternary Research* 47:216-228

Marchant RA, Taylor D (1998) Dynamics of montane forest in central Africa during the late Holocene: a pollen-based record from western Uganda. *The Holocene* 8:375-381

Marchitto TM, Curry WB, Oppo DW (1998) Millenial-scale changes in North Atlantic circulation since the last glaciation. *Nature* 393:557-561

Markgraf V (2001) *Interhemispheric Climate Linkages*. Academic Press,

Markgraf V, Diaz HF (2001) The past ENSO record: A synthesis. In: H.F. Diaz and Markgraf V (eds.) *El Niño and the Southern Oscillation: Multiscale variability and global impacts*. Cambridge University Press, Cambridge pp. 465-488

Markgraf V, McGlone M, Hope G (1995) Neogene paleoenvironmental and paleoclimatic change in southern temperate ecosystems - a southern perspective. *Trends in Ecology & Evolution* 10:143-147

Martin J, Fitzwater S, Gordon R (1990) Iron in Antarctic waters. *Nature* 345:156-158

Martin JH, Coale KH, Johnson KS, Fitzwater SE, Gordon RM, SJ Tanner, Hunter CN, Elrod VA (1994) Testing the iron hypothesis in ecosystems of the equatorial Pacific Ocean. *Nature* 371: 123-129.

Martin P, Steadman D (1999) Prehistoric extinctions on islands and continents. In: MacPhee (ed.) *Extinctions in Near Time*. Kluwer Academic/Plenum Publishers, New York pp. 17-55

Martin PS (1984) Pre-prehistoric overkill: The global model. In: Martin PS and Klein RG (eds.) *Quaternary Extinctions: A Prehistoric Revolution*. University of Arizone press, Tucson pp. 354 – 403

Martinerie P, Brasseur GP, Granier C (1995) The chemical composition of ancient atmospheres: A model study constrained by ice core data. *Journal of Geophysical Research* 100:14291-14304

Martínez-Cortizas A, Pontevedra-Pombal X, García-Rodeja E, Nóvoa-Muñoz JC, Shotyk W (1999) Mercury in a Spanish peat bog: Archive of climate change and atmospheric metal deposition. *Science* 284:939-942

Martinson DG, Pisias NG, Hays JD, Imbrie J, Moore TC, Shackleton NJ (1987) Age dating and the orbital theory of the ice ages: development of a high-resolution 0-300,000 year chronostratigraphy. *Quaternary Research* 27:1-30

Maslin MA, Burns SJ (2000) Reconstruction of the Amazon basin effective moisture available over the past 14000 years. *Science* 290:2285-2287

Maslin MA, Li XS, Loutre MF, Berger A (1998) The contribution of orbital forcing to the progressive intensification of Northern Hemisphere glaciation. *Quaternary Science Reviews* 17:411-426

Masson V, Vimeux F, Jouzel J, Morgan V, Delmotte M, Ciais P,

Hammer C, Johnsen S, Lipenkov V Ya, Mosley-Thompson E, Petit J-R, Steig EJ, Stievenard M, Vaikmae R (2001) Holocene climate variability in Antarctica based on 11 ice core isotopic records. *Quaternary Research* 54: 348-358

Mayewski PA, Meeker LD, Twickler MS, Whitlow S, Yang Q, Lyons WB, Prentice M (1997) Major features and forcing of high-latitude northern hemisphere atmospheric circulation using a 110,000-year-long glaciochemical series. *Journal of Geophysical Research* 102:26345-26366

Mayewski PA, Meeker LD, Whitlow S, Twickler MS, Morrison MC, Bloomfield P, Bond GC, Alley RB, Gow AJ, Grootes PM, Meese DA, Ram M, Taylor KC, Wumkes W (1994) Changes in atmospheric circulation and ocean ice cover over the North Atlantic during the last 41,000 years. *Science* 263:1747-1751

McDonald D, Pedersen TF, Crusius J (1999) Multiple late Quaternary episodes of exceptional diatom production in the Gulf of Alaska. *Deep-Sea Research* 46:2993-3017

McEvedy C, Jones R (1979) *Atlas of world population history* New York, 166-167

McGillicuddy DJ, Robinson AR, Siegel DJ, Jannasch HW, Johnson R, Dickey T, McNeil J, Michaels AF, Knap AH, NATURE, 1998 -J (1998) Influence of mesoscale eddies on new production in the Sargasso Sea. *Nature* 394:263-266

McGlone MS (1997) The response of New Zealand forest diversity to Quaternary climates. In: Huntley B, Cramer W, Morgan AV, Prentice HC and Allen JRM (eds.) *Past and future rapid environmental changes: The spatial and evolutionary responses of terrestrial biota*. Springer-Verlag, Berlin pp. 73-80

McGlone MS, Kershaw AP, Markgraf V (1992) El Niño/Southern Oscillation and climatic variability in Australasian and South American paleoenvironmental records. In: H.F. Diaz and Markgraf V (eds.) *El Niño: Historical and paleoclimatic aspects of the Southern Oscillation*. Cambridge University Press, Cambridge pp. 435-462

McGlone MS, Salinger MJ, Moar NT (1993) Paleovegetation studies of New Zealand's climate since the last glacial maximum. In: Wright HE, Kutzbach JE, Webb T, Ruddiman WF, Street-Perrott FA and Bartlein PJ (eds.) *Global Climates since the Last Glacial Maximum*. University of Minnesota Press, Minneapolis pp. 294-317

McIntosh RJ, Tainter JA, McIntosh SK (2000) *The Way the Wind Blows* Columbia University Press

McIntyre A, Molfino B (1996) Forcing of Atlantic equatorial and subpolar millenial cycles by precession. *Science* 274:1867-1870

McManus J, Berelson WM, Hammond DE, Klinkhammer GP (1999) Barium cycling in the North Pacific: Implications for the utility of Ba as a paleoproductivity and paleoalkalinity proxy. *Paleoceanography* 14:53-61

McManus JF, Oppo DW, Cullen JL (1999) A 0.5 million year record of millennial scale climate variability in the North Atlantic. *Science* 283:971-975

McManus JF, Anderson RF, Bond GC, Broecker WS, Fleisher MQ, Higgins SM, Leth P (1995) Radiometrically determined sedimentary fluxes in the sub-polar North Atlantic Ocean during the last 150,000 years. In: *5th International Conference on Paleoceanography*. Halifax, Canada pp. 105-106

Meehl GA, Branstator GW (1992) Coupled climate model simulation of El Niño/Southern Oscillation: implications for paleoclimate. In: H. Diaz and Markgraf V (eds.) *El Niño: Historical and paleoclimatic aspects of the Southern Oscillation*. Cambridge University Press, Cambridge pp. 69-91

Meese DA, Gow AJ, Grootes P, Mayewski PA, Ram M, Stuiver M, Taylor KC, Waddington ED, Zielinski GA (1994) The accumulation record from the GISP2 core as an indicator of climate change throughout the Holocene. *Science* 266:1680-1682

Meko DM, Cook ER, Stahle DW, Stockton CW, Hughes MK (1993) Spatial patterns of tree-growth anomalies in the United States and Southeastern Canada. *Journal of Climate* 6:1773-86

Mercone D, Thomson J, Croudace IW, Siani G, Paterne M, Troesla S (2000) Duration of S1, the most recent sapropel in the eastern Mediterranean Sea, as indicated by accelerator mass spectrometry radiocarbon and geochemical evidence. *Paleoceanography* 15:336-347

Messerli B, Grosjean M, Hofer T, Nuñez L, Pfister C (2000) From nature-dominated to human-dominated environmental changes. *Quaternary Science Reviews* 19:459-479

Meybeck M, Green P, Vörösmarty C (2001) A new typology for mountains and other relief classes. *Mountain Research and Development* 21:34-45

Michaelsen J, Thompson LG (1992) A comparison of proxy records of El Niño/Southern Oscillation. In: H.F. Diaz and Markgraf V (eds.) *El Niño: Historical and paleoclimatic aspects of the Southern Oscillation.* Cambridge University Press, Cambridge, UK pp. 323-348

Mikolajewicz U, Maier-Reimer E (1990) Internal secular variability in an ocean general circulation model. *Climate Dynamics* 4:145-156

Mikolajewicz U, Maier-Reimer E (1994) Mixed boundary conditions in ocean general circulation models and their influence on the stability of the model's conveyor belt. *Journal of Geophysical Research* 99:22633-22644

Milankovitch M (1930) Mathematische Klimalehre und Astronomische Theorie der Klimaschwankungen. *Gebrueder Borntraeger* Berlin

Milankovitch MM (1941) *Canon of insolation and the ice-age problem.* Koniglich Serbische Akademie, Belgrade

Millar CI, Woolfenden WB (1999) The role of climate change in interpreting historical variability. *Ecological Applications* 9:1207-1216

Millar CL (2000) Historical variability in ecosystem management. *PAGES Newsletter* 8:2-4

Miller SL (1969) Clathrate hydrates of air in Antarctic ice. *Science* 165:489-490

Millspaugh SH, Whitlock C, Bartlein PJ (2000) Variations in fire frequency and climate over the past 17,000 yr in central Yellowstone National Park. *Geology* 28:211-214

Minobe S (1997) A 50-70 year climatic oscillation over the North Pacific and North America. *Geophysical Research Letters* 24:683-686

Misra VN, Wadia S (1999) *Man and Environment.* Indian Society for Prehistoric and Quaternary Studies, Pune

Mitchell JFB, Johns TC, Gregory JM, Tett S (1995) Climate response to increasing levels of greenhouse gases and sulphate aerosols. *Nature* 376:501-504

Mitchell JM (1976) An overview of climatic variability and its causal mechanisms. *Quaternary Research* 6:481-494

Mitchell JM, Stockton CW, Meko DM (1979) Evidence of a 22-year rhythm of drought in the Western United States related to the Hale solar cycle since the 17th century. In: B. M. Seliga and McCormack TA (eds.) *Solar-Terrestrial Influences on Weather and Climate.* D.Reidel, Dordrecht pp. 125-144

Mitsch WJ, Gosseelink JG (1993) *Wetlands.* Van Nostrand Reinhold, New York

Mix AC, Mrey AE, Pisias NG (1999) Foraminiferal faunal estimates of paleotemperature: Circumventing the no-analog problem yelds cool ice age tropics. *Paleoceanography* 14:350-359

Mix AC, Bard E, Schneider RR (2002) Ice sheets and sea level of the Last Glacial Maximum. *Quaternary Science Reviews* 22:

Mix AC, Bard E, Schneider RR (in press) Environmental

processes of the Ice Age: Land, oceans, glaciers (EPILOG). *Quaternary Science Reviews* 20:

Molfino B, Heusser LH, Woillard GM (1984) Frequency components of a Grande Pile pollen record: evidence of precessional orbital forcing. In: Berger A, Imbrie J, Hays J, Kukla G and Salzman B (eds.) *Milankovitch and Climate.* D. Reidel, Dordrecht

Monnin E, Indermühle A, Dällenbach A, Flückiger J, Stauffer B, Stocker TF, Raynaud D, Barnola J-M (2001) Atmospheric $CO_2$ concentrations over the Last Glacial Termination. *Science* 291:112-114

Monserud RA, Leemans R (1992) Comparing global vegetation maps with the Kappa statistic. *Ecological Modelling* 62:275-293

Mooney HA, Billings WD (1961) Comparative physiological ecology of arctic and alpine populations of *Oxyria digyna*. *Ecological Monographs* 31:1-29

Moore GWK, Holdsworth G, Alverson K (2001) Extra-tropical response to ENSO 1736-1985 as expressed in an ice core from the Saint Elias mountain range in northwestern North America. *Geophysical Research Letters* 28:3457-3461

Moore PD (1973) The influence of prehistoric cultures upon the initiation and spread of blanket bog in Upland Wales. *Nature* 241:350 – 353

Moros M, Endler R, Lackschewitz KS, Wallrabe-Adams HJ, Mienert J, Lemke W (1997) Physical properties of Reykjanes Ridge sediments and their linkage to high-resolution Greenland Ice Sheet Project 2 ice core data. *Paleoceanography* 12:687-695

Morrill C, Overpeck JT, Cole JE (submitted) A synthesis of abrupt changes in the Asian monsoon since the last deglaciation. *The Holocene*

Mosley-Thompson E (1992) Paleoenvironmental conditions in Antarctica since A.D. 1500: ice core evidence. In: Bradley RS and Jones PD (eds.) *Climate Since A.D. 1500.* Routledge, London pp. 572-591

Muggli DL, Harrison PJ (1996) Effects of nitrogen source on the physiology and metal nutrition of *Emiliania huxleyi* grown under different iron and light conditions. *Marine Ecology Progress Series* 130:255-267

Muhs DR, Swinehart JB, Cowherd SD, Mahan SA, Bush CA, Madole RF, Maat PB (1997) Late Holocene eolian activity in the mineralogically mature Nebraska Sand Hills. *Quaternary Research* 48:162-176

Müller PJ, Kirst G, Ruhland G, Storch IV, Rosell-Melé A (1998) Calibration of the alkenone paleotemperature index Uk'37 based on core tops from the eastern South Atlantic and the global ocean (60°N-60°S). *Geochimica et Cosmochimica Acta* 62:1757-1772

Mulvaney R, Pasteur E, Peel DA, Saltzman ES, Whung PY (1992) The ratio of MSA to non-sea-salt sulphate in Antarctic Peninsula ice cores. *Tellus* 44B:295-303

Munk W, Wunsch C (1998) Abyssal recipes II: Energetics of tidal and wind mixing. *Deep-sea research, part 1: oceanographic papers* 45:1977-2010

Mworia-Maitima J (1997) Prehistoric fires and land-cover change in western Kenya: evidence from pollen, charcoal, grass cuticles and grass phytoliths. *The Holocene* 7:409-417

Myers N, Mittermeier R, Mittermeier C, da Fonseca G-A, Kent J (2000) Biodiversity hotspot for conservation priorities. *Nature* 403:853-858

Myneni RB, Keeling CD, Tucker CJ, Asrar G, Nemani RR (1997) Increased plant growth in the northern high latitudes from 1981 to 1991. *Nature* 386:698-702

Nakazawa T, Machida T, Tanaka M, Fujii Y, Aoki S, Watanabe O (1993) Differences of the atmospheric $CH_4$ concentration between the arctic and antarctic regions in pre-industrial/pre-agricultural era. *Geophysical Research Letters* 20:943-946

Nameroff TJ, Balistrieri LS, Murray JW (2002) Suboxic trace metal geochemistry in the eastern tropical North Pacific.

*Geochim. Cosmochim. Acta* 66: 1139-1158

National Research Council (2000) *Abrupt climate change: inevitable surprises*. National Academy Press, Washington DC, 182 pp.

Naqvi SWA, Jayakumar DA (2000) Ocean biogeochemistry and atmospheric composition: Significance of the Arabian Sea. *Current Science* 78:289-299

Neff U, Burns SJ, Mangini A, Mudelsee M, Fleitmann D, Matter A (2001) Strong coherence between solar variability and the monsoon in Oman between 9 and 6kyr ago. *Nature* 411:290-293

Neftel A, Oeschger H, Staffelbach T, Stauffer B (1988) $CO_2$ record in the Byrd ice core 50,000-5000 years BP. *Nature* 331:609-611

Nepstad DC, Verissimo A, Alencar A, Nobre C, Lima E, Lefebvre P, Schlesinger P, Potter C, Moutinho P, Mendoza E, Cochrane M, Brooks V (1999) Large-scale impoverishment of Amazonian forests by logging and fire. *Nature* 398:505-508

Nesje A, Kvamme M (1991) Holocene glacier and climatic variations in western Norway: evidence for early Holocene glacier demise and multiple neoglacial events. *Geology* 19:610-612

Nesje A, Matthews JA, Dahl SO, Berrisford MS, Andersson C (2001) Holocene glacier fluctuations of Flatebreen and winter precipitation changes in the Jostedalsbreen region, western Norway, based on glaciolacustrine sediment records. *The Holocene* 11:267-280

Neue HU, Sass RL (1994) Trace gas emissions from rice fields. In: Prinn RG (ed.) *Global Atmospheric-Biospheric Chemistry*. Plenum, New York

New M, Hulme M, Jones P (1999) Representing twentieth-century space–time climate variability. Part I: Development of a 1961–90 mean monthly terrestrial climatology. *Journal of Climate* 12:829-856

Nials FL, Gregory DA, Graybill DA (1989) Salt river stream flow and Hohokam irrigation systems. In: Graybill DA, Gregory DA, Nials FL, Gasser R, Miksicek C and Szuter C (eds.) *The 1982-1992 Excavations at Las Colinas: Environment and Subsistence 5*. Arizona State Museum Archaeological Series, pp. 59-78

Nicholson S (2000) Land surface processes and Sahel climate. *Reviews of Geophysics* 38:117-139

Nisbet EG (1990) The end of the ice age. *Canadian Journal of Earth Sciences* 27:148-157

Nobre C, Sellers P, Shukla J (1991) Amazonian deforestation and regional climate change. *Journal of Climate* 4:957-988

Nordli PØ (2001) Reconstruction of nineteenth century summer temperatures in Norway by proxy data from farmers' diaries, *Climatic Change* 48:201-218

Norton DA, Palmer JG (1992) Dendroclimatic evidence from Australasia. In: Bradley RS and Jones PD (eds.) *Climate Since AD 1500*. Routledge, London pp. 463-482

Norton S (1985) Geochemistry of selected Maine peat deposits. *Bulletin of the Maine Geological Survey* 34:1-38

Nriagu JO (1990) The rise and fall of leaded gasoline. *Science of the total environment* 92:13-28

Nriagu JO (1996) A history of global metal pollution. *Science* 272:223-224

Nuñez L, Grosjean M, Cartajena I (2001) Human dimensions of late Pleistocene/Holocene arid events in southern South America. In: Markgraf V (ed.) *Interhemispheric Paleoclimate of the Americas*. Academic Press, London pp. 105-117

Nürnberg D, Müller A, Schneider RR (2000) Paleo-sea surface temperature calculations in the equatorial east Atlantic from Mg/Ca ratios in planktic *Foraminifera*: A comparison to sea surface temperature estimates from Uk$^{37}$ oxygen isotopes and foraminiferal transfer function. *Paleoceanography* 15:124-134

Oba T, Pederson T (1999) Paleoclimate significance of eolian carbonates supplied to the Japan Sea during the last glacial maximum. *Paleoceanography* 14:34-41

Odén S (1968) The acidification of air precipitation and its consequences in the natural environment. *Energy Commission Bulletin* 1:225

Odgaard B (1999) Fossil pollen as a record of past biodiversity. *Journal of Biogeography* 26:7-18

Oeschger H, Neftel A, Staffelbach T, Stauffer B (1989) The dilemma of the rapid variations in $CO_2$ in Greenland ice cores.

Oeschger H, Stauffer B, Finkel R, Langway CC (1985) Variations of the $CO_2$ concentration of occluded air and of anions and dust in polar ice cores. In: E. Sundquist and Broecker WS (eds.) *The Carbon Cycle and Atmospheric $CO_2$: Natural Variations Archean to Present*. American Geophysical Union, Washington, D. C. pp. 132-142

Ogden J, Ahmed M (1989) Climate response function analyses of kauri (*Agathis australis*) tree-ring chronologies in northern New Zealand. *Journal of the Royal Society of New Zealand* 19:205-221

Ogilvie A, Farmer G (1997) Documenting the medieval climate. In: Hulme M and Barrow E (eds.) *Climates of the British Isles: present, past and future*. Routledge, London pp. 112-133

Oldfield F (1969) The ecological history of Blelham Bog National Nature Reserve. In: West RG and Walker D (eds.) *Studies in Vegetation History of the British Isles*. Cambridge University Press, pp. 141-157

Oldfield F (1977) Lakes and their drainage basins as units of sediment-based ecological study. *Progress in Physical Geography* 1:460-504

Oldfield F (1996) The PALICLAS project: synthesis and overview, *International Journal of Limnology* 55: 329-357

Oldfield F (1998) Past Global Changes Status Report and Implementation Plan. *International Geosphere Biosphere Program* 45:236

Oldfield F, Alverson K (in press) The human rationale for past global change research. In: Alverson K, Bradley R and Pedersen T (eds.) *Paleoclimate, Global Change and the Future*. Springer Verlag

Oldfield F, Appleby PG, Thompson R (1980) Paleo-ecological studies of three lakes in the Highlands of Papua New Guinea. I The chronology of sedimentation. *Journal of Ecology* 68:457-477

Oldfield F, Appleby PG, Worsley AT (1985) Evidence from lake sediments for recent erosion rates in the Highlands of Papua New Guinea. In: Douglas I and Spencer E (eds.) *Environmental Change and Tropical Geomorphology*. George Allen & Unwin, London pp. 185 – 195

Oldfield F, Richardson N, Appleby PG (1993) $^{241}$Am and $^{137}$Cs activity in fine grained sediments from parts of the N.E. Irish sea shoreline. *Journal of Environmental Radioactivity* 19:1-24

Oldfield F, Asioli A, Accorsi CA, Juggins S, Langone L, Rolph T, Trincardi F, Wolff G, Gibbs-Eggar Z, Vigliotti L, Frignani M (in prep) A high resolution Late-Holocene palaeo-environmental record from the Adriatic Sea: Core RF 93:30.

Ono Y, Naruse T (1997) Snowline elevation and eolian dust flux in the Japanese islands during isotope stages 2 and 4. *Quaternary International* 37:45-54

Ono Y, D. Liu, Zhou YM (1997) Paleoenvironments of the Tibetan Plateau viewed from glacial fluctuations at the northern foot of the West Kunlun Mountains. *Journal of Tokyo Geographical Society* 106:184-198

Opdyke BN, Walker JCG (1992) Return of the coral reef hypothesis: basin to shelf partitioning of $CaCO_3$ and its effect on atmospheric $CO_2$. *Geology* 20:733-736

Oppo DW, Fairbanks RG (1987) Variability in the deep and

intermediate water circulation of the Atlantic Ocean during the past 25,000 years: Northern Hemisphere modulation of the Southern Ocean. *Earth and Planetary Science Letters* 86:1-15

Oppo DW, Keigwin LD, McManus JF, Cullen JL (2001) Persistent suborbital climate variability in marine isotope stage 5 and Termination II. *Paleoceanography* 16:280-292

Osborn TJ, Hulme M (2001) Climate observations and GCM validation Report to UK Dept. Environment, Transport and the Regions, Climatic Research Unit, University of East Anglia

Osborn TJ, Briffa KR, Tett SFB, Jones PD, Trigo RM (1999) Evaluation of the North Atlantic Oscillation as simulated by a coupled climate model. *Climate Dynamics* 15:685-702

Osterkamp TE, Romanovsky VE (1999) Evidence for warming and thawing of discontinuous permafrost in Alaska. *Permafrost and Periglacial Processes* 10:17-37

Osterkamp TE, Viereck L, Shur Y, Jorgenson MT, Racine C, Doyle A, Boone RD (2000) Observations of thermokarst and its impact on boreal forests in Alaska, U.S.A. *Arctic, Antarctic and Alpine Research* 32:303-315

Otto-Bliesner B (1999) El Niño/La Niña and Sahel precipitation during the middle Holocene. *Geophysical Research Letters* 26:87-90

Overpeck J (1987) Pollen time series and Holocene climate variability of the midwest United States. In: Berger W and Labeyrie L (eds.) *Abrupt Climatic Change*. D. Reidel Publishing Co, pp. 137-143

Overpeck J (1996) Warm climate surprises. *Science* 271:1820-1821

Overpeck J, Webb R (2000) Nonglacial rapid climate events: past and future. *Proceedings of the National Academy of Sciences* 97:1335-1338

Overpeck J, Rind D, Goldberg R (1990) Climate-induced changes in forest disturbance and vegetation. *Nature* 51-53

Overpeck J, Bartlein P, Webb T III (1991) Potential magnitude of future vegetation change in eastern North America: comparisons with the past. *Science* 254:692-695

Overpeck J, Webb R, Webb T III (1992) Mapping eastern North American vegetation change of the past 18 ka: No-analogs and the future. *Geology* 20:1071-1074

Overpeck J, Rind D, Lacis A, Healy R (1996) Possible role of dust-induced regional warming in abrupt climate change during the last glacial period. *Nature* 447-449

Overpeck JT, Anderson D, Trumbore S, Prell W (1996) The southwest Indian monsoon over the last 18 000 years. *Climate Dynamics* 12:213-225

Overpeck JT, and 17 others (1997) Arctic environmental change of the last 4 centuries. *Science* 278:1251-1256

Oxburgh R (1998) Variations in the osmium isotope composition of sea water over the past 200,000 years. *Earth Planetary Science Letters* 159:183-191

Page MJ, Trustrum NA, DeRose RC (1994) A high resolution record of storm-induced erosion from lake sediments, New Zealand. *Journal of Palaeolimnology* 11:333-348

Paillard D (1998) The timing of Pleistocene glaciations from a simple multiple-state climate model. *Nature* 391:378-381

Paillard D (2001) Glacial hiccups. *Nature* 409:147-148

Paillard D, Labeyrie LD (1994) Role of the thermohaline circulation in the abrupt warming after Heinrich events. *Nature* 372:162-164

Paillard D, Cortijo E (1999) A simulation of the Atlantic meridional circulation during Heinrich event 4 using reconstructed sea surface temperatures and salinities. *Paleoceanography* 14:716-724

Paillard D, Labeyrie LD, Yiou P (1996) AnalySeries 1.0: a Macintosh software for the analysis of geophysical time-series. *E.O.S* 77:379

Parker DE, Legg TP, Folland CK (1992) A new daily Central England temperature series. *International Journal of Climatology* 12:317-342

Parker DE, Jones PD, Bevan A, Folland CK (1994) Interdecadal changes of surface temperature since the 19th century. *Journal of Geophysical Research* 99:14373-14399

Patterson DB, Farley KA, Norman MD (1999) [4]He as a tracer of continental dust: A 1.9 million year record of aeolian flux to the west equatorial Pacific Ocean. *Geochimicha Cosmochimica Acta* 63:615-625

Patterson WA, Edwards KJ, Macguire DJ (1987) Microsopic charcoal as a fossil indicator of fire. *Quaternary Science Reviews* 6:3-23

Pätzold J (1986) Temperature and $CO_2$ changes in tropical surface waters of the Phillippines during the last 120 years: record in the stable isotopes of hermatypic corals Geologische-Palaeontologische Institut, Christian-Albrechts-Universität, Kiel, Germany pp. 92

Pauli H, Gottfried M, Grabherr G (1996) Effects of climate change on mountain ecosystems – upward shifting of alpine plants. *World Resource Review* 8:382-390

Payette S, Fortin MJ, Morneau C (1996) The recent sugar maple decline in southern Québec: Probable causes deduced from tree rings. *Canadian Journal of Forest Research* 26:1069-1078

Pearman GI, Etheridge D, Silva FD, Fraser PJ (1986) Evidence of changing concentrations of atmospheric $CO_2$, $N_2O$ and $CH_4$ from air bubbles in Antarctic ice. *Nature* 320:248-250

Pedersen TF, Bertrand P (2000) Influences of oceanic rheostats and amplifiers on atmospheric $CO_2$ content during the Late Quaternary. *Quaternary Science Reviews* 19:273-283

Peel DA (1992) Ice core evidence from the Antarctic Peninsula region. In: Bradley RS and Jones PD (eds.) *Climate Since A.D. 1500*. Routledge, London pp. 549-571

Peglar S, Birks HJB (1993) The mid-Holocene *Ulmus* fall at Diss Mere, South-east England – disease and human impact? *Vegetation history and Archaeobotany* 2:61-68

Peglar SM (1993) The mid-Holocene *Ulmus* decline at Diss Mere, Norfolk, UK: a year-by-year pollen stratigraphy from annual laminations. *The Holocene* 3:1-13

Peltier WR (2002) On eustatic sea level history, Last Glacial Maximum to Holocene. *Quaternary Science Reviews* 21:377-396

Pendall E, Markgraf V, White JWC, Dreier M (2001) Multiproxy record of Late Pleistocene-Holocene climate and vegetation changes from a peat bog in Patagonia. *Quaternary Research* 55:168-178

Peng CH, Guiot J, Campo EV (1995) Reconstruction of the past terrestrial carbon storage of the northern hemisphere from the Osnabrück biosphere model and palaeodata. *Climate Research* 5:107-118

Penner J, Dickinson R, O'Neill C (1992) Effects of aerosol from biomass burning on the global radiation budget. *Science* 256:1432-1434

Pépin L, Raynaud D, Barnola J-M, Loutre MF (in press) Hemispheric roles of climate forcings during glacial-interglacial transitions as deduced from the Vostok record and LLN-2D model experiments. *Journal of Geophysical Research* 106:31885-31892

Peteet DM (1995) Global Younger-Dryas. *Quaternary International* 28:93-104

Peterson LC, Prell WL (1985) Carbonate dissolution in recent sediments of the eastern equatorial Indian Ocean: preservation patterns and carbonate loss above the lysocline. *Marine Geology* 64:259-290

Peterson LC, Haug GH, Hughen KA, Röhl U (2000) Rapid changes in the hydrologic cycle of the tropical Atlantic during the Last Glacial. *Science* 290:1947-1951

Petit JR, L. Mounier, J. Jouzel, Y.S. Korotkevich, V.I. Kotlyakov, Lorius C (1990) Palaeoclimatological and chronological implications of the Vostok core dust record. *Nature* 343:56-58

Petit JR, Jouzel J, Raynaud D, Barkov NI, Barnola JM, Basile I, Bender M, Chappellaz J, Davis M, Delaygue G, Delmotte M, Kotlyakov VM, Legrand M, Lipenkov VY, Lorius C, Pepin L, Ritz C, Saltzman E, Stievenard M (1999) Climate and atmospheric history of the past 420,000 years from the Vostok ice core, Antarctica. *Nature* 399:429-436

Petit-Maire N (1999) Variabilité naturelle des enviornnements terrestres: les deux extremes climatiques (18 000± 2 000 and 8 000± 1 000 yrs BP). *Earth and Planetary Sciences* 328: 273-279

Petit-Maire N, Riser J (1983) *Sahara ou Sahel? Quaternaire Récent du Bassin de Taoudenni* Luminy, Marseille, 473pp

Petit-Maire N, Fontugne M, Rouland C (1991) Atmospheric methane ratio and environmental changes in the Sahara and Sahel during the last 130 kyrs. *Palaeogeography, Palaeoclimatology, Palaeoecology* 86:197-204

Petoukhov V, Ganopolski A, Brovkin V, Claussen M, Eliseev A, Kubatzki C, Rahmstorf S (2000) CLIMBER-2: A climate system model of intermediate complexity. Part I: Model description and performance for present climate. *Climate Dynamics* 16:1-17

Peyron O, Guiot J, Cheddadi R, Tarasov P, Reille M, de Beaulieu JL, Andrieu V (1998) Climatic reconstruction in Europe for 18,000 years B.P. from pollen data. *Quaternary Research* 49:183-196

Pfister C (1992) Monthly temperature and precipitation in central Europe 1525-1979: Quantifying documentary evidence on weather and its effects. In: Bradley RS and Jones PD (eds.) *Climate Since AD 1500.* Routledge, London pp. 118-142

Pfister C, Luerbacher J, Schwartz-Zanetti G, Wegmann M (1998) Winter air temperature variations in western Europe during the Early and High Middle Ages (A.D. 750-1300. *The Holocene* 8:535-552

Pfister C, Brázdil R, Glaser R, Bowka A, Holawe F, Limanowka D, Kotyza O, Munzar J, Rácz L, Strommer E, Schwartz-Zanetti G (1999) Daily weather observations in sixteenth-century Europe. *Climatic Change* 43:111-150

Pfister C, Brázdil R, Glaser R, Barriendos M, Camuffo D, Deutsch M, Dobrovoln P, Enzi S, Guidoboni E, Kotyza O, Militzer S, Rácz L, Rodrigo FS (1999) Documentary evidence on climate in sixteenth century Europe. *Climatic Change* 43:55-110

Pflaumann U, Duprat J, Pujol C, Labeyrie L (1996) SIMMAX, a modern analog technique to deduce Atlantic sea surface temperatures from planktonic *Foraminifera* in deep sea sediments. *Paleoceanography* 11:15-35

Philander SGH (1990) *El Niño, La Niña, and the Southern Oscillation* Academic Press, San Diego, 293 pp.

Pierce DW, Barnett TP, Mikolajewicz U (1995) Competing roles of heat and freshwater flux in forcing the thermohaline circulation. *Journal of Physical Oceanography* 25:2046-2064

Pinot S, Ramstein G, Marsiat I, Vernal Ad, Peyron O, Duplessy JC, Weinelt M (1999) Sensitivity of the European LGM climate to North Atlantic sea-surface temperature. *Geophysical Research Letters* 26:1893-1896

Pitelka LF, Plant Migration Working Group (1997) Plant migration and climate change. *American Scientist* 85:464-473

PMIP (2000) Paleoclimate Modelling Intercomparison Project (PMIP) Proceedings of the third PMIP Workshop

Pollack HN, Huang S, Shen P (1998) Climate change record in subsurface temperatures: a global perspective. *Science* 282:279-281

Polley HW, Johnson HB, Marino BD, Mayeux HS (1993) Increase in C3 plant water-use efficiency and biomass over glacial to present $CO_2$ concentrations. *Nature* 361:61-64

Polzin KL, Toole JM. Ledwell JR, Schmilt RW (1997) Spatial variability of turbulent mixing in the abyssal ocean. *Science* 276: 93-96

Porter SC (2001) Snowline depression in the tropics during the Last Glaciation. *Quarternary Science Reviews* 20:1067-1091

Porter SC, Denton GH (1967) Chronology of neoglaciation in the North American Cordillera. *American Journal of Science* 265:177-210

Porter SC, An ZS (1995) Correlation between climate events in the North Atlantic and China during the last glaciation. *Nature* 375:305-308

Powell JM (1982) The history of plant use and man's impact on the vegetation. In: Gressitt L (ed.) *Biogeography and Ecology in New Guinea.* Junk, The Hague pp. 207-277

Power S, Casey T, Folland C, Colman A, Mehta V (1999) Inter-decadal modulation of the impact of ENSO on Australia. *Climate Dynamics* 15:319-324

Prahl FG, Wakeham SG (1987) Calibration of unsaturated patterns in long chain ketone compositions for paleotemperature assessment. *Nature* 330:367-369

Prather M, Derwent R, Erhlt D, Fraser P, Sanhueza E, Zhou X (1995) Other trace gases and atmospheric chemistry. In: Houghton JT, Filho LGM, Callander BA, Harris N, Kattenberg A and Maskell K (eds.) *Climate Change 1995: The Science of Climate Change.* CUP, Cambridge

Preece RC (1997) The spatial response of non-marine *Mollusca* to past climate changes. In: Huntley B, Cramer W, Morgan AV, Prentice HC and Allen JRM (eds.) *Past and future rapid environmental changes: The spatial and evolutionary responses of terrestrial biota.* Springer-Verlag, Berlin pp. 163-177

Prell WL (1984) Variation of monsoonal upwelling: a response to changing solar radiation. In: Hansen J and Takahashi T (eds.) *Climate Processes and Climate Sensitivity.* American Geophysical Union, Washington DC pp. 48-57

Prell WL (1985) The stability of low-latitude sea-surface temperatures: an evaluation of the CLIMAP reconstruction with emphasis on the positive SST anomalies U.S. Department of Energy

Prell WL, van Campo E (1986) Coherent response of Arabian Sea upwelling and pollen transport to late Quaternary monsoonal winds. *Nature* 323:526-528

Prell WL, Kutzbach JE (1987) Monsoon variability over the past 150,000 years. *Journal of Geophysical Research* 92:8411-8425

Prentice I (1986) Vegetation responses to past climatic changes. *Vegetation History and Archaebotany* 67:131

Prentice I, Bartlein P, Webb T III (1991) Vegetation and climate change in Eastern North America since the last glacial maximum. *Ecology* 72:2038-56

Prentice IC (1992) Climate change and long-term vegetation dynamics. In: Glenn–Lewin DC, Peet RK and Veblen TT (eds.) *Plant succession: Theory and prediction.* Chapman & Hall, London

Prentice IC, Guiot J, Huntley B, Jolly D, Cheddadi R (1996) Reconstructing biomes from palaeoecological data: a general method and its application to European pollen data at 0 and 6 ka. *Climate Dynamics* 12:185-194

Prentice IC, Cramer W, Harrison SP, Leemans R, Monserud RA, Solomon AM (1992) A global biome model based on plant physiology and dominance, soil properties and climate. *Journal of Biogeography* 19:117-134

Prentice IC, Sykes MT, Lautenschlager M, Harrison SP, Dennissenko O, Bartlein PJ (1993) Modelling the global vegetation patterns and terrestrial carbon storage at the last glacial maximum. *Global Ecology and Biogeography Letters* 3:67-76

Prentice KC, Fung IY (1990) The sensitivity of terrestrial carbon storage to climate change. *Science* 346:48-51

Pringle H (1997) Death in Norse Greenland. *Science* 275:924-926

Prinn R, D. Cunnold, R. Rasmussen, P. Simmonds, F. Alyea, A. Crawford, P. Fraser, Rosen R (1990) Atmospheric emissions and trends of nitrous-oxide deduced from 10 years of Ale-Gauge data. *Journal of Geophysical Research-Atmospheres* 95:18369-18385

Pye K (1987) *Aeolian dust and dust deposits* Academic Press Inc., Orlando, FL, 330 pp.

Pyne SJ (1998) Forged in Fire: History, Land and Anthropogenic Fire. In: Alén W (ed.) *Advances in Historical Ecology.* Columbia University Press, New York pp. 64 – 103

Quade J, Forester RM, Pratt WL, Carter C (1998) Black mats, spring-fed streams, and Late-Glacial-Age recharge in the southern Great Basin. *Quaternary Research* 49:129-148

Quinn TM, Crowley TJ, Taylor FW (1996) New stable isotope records from a 173-year coral record from Espiritu Santo, Vanuatu. *Geophysical Reserch Letters* 23:3413-3416

Quinn TM, Crowley TJ, Taylor FW, Henin C, Joannot P, Join Y (1998) A multicentury stable isotope record from a New Caledonia coral: Interannual and decadal see surface temperature variability in the southwest Pacific since 1657 A.D. *Paleoceanography* 13:412-426

Rahmstorf S (1995) Bifurcations of the Atlantic thermohaline circulation in response to changes in the hydrological cycle. *Nature* 378:145-149

Rajagopalan B, Lall U, Cane MA (1997) Anomalous ENSO occurrences: an alternate view. *Journal of Climate* 10:2351-2357

Rajagopalan B, Kushnir Y, Tourre YM (1998) Observed decadal midlatitude and tropical Atlantic climate variability. *Geophysical Research Letters* 25:3967-3970

Ramankutty N, Foley JA (in press) Estimating historical changes in global land cover: Croplands from 1700 to 1992. *Global Biogeochemical Cycles*

Ramrath A, Sadori L, Negendank JFW (2000) Sediments from Lago di Mezzano, central Italy: a record of Late glacial/Holocene climatic variations and anthropogenic impact. *The Holocene* 10:87-95

Raper SCB, Cubasch U (1996) Emulation of the results from a coupled general circulation model using a simple climate model, *Geophysical Research Letters* 23:1107-1110

Rappaport RA (1978) Maladaptations in social systems. In: Friedman J and Rowlands MJ (eds.) *The Evolution of Social Systems.* University of Pittsburgh Press, Pittsburgh pp. 49 –87

Rasmussen RA, Khalil MAK (1984) Atmospheric methane in the recent and ancient atmospheres: concentrations, trends, and interhemispheric gradient. *Journal of Geophysical Research* 89:11599-11605

Rasmussen TL, Thomsen E, Labeyrie L, van Weering TCE (1996) Circulation changes in the Faeroe-Shetland channel correlating with cold events during the last glacial period. *Geology* 24:937-940

Rasmussen TL, Thomsen E, van Weering TCE, Labeyrie L (1996) Rapid changes in surface and deep water conditions at the Faeroe margin during the last 58,000 years. *Paleoceanography* 11:757-771

Raup DM, Sepkoski JJ (1984) Periodicity of extinctions in the geologic past. *Proceedings of the National Academy of Sciences of the U.S.A.* 81: 801-805

Raymo ME, Ganley K, Carter S, Oppo DW, McManus J (1998) Millennial-scale climate instability during the early Pleistocene epoch. *Nature* 392:699-702

Raynaud D, Jouzel J, Barnola JM, Chappellaz J, Delmas RJ, Lorius C (1993) The ice record of greenhouse gases. *Science* 259:926-933

Rayner PJ, Law RM, Dargaville R (1999) The relationship between tropical $CO_2$ fluxes and the El Niño-Southern Oscillation. *Geophysical Research Letters* 26:493-496

Redman CL (1999) *Human Impact on Ancient Environments.* The University of Arizona Press

Reille M, DeBeaulieu J-L (1995) Long Pleistocene pollen records from the Praclaux Crater, South-Central France.

*Quaternary Research* 44:205-215

Ren G, Zhang L (1998) A preliminary mapped summary of Holocene pollen data for Northeast China. *Quaternary Science Reviews* 17:669 – 688

Renberg I (1990) A 12,6000 year perspective of the acidification of Lille Oresjon, southwest Sweden. *Philosophical Transactions of the Royal Society of London Series B* 327:357-361

Renberg I, Persson MW, Emteryd O (1994) Pre-industrial atmospheric lead contamination detected in Swedish lake sediments. *Nature* 368:323-326

Renberg I, Bindler R, Brännvall M-L (2001) Using the historical atmospheric lead-deposition record as a chronological marker in sediment deposits in Europe. *The Holocene* 11:511-516

Revel M, Sinko JA, Grousset FE, Biscaye PE (1996) Sr and Nd isotopes as tracers of North Atlantic lithic particles: paleoclimatic implications. *Paleoceanography* 11:95-113

Rice DS (1994) The human impact on lowland mesoamerican environments. Society for American Archaeology, 59th Annual Meeting Anaheim, California

Rind D (1984) The influence of vegetation on the hydrological cycle in a global climate model. In: Hansen J and Takahashi T (eds.) *Climate Processes and Climate Sensitivity.* American Geophysical Union, Washington, DC pp. 73-91

Rind D (2000) Relating paleoclimate data and past temperature gradients: some suggestive rules. *Quaternary Science Reviews* 19:381-390

Rind D, Overpeck JT (1993) Hypothesized causes of decade-to-century-scale climate variability: climate model results. *Quaternary Science Reviews* 12:357-374

Rind D, Lean J, Healy R (1999) Simulated time-dependent climate response to solar radiative forcing since 1600. *Journal of Geophysical Research-Atmospheres* 104:1973-1990

Rippey B, Murphy RJ, Kyle SW (1982) Anthropogenically derived changes in the sedimentary flux of Mg, Ni, Cu, Zn, Hg, Pb and P in Louogh Neagh, Northern Ireland. *Environmental Science and Technology* 16:23-30

Ritchie JC (1987) *Postglacial Vegetation of Canada* Cambridge University Press, Cambridge, 178 pp

Ritchie JC, MacDonald GM (1986) The patterns of post-glacial spread of white spruce. *Journal of Biogeography* 13:527-540

Robertson AD, Overpeck JT, Rind D, Healy R (submitted) Simulated and observed climate variability of the last 500 years. *Science*

Robertson AD, Overpeck JT, Rind D, Mosley-Thompson E, Zielinski GA, Lean J, Koch D, Penner JE, Tegen I, Healy R (2001) Hypothesized climate forcing time series for the last 500 years. *Journal of Geophysical Research-Atmospheres* 106:14,783-14,803

Roelandt C (1998) Modelisation de la biosphere continentale a l'echelle globale: cycle du carbone et albedo, Faculte des Sciences, Universite Catholique de Louvain, Louvain-La-Neuve, Belgique

Robock A (2000) Volcanic eruptions and climate. *Reviews of Geophysics* 38:191-219

Rodbell D, Seltzer GO, Anderson DM, Enfield DB, Abbott MB, Newman JH (1999) A high-resolution 15,000 year record of El Niño driven alluviation in southwestern Ecuador. *Science* 283:516-520

Rohling EJ, Fenton M, Jorissen FJ, Bertrand P, Ganssen G, Caulet JP (1998) Magnitudes of sea-level lowstands of the past 500,000 years. *Nature* 394:162-165

Rommelaere V, Arnaud L, Barnola J-M (1997) Reconstructing recent atmospheric trace gas concentrations from polar firn and bubbly ice data by inverse methods. *Journal of Geophysical Research* 102:30069-30083

Rose NL (1994) Characterization of carbonaceous particles from lake sediments. *Hydrobiologia* 274:127-132

Rosman KJR, Ly C, Van de Velde K, Boutron CF (2000) A two century record of lead isotopes in high altitude alpine snow and ice. *Earth And Planetary Science Letters* 176: 413-424

Rosman KJR, Chisholm W, Boutron CF, Candelone JP, Gorlach U (1993) Isotopic evidence for the source of lead in Greenland snows since the late 1960s. *Nature* 362:333-334

Rossignol-Strick M (1983) African monsoons, an immediate climate response to orbital insolation. *Nature* 303:46-49

Rossignol-Strick M (1985) Mediterranean Quaternary sapropels, an immediate response of the African Monsoon to variation of insolation. *Palaeogeography, Palaeoclimatology, Palaeoecology* 49:237-263

Rossignol-Strick M, Paterne M, Bassinot F, Emeis C, Lange GJD (1998) An unusual mid-Pleistocene monsoon period over Africa and Asia. *Nature* 392:269-272

Rothrock DA, Yu Y, Maykut GA (1999) Thinning of the Arctic sea-ice cover. *Geophysical Research Letters* 26:3469-3472

Rousseau D-D (1997) The weight of internal and external constraints on *Pupilla muscorum* L. (*Gastropoda*: *Stylommatophora*) during the Quaternary in Europe. In: Huntley B, Cramer W, Morgan AV, Prentice HC and Allen JRM (eds.) *Past and future rapid environmental changes: The spatial and evolutionary responses of terrestrial biota.* Springer-Verlag, Berlin pp. 303-318

Ruddiman W, William F (1997) Tectonic uplift and climate change. *Science*

Ruddiman WF, McIntyre A (1979) Warmth of the subpolar north Atlantic Ocean during northern hemisphere ice-sheet growth. *Science* 204:173-175

Rühlemann C, Mulitza S, Müller PJ, Wefer G, Zahn R (1999) Warming of the tropical Atlantic Ocean and slow down of thermocline circulation during the last deglaciation. *Nature* 402:511-514

Sachs JP, Lehman SJ (1999) Subtropical North Atlantic temperatures 60,000 to 30,000 years ago. *Science* 286:756-759

Saigne C, Legrand M (1987) Measurements of methanesulphonic acid in Antarctic ice. *Nature* 330:240-242

Salgado-Labouriau ML, Barberi M, Ferraz-Vicentini KR, Parizzi MG (1998) A dry climatic event during the late Quaternary of tropical Brazil. *Review of Palaeobotany and Palynology* 99:115-129

Salinger MJ, Palmer JG, Jones PD, Briffa KR (1994) Reconstruction of New Zealand climate indices back to AD 1731 using dendroclimatic techniques: some preliminary results. *International Journal of Climatology* 14:1135-1149

Saltzman B, Sutera A, Hansen AR (1984) Earth-Orbital eccentricity variations and climatic change. In: Berger A (ed.) *Milankovitch and Climate, part 2.* Reidel, Hingham, Mass. pp. 615-636

Salvignac ME (1998) Variabilite hydrologique et climatique dans l'ocean Austral au cours du Quaternaire terminal. Essai de correlation inter-hemispherique. Universite de Bordaux 1.

Sandor JA, Gesper PL (1988) Evaluation of soil fertility in some prehistoric agricultural terraces in New Mexico. *Agronomy Journal* 80: 846-850

Sandor JA, Eash NS (1991) Significance of ancient agricultural soils for long-term agronomic studies and sustainable agricultural research. *Agronomy Journal* 83: 29-37

Sandweiss DH, Richardson JB, Reitz EJ, Rollins HB, Maasch KA (1996) Geoarcheological evidence from Peru for a 5,000 years B.P. onset of El Niño. *Science* 273:1531-1533

Santer BD, Taylor KE, Wigley TML, Penner JE, Jones PD, Cubasch U (1995) Towards the detection and attribution of an anthropogenic effect on climate. *Climate Dynamics* 12:77-100

Sanudo-Wilhelmy SA, Kustka AB, Gobler CJ, Hutchins DA, Yang M, Lwiza K, Burns J, Capone DG, Raven JA, Carpenter EJ (2001) Phosphorus limitation of nitrogen fixation by *Trichodesmium* in the central Atlantic Ocean. *Nature* 411:66-69

Sarmiento JS, Toggweiler JR (1984) A new model for the role of the oceans in determining atmospheric pCO$_2$. *Nature* 308:621-624

Sarnthein M (1978) Sand deserts during glacial maximum and climatic optimum. *Nature* 272:43-46

Sarnthein M, Winn K, Jung SJA, Duplessy JC, Labeyrie LD, Erlenkeuser H, Ganssen G (1994) Changes in East Atlantic deep water circulation over the last 30,000 years : Eight time slice reconstructions. *Paleoceanography* 9:209-267

Sarnthein M, Jansen E, Weinelt M, Arnold M, Duplessy JC, Erlenkeuser H, Flatoy A, Johannessen G, Johannessen T, Jung SJA, Koc N, Labeyrie L, Maslin M, Pflaumann U, Schultz H (1995) Variations in Atlantic surface ocean paleoceanography, 50°-80°N: A time-slice record of the last 30,000 years. *Paleoceanography* 10:1063-1094

Sato M, Hansen JE, McCormick MP, Pollack JB (1993) Stratospheric aerosol optical depths, 1850-1990. *Jounal of Geophysic Research- Atmosphere* 98:22987-22994

Scambos TA, C. Hulbe, Fahnestock MA (2000) The link between climate warming and ice shelf breakup in the Antarctic Peninsula. *Annals of Glaciology* 46:516-530

Scarborough VL (1993) Water management systems in the Southern maya Lowlands: An accretive model for the engineered landscape. In: Scarborough VL and Isaacs BL (eds.) *Economic Aspects of Water management in the Prehispanic New World; Research in Economic Anthropology.* JAI Press, pp. 17 – 69

Scarborough VL (1994) Maya water management. *National Geographic Research and Exploration* 10:184 – 199

Scheffer, 2001. Catastrophic shifts in ecosystems. Nature 413: 591-596

Schiller A, Mikolajewicz U, Voss R (1997) The stability of the North Atlantic thermohaline circulation in a coupled ocean-atmosphere general circulation model. *Climate Dynamics* 13:325-347

Schimel D, Alves D, Enting I, Heimann T, Joos F, Raynbaud D, Wigley T, Prather M, Derwent R, Ehhalt D, Fraser P, Sanhueza E, Zhou X, Jonas P, Charlson R, Rodhe H, Sadasivan S, Shine KP, Fouguart Y, Ramaswamy V, Solomon S, Srinivasan J, Albritton D, Isaksen I, Lal M, Wuebbles D (1996) Radiative forcing of climate change, in Climate Change 1995. The Science of Climate Change. In: Houghton JT, Meira-Filho LG, Callander BA, Harris N, Kattenberg A and Maskell K (eds.) *Contribution of Working Group I to the Second Assessment Report of the Intergoovernmental Panel on Climate Change.* Cambridge University Press, Cambridge pp. 65-131

Schimmelmann A, Zhao M, Harvey CC, Lange CB (1998) A large California flood and correlative global climatic events 400 years ago. *Quaternary Research* 49:51-61

Schindler DW (2000) The cumulative effects of climate warming and other human stresses on Canadian freshwaters in the new millennium. *Canadian Journal of Fisheries and Aquatic Sciences* 58:18-29

Schlesinger M, Zhao Z-C (1989) Seasonal climatic changes induced by doubled CO$_2$ as simulated by the OSU atmospheric GCM/mixed-layer ocean model. *Journal of Climate* 2:459-495

Schlesinger ME, Ramankutty N (1992) Implications for global warming of intercycle solar irradiance variations. *Nature* 360:330-333

Schmittner A, Appenzeller C, Stocker TF (2000) Enhanced Atlantic freshwater export during El Niño. *Geophysical Research Letters* 27:1163-1166

Schmutz C, Luterbacher J, Gyalistras D, Xoplaki E, Wanner H (2000) Can we trust proxy-based NAO index reconstructions? *Geophysical Research Letters* 27:1135-1138

Schrag DP, Hampt G, Murray DW (1996) Pore fluid constraints

on the temperature and oxygen isotopic composition of the glacial ocean. *Science* 272:1930-1932

Schrott L, Pasuto A (1999) Temporal stability and activity of landslides in Europe with respect to climate change. *Geomorphology Special Issue* 30:1-211

Schulz H, von Rad S, Erlenkeuser H (1998) Correlation between Arabian Sea and Greenland climate oscillations of the past 110,000 years. *Nature* 393:54-57

Schwander J (1996) Gas diffusion in firn. In: Wolff EW and Bales RC (eds.) *Chemical Exchange Between the Atmosphere and Polar Snow.* pp. 527-540

Schwander J, Stauffer B (1984) Age difference between polar ice and the air trapped in its bubbles. *Nature* 311:45-47

Schwander J, Stauffer B (1989) The transformation of snow to ice and the occlusion of gases. In: Oeschger H and Langway CC (eds.) *The Environmental Record in Glaciers and Ice Sheets.* John Wiley, New York pp. 53-67

Schwander J, Barnola J-M, Andrié C, Leuenberger M, Ludin A, Raynaud D, Stauffer B (1993) The age of the air in the firn and the ice at Summit, Greenland. *Journal of Geophysical Research* 98:2831-2838

Schwikowski M, Brutsch S, Gäggeler HW, Schotterer U (1999) A high-resolution air chemistry record from an Alpine ice-core; Fiescherhorn glacier, Swiss Alps. *Journal of Geophysical Research – Atmospheres* 104:13709-13719

Seltzer GO (2001) Late Quaternary glaciation in the tropics: future research directions. *Quaternary Science Reviews* 20:1063-1066

Seppä H, Birks HJB (2001) July mean temperature and annual precipitation trends during the Holocene in the Fennoscandian tree-line area: pollen-based climate reconstructions. *The Holocene* 11:527-539

Servain J (1991) Simple climatic indices for the tropical Atlantic and some applications. *Journal of Geophysical Research* 96:15,137-15,146

Severinghaus JP, Brook EJ (1999) Abrupt climate change at the end of the Last Glacial period inferred from trapped air in polar ice. *Science* 286:930-934

Severinghaus JP, Sowers T, Brook EJ, Alley RB, Bender ML (1998) Timing of abrupt climate change at the end of the Younger Dryas interval from thermally fractionated gases in polar ice. *Nature* 391:141-146

Shackleton NJ (2000) The 100,000-year ice-age cycle identified and found to lag temperature, carbon dioxide, and orbital eccentricity. *Science* 289:1897-1902

Shackleton NJ, Opdyke ND (1973) Oxygen isotope and paleomagnetic stratigraphy of equatorial Pacific core V28-238: oxygen isotope temperatures and ice volumes on a 105 and 106 year scale. *Quaternary Research* 3:39-55

Shackleton NJ, Imbrie J, Hall MA (1983) Oxygen and carbon isotope record of East Pacific core V19-30: implications for the formation of deep water in the late Pleistocene North Atlantic. *Earth Planetary Science Letters* 65:233-244

Shackleton NJ, Berger A, Peltier WR (1990) An alternative astronomical calibration of the lower Pleistocene timescale based on ODP Site 677. *Transactions of the Royal Society of Edinburgh: Earth Scicences* 81:251-261

Shackleton NJ, Hall MA, Vincent E (2000) Phase relationships between millenial-scale events 64,000-24,000 years ago. *Paleoceanography* 15:565-569

Shackleton NJ, deAbreu L, Hall MA, Vincent E (1999) Phase relationships during millennial scale events. *EOS* 80:F10

Shafer S, Bartlein PJ, Thompson RS (2001) Potential changes in the distributions of western North America tree and shrub taxa under future climate scenarios. *Ecosystems* 4:200-215

Sher AV (1997) Late-Quaternary extinction of large mammals in northern Eurasia: A new look at the Siberian contribution. In: Huntley B, Cramer W, Morgan AV, Prentice HC and Allen JRM (eds.) *Past and future rapid environmental changes: The spatial and evolutionary responses of*

*terrestrial biota.* Springer-Verlag, Berlin pp. 319-339

Shinn EA, Smith GW, Prospero JM, Betzer P, Hayes ML, Garrison V, Barber RT (2000) African dust and the demise of Caribbean coral reefs. *Geophysical Research Letters* 27:3029-3032

Shoji H, Miyamoto A, Kipfstuhl J, Langway CC (2000) Microscopic observations of air hydrate inclusions in deep ice core samples. In: Hondoh T (ed.) *Physics of ice core records.* Hokkaido University Press, Sapporo

Shotyk W, Cheburkin AK, Appleby PG, Frankhauser A, Kramers JD (1996) Two thousand years of atmospheric arsenic, antimony, and lead deposition recorded in an ombrotrophic peat bog profile, Jura Mountains, Switzerland. *Earth Planetary Science Letters* 145:E1-E7

Shotyk W, Weiss D, Appleby PG, Cheburkin AK, Frei R, Gloor M, Kramers JD, Reese S, Knaap WOVD (1998) History of atmospheric lead deposition since 12,370 $^{14}$C yr BP from a peat bog, Jura mountains, Switzerland. *Science* 281:1635-1640

Shulmeister J, Lees BG (1995) Pollen evidence from tropical Australia for the onset of an ENSO-dominated climate at c. 4000 BP. *The Holocene* 5:10-18

Siani G, Paterne M, Michel E, Sulpizio R, Sbrana A, Arnold M, Haddad G (2001) Mediterranean sea surface radiocarbon reservoir age changes since the Last Glacial Maximum. *Science* 294:1917-1920

Siegel DA, McGillicuddy DJ, Fields EA (1999) Mesoscale eddies, satellite altimetry, and new production in the Sargasso Sea. *Journal of Geophysics Research Oceans* 104:13359-13379

Siegenthaler U, Wenk T (1984) Rapid atmospheric $CO_2$ variations and ocean circulation. *Nature* 308:624-626

Sifeddine A, Martin L, Turcq B, Volkmer-Ribeiro C, Soubies F, Cordeiro RC, Suguio K (2001) Variations of the Amazonian rainforest environment: a sedimentological record covering 30,000 years. *Palaeogeography Palaeoclimatology Palaeoecology* 168:221-235

Sigman DM, Boyle EA (2000) Glacial/interglacial variations in atmospheric carbon dioxide. *Nature* 407:859-869

Sigman DM, McCorkle DC, Martin WR (1998) The calcite lysocline as a constraint on glacial/interglacial low-latitude production changes. *Global Biogeochemical Cycles* 12:409-427

Sigman DM, Altabet MA, Francois R, McCorkle DC, Gaillard JF (2000) The $\delta N15$ of nitrate in the Southern Ocean: Nitrogen cycling and circulation in the ocean interior. *Journal of Geophysical Research* 105(C8): 599-614

Sikes EL, Ramson CR, Guilderson TP, Howard W (2000) Old radiocarbon ages in the southwest Pacific Ocean during the Last Glacial period and deglaciation. *Nature* 405:555-559

Simmonds I, Jacka TH (1995) Relationships between the interannual variability of Antarctic sea-ice and the Southern Oscillation. *Journal of Climate* 8:637-647

Simmons IG (1969) Evidence for vegetation change associated with mesolithic man in Britain. In: Ucko PJ and Dimbleby GW (eds.) *The Domestication and Exploitation of Plants and Animals.* Duckworth, London pp. 111-119

Sirocko F, Garbe-Schonberg D, McIntyre A, Molfino B (1996) Teleconnections between the subtropical monsoons and high-latitude climates during the last deglaciation. *Science* 272:526-529

Skinner WR, Majorovicz JA (1999) Regional climatic warming and associated twentieth century land-cover changes in northwestern North America. *Climate Research* 12:39-52

Slowey NC, Curry WB (1987) Structure of the glacial thermocline at Little Bahama Bank. *Nature* 328:54-58

Smith (1872) *Air and Rain: The Beginning of a Chemical Climatology* Longman, London

Smith FA, Betancourt JL, Brown JH (1995) Evolution of body size in the woodrat over the past 25,000 years of climate change. *Science* 270:2012-2014

Smith SV, Hollibaugh JT (1993) Coastal metabolism and the organic carbon balance. *Review of Geophysics* 31: 75-89

Smol JP, Cumming BF, Dixit AS, Dixit SS (1998) Tracking recovery in acidified lakes: A paleolimnological perspective. *Restoration Ecology* 6:318 – 326

Solomon A, Shugart H (1984) Integrating forest-stand simulations with palaeoecological records to examine long-term forest dynamics. In: Agren G (ed.) *State and change of forest ecosystems-indicators in current research.* Swedish University of Agricultural Sciences, Uppsala pp. 333-356

Solomon AM (1986) Transient response of forests to $CO_2$-induced climate change: Simulation modeling experiments in eastern North America. *Oecologia* 68:567-579

Solomon AM, West DC, Solomon JA (1981) Simulating the role of climate change and species immigration in forest succession. In: West DC, Shugart HH and Botkin DB (eds.) *Forest Succession. Concepts and Application.* Springer-Verlag, New York pp. 154-177

Sowers T (2001) The $N_2O$ record spanning the penultimate deglaciation from the Vostok ice core. *Journal of Geophysical Research-Atmospheres*

Sowers T, Bender M, Labeyrie L, Martinson D, Jouzel J, Raynaud D, Pichon JJ, Korotkevich YS (1993) A 135,000-year Vostok-Specmap common temporal framework. *Paleoceanography* 8:737-766

Spero HJ, Bijma J, Lea DW, Bemis BE (1997) Effect of seawater carbonate concentration on foraminiferal carbon and oxygen isotopes. *Nature* 390:497-500.

Stahle DW, Cleaveland MK (1997) Development of a rainfall-sensitive tree-ring chronology in Zimbabwe. *Eight Symposium on Global Change Studies,* American Meteorological Society, Boston, Massachussets

Stahle DW, Cleaveland MK, Blanton DB, Therrell MD, Gay DA (1998a) The Lost Colony and Jamestown droughts. *Science* 280:564-567

Stahle DW, Cook ER, Cleaveland MK, Therrell MD, Meko DM, Grissino-Mayer HD, Watson E (2000) Tree-ring data document 16th century megadrought over North America. *EOS* 81:121 and 125

Stahle DW, D'Arrigo RD, Krusic PJ, Cleaveland MK, Cook ER, Allan RJ, Cole JE, Dunbar RB, Therrell MD, Gay DA, Moore MD, Stokes MA, Burns BT, Villanueva-Diaz J, Thompson LG (1998b) Experimental dendroclimatic reconstruction of the Southern Oscillation. *Bulletin of the American Meteorological Society* 79:2137-2152

Starkel L (1987) Man as a cause of sedimentological changes in the Holocene: anthropogenic sedimentological changes during the Holocene. *Striae* 26:5-12

Stauffer B, Fischer G, Neftel A, Oeschger H (1985) Increase of atmospheric methane recorded in Antarctic ice core. *Science* 229:1386-1388

Stauffer B, Hofer H, Oeschger H, Schwander J, Siegenthaler U (1984) Atmospheric $CO_2$ concentration during the last glaciation. *Annals of Glaciology* 5:160-164

Stauffer B, Blunier T, Dällenbach A, Indermühle A, Schwander J, Stocker TF, Tschumi J, Chappellaz J, Raynaud D, Hammer CU, Clausen HB (1998) Atmospheric $CO_2$ concentration and millennial-scale climate change during the last glacial period. *Nature* 392:59-62

Stein M, Wasserburg GJ, Aharon P, Chen JH, Zhu ZR, Bloom A, Chappell J (1993) TIMS U-series dating and stable isotopes of the last interglacial event in Papua New Guinea. *Geochemica et Cosmochemica Acta* 57:2541-2554

Stephens BB, Keeling RF (2000) The influence of Antarctic sea ice on glacial-interglacial $CO_2$ variations. *Nature* 404:171-174

Stine S (1994) Extreme and persistent drought in California and Patagonia during Medieval time. *Nature* 269:546-549

Stine S (1998) Medieval climatic anomaly in the Americas. In:

Issar AS and Brown N (eds.) *Water, Environment and Society in Times of Climatic Change.* Kluwer, Dordrecht pp. 43-67

Stirling CH, Esat TM, McCulloch MT, Lambeck K (1995) High-precision U-series dating of corals from Western Australia and implications for the timing and duration of the Last Interglacial. *Earth and Planetary Science Letters* 135:115-130

Stirling CH, Esat TM, Lambeck K, McCulloch MT (1998) Timing and duration of the last interglacial: evidence for a restricted interval of widespread coral reef growth. *Earth and Planetary Science Letters* 160:745-762

Stocker TF (1998) The Seesaw effect. *Science* 282:61-62

Stocker TF (2000) Past and future reorganization in the climate system. *Quaternary Science Reviews* 19:301-319

Stocker TF, Marchal O (2000) Abrupt climate change in the computer: is it real? *PNAS* 97:1362-1365

Stocker TF, Wright DG (1991). Rapid transitions of the ocean's deep circulation induced by changes in surface water fluxes. *Nature* 351: 729-732

Stocker TF, Wright DG (1998) The effect of a succession of ocean ventilation changes on radiocarbon. *Radiocarbon* 40:359-366

Stocker, T.F., Knutti, R., Plattner, G.-K. (2001). The future of the thermohaline circulation - a perspective. In: Seidov, D., Haupt, B., Maslin, M. (eds.) *The Oceans and Rapid Climate Change: Past, Present and Future.* Geophysical Monograph 126, American Geophysical Union, Washington DC, pp. 277-293

Stockton CW, Jacoby GC (1976) Long-term surface water supply and streamflow levels in the Upper Colorado River Basin. Lake Powell Research Project Bulletin pp. 70

Stott PA, Tett SFB, Jones GS, Allen MR, Mitchell JFB, Jenkins GJ (2000) External control of 20th century temperature by natural and anthropogenic forcings. *Science* 290:2133-2137

Stott PA, Tett SFB, Jones GS, Allen MR, Ingram WJ, Mitchell JFB (2001) Attribution of twentieth century temperature change to natural and anthropogenic causes. *Climate Dynamics* 17:1-21

Street FA, Grove AT (1979) Global maps of lake-level fluctuations since 30,000 yr BP. *Quaternary Research* 12:83-118

Street-Perrott F, Mitchell J, Marchand D, Brunner J (1990) Milankovitch and albedo forcing of the tropical monsoons: A comparison of geological evidence and numerical simulations for 9000 y BP. *Transactions of the Royal Society of Edinburgh: Earth Scicences* 81:407-427

Street-Perrott F, Huang Y, Perrott R, Eglinton G, Barker P, Khelifa L, Harkness D, Olago D (1997) Impact of lower atmospheric carbon dioxide on tropical mountain ecosystems. *Science* 278:1422-1426

Street-Perrott FA (1992) Tropical wetland sources. *Nature* 355:23-24

Stuart AJ (1993) The failure of evolution: Late Quaternary mammalian extinctions in the Holarctic. *Quaternary International* 19:101-107

Stuijts I-LM (1993) Late Pleistocene and Holocene vegetation of west Java, Indonesia. *Modern Quaternary Research in Southeast Asia* 12:1-173

Stuiver M, Braziunas TF (1993) Sun, ocean, climate and atmospheric $^{14}CO_2$: an evaluation of causal and spectral relationships. *The Holocene* 3:289-305

Stuiver M, Reimer PJ (1993) Extended $^{14}C$ data base and revised Calib 3.0 $^{14}C$ age calibration program. *Radiocarbon* 35:215-230

Stuiver M, Grootes PM (2000) GISP2 oxygen isotope ratios. *Quaternary Research* 53:277-283

Stuiver M, Grootes PM, Braziunas TF (1995) The GISP2 $_{-}{}^{18}O$ climate record of the past 16,500 years and the role of sun, ocean, and volcanoes. *Quaternary Research* 44:341-354

Stuiver M, Braziunas TF, Becker B, Kromer B (1991) Climatic, solar, oceanic and geomagnetic influences on late glacial and Holocene atmospheric $^{14}C/^{12}C$ change. *Quaternary Research* 35:1-24

Stute M, Talma S (1998) Glacial temperatures and moisture transport regimes reconstructed from noble gas and $_{-}^{18}O$, Stampriert aquifer, Namibia. In: *Isotope Techniques in the study of past and current Environmental Changes in the Hydrosphere and the Atmosphere*. Proceedings of Vienna Symposium, IAEA, Vienna pp. 307-328

Stute M, Forster M, Frischkorn H, Serejo A, Clark JF, Schlosser P, Broecker WS, Bonani G (1995) Cooling of tropical Brazil (5°C) during the last glacial maximum. *Science* 269:379-383

Subak S (1994) Methane from the house of Tudor and the Ming dynasty: Anthropogenic emissions in the sixteenth century. *Chemosphere* 29:843-854

Sunda WG, Huntsman SA (1995) Cobalt and zinc interreplacement in marine phytoplankton: Biological and geochemical implications. *Limnology and Oceanography* 40:1404-1417

Swetnam T, Allen C, Betancourt J (1999) Applied historical ecology: Using the past to manage for the future. In: *Ecological Applications*. pp. 1189-1206

Swetnam TW (1993) Fire history and climate change in Giant Sequoia groves. *Science* 262:885-889

Swetnam TW, Betancourt JL (1998) Mesoscale disturbance and ecological response to decadal climatic variability in the American Southwest. *Journal of Climate* 11:3128-3147

Sykes MT (1997) The biogeographic consequences of forecast changes in the global environment: Individual species' potential range changes. In: Huntley B, Cramer W, Morgan AV, Prentice HC and Allen JRM (eds.) *Past and future rapid environmental changes: The spatial and evolutionary responses of terrestrial biota*. Springer-Verlag, Berlin pp. 427-440

Sykes MT, Prentice IC (1996) Climate change, tree species distributions and forest dynamics: A case study in the mixed conifer northern hardwoods zone of northern Europe. *Climatic Change* 34:161-177

Sykes MT, Prentice IC, Cramer W (1996) A bioclimatic model for the potential distributions of north European tree species under present and future climates. *Journal of Biogeography* 23:203-233

Sykes MT, Prentice IC, Laarif F (1999) Quantifying the impact of global climate change on potential natural vegetation. *Climatic Change* 41:37-52

Tada R, Irino T, Koizumi I (1999) Land-ocean linkage over orbital and millenial timescales recorded in the late Quaternary sediments of the Japan Sea. *Paleoceanography* 14:236-247

Takeda S (1998) Influence of iron availability on nutrient consumption ratio of diatoms in oceanic waters. *Nature* 393:774-777

Tarasov L, Peltier WR (1997) A high-resolution model of the 100 ka ice-age cycle. *Annals of Glaciology* 25:58-65

Taylor D (1990) Late Quaternary pollen diagrams from two Ugandan mires: evidence for environmental change in the Rukiga Highlands of southwest Uganda. *Palaeogeography, Palaeoclimatology, Palaeoecology* 80:283-300

Taylor D, Marchant RA, Robertshaw P (1999) A sediment-based history of medium altitude forest in central Africa: a record from Kabata Swamp, Ndale volcanic field, Uganda. *Journal of Ecology* 87:303-315

Taylor KC, Mayewski PA, Alley RB, Broook EJ, Gow AJ, Grootes PM, Meese DA, Saltzman ES, Severinghaus JP (1997) The Holocene/Younger Dryas transition recorded at Summit Greenland. *Science* 278:825-827

TEMPO M (1996) Potential role of vegetation feedback in the climate sensitivity of high-latitude regions: a case study at 6000 years B.P. *Global Biogeochemical Cycles* 10:727-737

Terray P (1995) Space-time structure of interannual monsoon variability. *Journal of Climate* 8:2595-2619

Tett SFB, Johns TC, Mitchell JFB (1997) Global and regional variability in a coupled AOGCM. *Climate Dynamics* 13:303-323

Tett SFB, Stott PA, Allen MR, Ingram WJ, Mitchell JFB (1999) Causes of twentieth-century temperature change near the Earth's surface. *Nature* 399:569-572

Texier D, de Noblet N, Harrison S, Haxeltine A (1997) Quantifying the role of biosphere-atmosphere feedbacks in climate change: coupled model simulations for 6000 years BP and comparison with palaeodata for northern Eurasia and northern Africa. *Climate Dynamics* 13:865-882

Thomas JA (1991) Rare species conservation: case studies of European butterflies. In: Spellerberg IF, Goldsmith FB and Morris MG (eds.) *The Scientific Management of Temperate Communities for Conservation*. Blackwell Scientific Publications, Oxford pp. 149-197

Thompson AM (1992) The oxidizing capacity of the Earth's atmosphere: probable past and future changes. *Science* 256:1157-1168

Thompson AM, Chappellaz JA, Fung IY, Kucsera TL (1993) Atmospheric methane increase since the Last Glacial Maximum. 2. Interactions with oxidants. *Tellus* 45B:242-257

Thompson DWJ, Wallace JM (1998) The Arctic Oscillation signature in the wintertime geopotential height and temperature fields. *Geophysical Research Letters* 25:1297-1300

Thompson K (1994) Predicting the fate of temperate species in response to human disturbance and global change. In: Boyle TJB and Boyle CEB (eds.) *Biodiversity, Temperate Ecosystems and Global Change*. Springer-Verlag, Berlin, Heidelberg pp. 61-76

Thompson LG (1995) Late Glacial Stage and Holocene tropical ice core records from Huascaran, Peru. *Science* 269:46-50

Thompson LG (2000) Ice core evidence for climate change in the Tropics: implications for our future. *Quaternary Science Reviews* 19:19-35

Thompson LG (2001) Stable isotopes and their relationship to temperature as recorded in low latitude ice cores. In: Gerhard LC, Harrison WE, Hanson BM (eds). *Geological Perspectives of Global Climate Change* pp. 99-119

Thompson LG, Mosley-Thompson E (1981) Microparticle concentration variations linked with climatic change: Evidence from polar ice cores. *Science* 212:812-815

Thompson LG, Mosley-Thompson E, Bolzan JF, Koci BR (1985) A 1500 year record of tropical precipitation in ice cores from the Quelccaya Ice Cap, Peru. *Science* 229:971-973

Thompson LG, Davis ME, Mosley-Thompson E, Liu K-B (1988) Pre-Incan agricultural activity recorded in dust layers in two tropical ice cores. *Nature* 336:763-765

Thompson LG, Yao T, Mosley-Thompson E, Davis ME, Henderson KA, Lin P-N (2000) A high-resolution millennial record of the South Asian monsoon from Himalayan ice cores. *Science* 289:1916-1919

Thompson LG, Mosley-Thompson E, Davis ME, Lin N, Yao T, Dyurgerov M, Dai J (1993) "Recent warming": ice core evidence from tropical ice cores, with emphasis on central Asia. *Global and Planetary Change* 7:145-156

Thompson LG, Mosley-Thompson E, Davis ME, Lin PN, Dai J, Bolzan JF, Yao T (1995) A 1000 year climate ice-core record from the Guliya ice cap, China: its relationship to global climate variability. *Annals of Glaciology* 21:175-181

Thompson LG, Davies ME, Mosley-Thompson E, Sowers TA, Henderson KA, Zagoronov VS, Lin P-N, Mikhalenko VN, Campen RK, Bolzan JF, Cole-Dai J, Francou B (1998) A 25,000-year tropical climate history from Bolivian ice cores. *Science* 282:1858-1864

Thompson RS (1988) Glacial and Holocene vegetation history: Western North America. In: Huntley B and WebbIII T (eds.) *Vegetation History*. Kluwer Academic Publishers, Dordrecht pp. 415-458

Thompson RS (1992) Late Quaternary environments in Ruby Valley, Nevada. *Quaternary Research* 37:1-15

Thorndycraft V, Hu Y, Oldfield F, Crooks PRJ, Appleby PG (1998) Individual flood events detected in the recent sediments of the Petit Lac d'Annecy, eastern France. *The Holocene* 8:741-746

Tian HQ, Melillo JM, Kicklighter DW, McGuire AD, Helfrich JVK, Moore B, Vorosmarty CJ (1998) Effect of interannual climate variability on carbon storage in Amazonian ecosystems. *Nature* 396:664-667

Tiedemann RM, Sarnthein M, Shackleton NJ (1994) Astronomical timescale for the Pliocene Atlantic $\delta^{18}O$ and dust records of Ocean Drilling Program site 659. *Paleoceanography* 9:619-638

Tinner W, Lotter AF (2001) Central European vegetation response to abrupt climate change at 8.2 ka. *Geology* 29:551-554

Tinner W, Hubschmid P, Wehrli M, Ammann B, Conedera M (1999) Long-term forest fire ecology and dynamics in southern Switzerland. *Journal of Ecology* 87:273-289

Tinner W, Conedera M, Ammann B, Gäggeler HW, Gedye S, Jones R, Sägesser B (1998) Pollen and charcoal in lake sediments compared with historically documented forest fires in southern Switzerland since AD 1920. *The Holocene* 8:43-53

Tolman T (1998) *Butterflies of Britain and Europe* Harper Collins, London, 320 pp.

Touchan R, Meko D, Hughes MK (1999) A 396-year reconstruction of precipitation in southern Jordan. *Journal of the American Water Resources Association* 35:49-59

Trenberth KE (1998) Atmospheric moisture residence times and cycling: Implications for rainfall rates and climate change. *Climatic Change* 39:667-694

Trenberth KE, Shea DJ (1987) On the evolution of the Southern Oscillation. *Monthly Weather Review* 115:3078-3096

Trenberth KE, Hurrell JW (1994) Decadal atmosphere-ocean variations in the Pacific. *Climate Dynamics* 9:303-319

Trenberth KE, Hoar TW (1996) The 1990-1995 El Niño-Southern Oscillation event: Longest on record. *Geophysical Research Letters* 23:57-60

Trenberth KE, Branstator GW, Karoly D, Kumar A, Lau N-C, Ropelewski CF (1998) Progress during TOGA in understanding and modeling global teleconnections associated with tropical sea surface temperatures. *Journal of Geophysical Research* 103:14,291-14,324

Tschumi J, Stauffer B (2000) Reconstructing past atmospheric $CO_2$ concentration based on ice-core analyses: open questions due to in situ production of $CO_2$ in the ice. *Journal of Glaciology* 46:45-53

Tudhope AW, Shimmield GB, Chilcott CP, Jebb M, Fallick AE (1995) Recent changes in climate in the far western equatorial Pacific and their relationship to the Southern Oscillation: oxygen isotope records from massive corals, Papua New Guinea. *Earth and Planetary Science Letters* 136:575-590

Tudhope AW, Chilcott CP, McCulloch MT, Cook ER, Chappell J, Ellam RM, Lea DW, Lough JM, Shimmield GB (2001) Variability in the El Niño-Southern Oscillation through a Glacial-Interglacial cycle. *Science* 291:1511-1517

Turner BL, Kasperson RE, Meyer WB, Dow KM, Golding D, Kasperson JX, Mitchell RC, Ratick SJ (1990) Two types of global environmental change. *Global Environmental Change* 15-22

Tyrrell T (1999) The relative influences of nitrogen and phophorus on oceanic primary production. *Nature* 400:525-531

Tyson PD, Lindesay JA (1992) The climate of the last 2000 years in southern Africa. *The Holocene* 2:271-278

Urban FE, Cole JE, Overpeck JT (2000) Influence of mean climate change on climate variability from a 155-year tropical Pacific coral record. *Nature* 407:989-993

Vaganov EA, Hughes MK, Kidyanov AV, Schweingruber FH, Silkin PP (1999) Influence of snowfall and melt timing on tree growth in subarctic Eurasia. *Nature* 400:149-151

Van Andel TH, Runnels CN, Pope KO (1986) Five thousand years of land use and abuse in the Southern Argolid, Greece. *Hesperia* 55:103-128

Van Andel TH, Zangger E, Demitrack A (1990) Land use and soil erosion in prehistoric and historical Greece. *Journal of Field Archaeology* 17:379-396

Van Campo E, Guiot J, C. Peng (1993) A data-based re-appraisal of the terrestrial carbon budget at the last glacial maximum. *Global Planetary Change* 8:189-201

Van de Water PK, Leavitt SW, Betancourt JL (1994) Trends in stomatal density a $^{13}C/^{12}C$ ratios of *Pinus flexilis* needles during last glacial-interglacial cycle. *Science* 264:239-243

Van den Bosch F, Hengeveld R, Metz JAJ (1992) Analysing the velocity of animal range extension. *Journal of Biogeography* 19:135-150

van der Hammen T, Hooghiemstra H (2000) Neogene and Quaternary history of vegetation, climate, and plant diversity in Amazonia. *Quarternary Science Reviews* 19:725-742

Van der Kaars WA, Dam MAC (1995) A 135,000-year record of vegetational and climatic change from the Bandung area, West-Java, Indonesia. *Palaeogeography, Palaeoclimatology, Palaeoecology* 117:55-72

Van der Post KD, Oldfield F, Haworth EY, Crooks PRJ, Appleby PJ (1997) A record of accelerated erosion in the recent sediments of Blelham Tarn in the English Lake District. *Journal of Paleolimnology* 18:103-120

van Engelen AFV, Buisman J, Ijnsen F (2001) A millennium of weather, winds and water in the low countries. In: Jones PD, Davies TD, Ogilvie AEJ and Briffa KR (eds.) *History and Climate: Memories of the Future?* Kluwer/Plenum, pp. 101-124

van Krevelt S, Sarnthein M, Erlenkeuser H, Grootes P, Jung S, Nadeau MJ, Pflaumann U, Voelker A (2000) Potential links between surging ice sheets, circulation changes, and the Dansgaard-Oeschger cycles in the Irminger sea, 60-18 kyr. *Paleoceanography* 15:425-442

van Loon H, Rogers JC (1978) The seesaw in winter temperatures between Greenland and Northern Europe, part I: general description. *Monthly Weather Review* 106:296-310

Vardy SR, Warner BG, Turunen J, Aravena R (2000) Carbon accumulation in permafrost peatlands in the Northwest Territories and Nunavut, Canada. *Holocene* 10:273-280

Varela DE, Harrison PJ (1999) Seasonal variability in nitrogenous nutrition of phytoplankton assemblages in the northeastern subarctic Pacific Ocean. *Deep-Sea Research* 46:2505-2538

Verba ES, Denton GH, Partridge TC, Buckle LH (1995) *Palaeoclimate and evolution, with emphasis on human origins*. Yale University Press, New Haven, 547 pp.

Verschuren D, Laird KR, Cumming BF (2000) Rainfall and drought in equatorial east Africa during the past 1,100 years. *Nature* 403:410-414

Vidal L, Labeyrie L, vanWeering TCE (1998) Benthic $\delta^{18}O$ records in the North Atlantic over the last glacial period (60-10 kyr): Evidence for brine formation. *Paleoceanography* 13:245-251

Vidal L, Schneider RR, Marchal O, Bickert T, Stocker T, Wefer G (1999) Link between the North and South Atlantic during the Heinrich events of the last glacial period. *Climate Dynamics* 15:909-919

Vidal L, Labeyrie L, Cortijo E, Arnold M, Duplessy JC, Michel

E, Becqué S, vanWeering TCE (1997) Evidence for changes in the North Atlantic deep water linked to meltwater surges during the Heinrich events. *Earth and Planetary Science Letters* 146:13-26

Villalba R, D'Arrigo RD, Cook ER, Jacoby GC, Wiles G (2001) Decadal-scale climatic variability along the extra-tropical western coast of the Americas: Evidence from tree-ring records. In: Markgraf V (ed.) *Interhemispheric Climate Linkages*. Academic Press, pp. 155-172

Villalba R, Boninsegna JA, Lara A, Veblen TT, Roig FA, Aravena JC, Ripalta A (1996) Interdecadal climatic variations in millennial temperature reconstructions from southern South America. In: Jones PD, Bradley RS and Jouzel J (eds.) *Climatic Variations and Forcing Mechanisms of the Last 2000 Years*. Springer, Berlin pp. 161-192

Villalba R, Cook ER, D'Arrigo RD, Jacoby GC, Jones PD, Salinger MJ, Palmer J (1997) Sea-level pressure variability around Antarctica since A.D.1750 inferred from Subantarctic tree-ring records. *Climate Dynamics* 375-390

Villalba R, Holmes RL, Boninsegan JA (1992) Spatial patterns of climate and tree growth variations in subtropical northwestern Argentina. *Journal of Biogeography* 19: 631-649

Vimeux F, Masson V, Jouzel J, Petit JR, Steig EJ, Stievenard M, Vaikmae R, White JWC (2001) Holocene hydrological cycle changes in the southern hemisphere documented in East Antarctic deuterium excess records. *Climate Dynamics* 17:503-513

Vinnikov K, Robock A, Stouffer RJ, Walsh J, Parkinson CL, Cavalieri DJ, Mitchell JFB, Garrett D, Zakharov VF (1999) Global warming and Northern Hemisphere sea ice extent. *Science* 286:1934-1937

Vita-Finzi C (1969) *The Mediterranean Valleys: Geological Changes in Historical Times* Cambridge University Press, Cambridge

Vitousek PM, Mooney HA, Lubchenco J, Melillo JM (1997) Human domination of Earth's ecosystems. *Science* 277:494

von Grafenstein U, Erlenkeuser H, Müller J, Jouzel J, Johnsen S (1998) The cold event 8200 years ago documented in oxygen isotope records of precipitation in Europe and Greenland. *Climate Dynamics* 14:73-81

Vuille M, Bradley RS (2000) Mean annual temperature trends and their vertical structure in the tropical Andes. *Geophysical Research Letters* 27:3885-3888

Wadhams P, Davis NR (2001) Further evidence of ice thinning in the Arctic Ocean. *Geophysical Research Letters* 27:3973-3975

Waelbroeck C, Duplessy J-C, Michel E, Labeyrie L, Paillard D, Duprat J (2001) The timing of the last deglaciation in North Atlantic climate records. *Nature* 412:724-727

Waelbroek C, Labeyrie L, Michel E, Duplessy JC, McManus JF, Lambeck K, Balbon E, Labracherie M (2002) Sea-level and deep water temperature changes derived from benthic Foraminifera isotopic records. *Quaternary Science Reviews* 21:295-306

Wagner G, Beer J, Laj C, Kissel C, Masarik J, Muscheler R, Synal H-A (2000) Chlorine-36 evidence for the Mono Lake event in the Summit GRIP ice core. *Earth and Planetary Science Letters* 181:1-6

Wagnon P, Delmas RJ, Legrand M (1999) Loss of volatile acid species from upper firn layers at Vostok, Antarctica. *Journal of Geophysical Research* 104:3423-3431

Walker D, Singh G (1993) Earliest palynological records of human impact on the world's vegetation. In: Chambers FM (ed.) *Climate Change and Human Impact on the Landscape*. Chapman and Hall, London

Walker GT (1924) Correlations in seasonal variations of weather IX. *Memoirs of the Indian Meteorological Department* 24:275-332

Walker JCG, Kasting JF (1992) Effects of fuel and forest conservation on future levels of atmospheric carbon dioxide. *Global Planetary Change* 97:151-189

Wallace JM, Rasmusson EM, Mitchell TP, Kousky VE, Sarachik ES, vonStorch H (1998) The structure and evolution of ENSO-related climate variability in the tropical Pacific: Lessons from TOGA. *Journal of Geophysical Research-Oceans* 103:14241-14259

Walling DE, Quine TA (1990) Use of Caesium-137 to investigate patterns and rates of soil erosion on arable fields. In: Boardman J, Foster IDL and Dearing JA (eds.) *Soil erosion on agricultural land*. Wiley, Chichester

Wang G, Elfatih A-E (2000) Ecosystem dynamics and the Sahel drought. *Geophysical Research Letters* 27:795-798

Wang L, Oba T (1998) Tele-connections between east Asian monsoon and the high-latitude climate: A comparison between the GISP2 ice core record and the high resolution marine records from the Japan and South China Seas. *The Quaternary Research (Daiyonki-Kenkyu)* 37:211-219

Wang YJ, Cheng H, Edwards RL, An Z, Wu J, Shen C-C, Dorale JA (2001) A high-resolution absolute-dated late Pleistocene monsoon record from Hulu Cave, China. *Science* 294:2345-2348

Wania F, Mackay D (1993) Global fractionation and cold condensation of low volatility organochlorine compounds in polar regions. *Ambio* 22:10-18

Wanner H, Pfister C, Brazdil R, Frich P, Frydendahl K, Jónsson T, Kington J, Rosenørn S, Wishman E (1995) Wintertime European circulation patterns during the Late Maunder Minimum cooling period (1675-1704). *Theoretical and Applied Climatology* 51:167-175

Wanninkhof R (1992) Relationship between wind speed and gas exchange over the ocean. *Journal of Geophysical Research* 97:7373-7382

Ware DM, Thomson RE (2000) Interannual to multidecadal timescale climate variations in the northeast Pacific. *Journal of Climate* 13:3209-3220

Warnant P, Francois L, Strivay D, Gerard JC (1994) CARAIB: A global model of terrestrial biological productivity. *Global Biogeochemical Cycles* 8:255-270

Warrick R, Oerlemans J (1990) Sea-level rise. In: Houghton JT, Jenkins GJ and Ephraums JJ (eds.) *Climate Change. The IPCC Scientific Assessment*. Cambridge University Press, Cambridge pp. 257-282

Wasson RJ (1996) *Land and Climate impacts on Fluvial Systems during the Period of Agriculture. Recommendations for a research project and its implementations*. PAGES report

Wasson RJ, Mazari RK, Starr B, Clifton G (1998) The recent history of erosion and sedimentation on the Southern tablelands of southeastern Australia: sediment flux dominated by channel incision. *Geomorphology* 24:291-308

Watson AJ, Liss PS (1998) Marine biological controls on climate via the carbon and sulphur geochemical cycles. *Philosohpical Transactions of the Royal Society, London* 353:41-51

Watson AJ, Bakker DCE, Ridgewell AJ, Boyd PW, Law CS (2000) Effect of iron supply on Southern Ocean $CO_2$ uptake and implications for glacial atmospheric $CO_2$. *Nature* 407:730-733

Watts W, Allen J, Huntley B (1996) Vegetation history and palaeoclimate of the Last Glacial Period at Lago Grande di Monticchio, southern Italy. *Quaternary Science Reviews* 15:113-132

Weaver AJ, Eby M, Fanning AF, Wiebe EC (1998) Simulated influence of carbon dioxide, orbital forcing and ice sheets on the climate of the Last Glacial Maximum. *Nature* 394:847-853

Webb RS, Anderson KH, Webb T (1993) Pollen response-surface estimates of Late-Quaternary changes in the moisture balance of the northeastern United-States. *Quaternary Research* 40:213-227

Webb RS, Rind DH, Lehman SJ, Healy RJ, Sigman D (1997) Influence of ocean heat transport on the climate of the Last Glacial Maximum. *Nature* 385:695-699

Webb T, III (1981) The past 11,000 years of vegetational change in eastern North America. *Bioscience* 31:501-508

Webb T, III (1986) Is vegetation in equilibrium with climate? How to interpret late-Quaternary pollen data. *Vegetation History and Archaebotany* 67:75

Webb T, III (1988) Glacial and Holocene vegetation history: Eastern North America. In: Huntley B and WebbIII T (eds.) *Vegetation History.* Kluwer Academic Publishers, pp. 385-414

Webb T, III (1992) Past changes in vegetation and climate: lessons for the future. In: Peters R and Lovejoy T (eds.) *Global warming and biological diversity.* Yale University Press, New Haven, CT pp. 59-75

Webb T, III, Bartlein PJ (1992) Global changes during the last 3 million years: climatic controls and biotic responses. *Annual Review of Ecology and Systematics* 23:141-173

Webb T, III, Anderson KH, Bartlein PJ, Webb RS (1998) Late Quaternary climate change in eastern North America: A comparison of pollen-derived estimates with climate model results. *Quaternary Science Reviews* 17:587-606

Weeks RJ, Laj C, Endignoux L, Mazaud A, Labeyrie L, Roberts AP, Kissel C, Blanchard E (1995) Normalized natural remanent magnetisation intensity during the last 240,000 years in piston cores from the central North Atlantic Ocean : geomagnetic field intensity or environmental signal? *Physics of the Earth and Planetary Interior* 87:213-229

Weihenmeyer CE, Burns SJ, Waber HN, Aeschbach-Hertig W, Kipfer R, Loosli HH, Matter A (2000) Cool glacial temperatures and changes in moisture source recorded in Oman groundwaters. *Science* 287:842-845

Weiss H (1997) Late third millennium abrupt climate change and social collapse in West Asia and Egypt. In: Dalfes HN, Kukla G and Weiss H (eds.) *Third Millennium BC Climate Change and Old World Collapse.* NATO ASI Series, pp. 711-722

Weiss H, Bradley RS (2001) What drives societal collapse? *Science* 291:609-610

Weiss H, Courtney M-A, Wetterstrom W, Guichard F, Senior L, Meadow R, Curnow A (1993) The genesis and collapse of third millennium north Mesopotamian civilization. *Science* 261:995-1004

Weiss D, Shotyk W, Appleby PG, Kramers JG, Cherbukin AK (1999) Atmospheric Pb deposition since the industrial revolution recorded by five Swiss peat profiles: enrichment factors, fluxes, isotopic composition and sources. *Environmental Science and Technology* 33: 1340-1352

White JWC, Steig EJ, Cole JE, Cook ER, Johnsen SJ (1999) Recent, annually resolved climate as recorded in stable isotope ratios in ice cores from Greenland and Antarctica. In: Karl TR (ed.) *The ENSO Experiment research activities; exploring the linkages between the El Nino-Southern Oscillation (ENSO) and human health.* American Meteorological Society, pp. 300-302

White JWC, Barlow LK, Fisher DA, Grootes P, Jouzel J, Johnsen S, Mayewski PA (1997) The climate signal in the stable isotopes of snow from Summit Greenland: results of comparisons with modern climate observations. *Journal of Geophysical Research* 102:26425-26440

Whitehead DR, Charles DF, Goldstein RA (1990) The PIRLA project (Paleoecological Investigation of Recent lake Acidification): an introduction to the synthesis of the project. *Journal of Paleolimnology* 3:187-194

Whitlock C, Millspaugh SH (1996) Testing the assumptions of fire-history studies: an examination of modern charcoal accumulation in Yellowstone National Park, USA. *The Holocene* 6:7-15

Whitlock C, Bartlein PJ (1997) Vegetation and climate change in

northwest America during the past 125kyr. *Nature* 57-61

Whitlock C, Grigg L (1999) Paleoecological evidence of Milankovitch and sub-Milankovitch climate variations in the Western U.S. during the late Quaternary. In: Webb R and Clark P (eds) *Mechanisms of millennial-scale global climate change.* American Geophysical Union Geophysical Monograph, pp. 227-241

Whitlock C, Sarna-Wojcicki AM, Bartlein PJ, Nickmann RJ (2000) Environmental history and tephrostratigraphy at Carp lake, southwestern Columbia basin, Washington, U.S.A. *Palaeogeography, Palaeoclimatology, Palaeoecology* 155: 7-29

Wik M, Renberg I (1991) Spheroidal carbonaceous particles as a marker for recent sediment distribution. *Hydrobiologia* 214:85-90

Williams JW, Shuman BN, Webb T III (in press) Dissimilarity analysis of Late Pleistocene and Holocene vegetation and climate in eastern North America. *Ecology*

Winkler M, Wang PK (1993) The late-Quaternary vegetation and climate of China. In: Wright HE, Kutzbach JE, Ruddiman WF, Street-Perrott FA, Webb T III and Bartlein PJ (eds.) *in Global Climate Since the Last Glacial Maximum.* University of Minnesota, Minneapolis, MN

Winograd IJ, Szabo BJ, Coplen TB, Riggs AC (1988) A 250,000 climatic record from Great Basin vein calcite: implications for Milankovitch theory. *Science* 242:1275-1280

Wohlfarth B, Holmquist B, Cato I, Linderson H (1998) The climatic significance of clastic varves in the Angermanälven Estuary, northern Sweden, AD 1860 to 1950. *The Holocene* 8:521-534

Woillard GM (1978) Grande Pile peat bog - continuous pollen record for last 140,000 years. *Quaternary Research* 9:1-21

Wolff EW (1996) Location, movement and reactions of impurities in polar ice. In: Wolff EW and Bales RC (eds.) *Chemical Exchange between the Atmosphere and Polar Snow.* Springer, Berlin pp. 541-560

Wong CS, Chan Y-H (1991) Temporal variations in the partial pressure and flux of $CO_2$ at ocean station P in the subarctic northeast Pacific Ocean. *Tellus* 43B:206-223

Wong CS, F.A W, Crawford DW, Iseki K, Matear RJ, Johnson WK, Page JS, Timothy D (1999) Seasonal and interannual variability in particle fluxes of carbon, nitrogen and silicon from time series of sediment traps at Ocean Station P, 1982-1993: relationship to changes in subarctic primary productivity. *Deep-Sea Research* 46:2735-2760

Woodhouse CA, Overpeck JT (1998) 2000 years of drought variability in the central United States. *Bulletin of the American Meteorological Society* 79:2693-2714

Woodward F, Lomas M, Betts R (1998) Vegetation-climate feedbacks in a greenhouse world. *Philosohpical Transactions of the Royal Society, London* 353:29-39

Wright DG, Stocker TF (1993) Younger Dryas experiments. In: Peltier WR (ed.) *Ice in the Climate System.* Springer-Verlag, Heidelberg pp. 395-416

Wu J, Sunda W, Boyle EA, Karl DM (2000) Phosphate depletion in the western North Atlantic Ocean. *Science* 289:759-762

Wunsch C (1996) *The Ocean Circulation Inverse Problem* Cambridge University Press, 442 pp.

Wunsch C (1999) The interpretation of short climate records, with comments on the North Atlantic and Southern Oscillations. *Bulletin of the American Meteorological Society* 80:245-255

Xiao JL, Porter SC, An ZS, Kumai H, Yoshikawa S (1995) Grain size of quartz as an indicator of winter monsoon strength on the Loess Plateau of central China during the last 130,000 years. *Quaternary Research* 43:22-29

Xiao JL, An ZS, Liu TS, Inouchi Y, Kumai H, Yoshikawa S, Kondo Y (1999) East Asian monsoon variation during the last 130,000 years: evidence from the Loess Plateau of central China and Lake Biwa of Japan. *Quaternary Science*

*Reviews* 18:147-157

Xoplaki E, Maheras P, Luterbacher J (2001) Variability of climate in meridional Balkans during the periods 1675-1715 and 1780-1830 and its impact on human life. *Climatic Change* 48:581-614

Xue Y (1997) Biosphere feedback on regional climate in tropical north Africa. *Quarterly Journal of the Royal Meteorological Society* 123:1483-1515

Yokohama Y, Lambeck K, Dekker Pd, Johnston P, Fifleds KL (2000) Timing of the Last Glacial Maximum from observed sea-level minima. *Nature* 406:713-716

Yokohama Y, Deckker PD, Lambeck K, Johnston P, Fifield LK (2001) Sea-Level at the Last Glacial Maximum: evidence from northwestern Australia to constrain ice volumes for oxygen isotope stage 2. *Palaeogeography, Palaeoclimatology, Palaeoecology* 165:281-297

Young M, Bradley RS (1984) Insolation gradients and the paleoclimatic record. In: Berger AL, Imbrie J, Hays J, Kukla G and Saltzman B (eds.) *Milankovitch and climate.* D. Reidel, Dordrecht pp. 707-13

Yu E-F, Francois R, Bacon M (1996) Similar rates of modern and last-glacial ocean thermohaline circulation inferred from radiochemical data. *Nature* 379:689-694

Yu EF, Francois R, Bacon MP, Honjo S, Fleer AP, Manganini SJ, van der Loeff MMR, Ittekot V, PAPERS D-SRPI-OR, 2001 -M (2001) Trapping efficiency of bottom-tethered sediment traps estimated from the intercepted fluxes of Th-230 and Pa-231. *Deep-Sea Research* 48:865-889

Yu Z, Ito E (1999) Possible solar forcing of century-scale drought frequency in the northern Great Plains, *Geology* 27:263-266

Zachos J, Pagani M, Sloan L, Thomas E, Billups K (2001) Trends, rhythms and aberrations in global climate 65 ma to present. *Science* 292:686-693

Zangger E (1992) Neolithic to present soil erosion in Greece. In: Bell M and Boardman J (eds.) *Past and Present Soil Erosion. Archaeological and Geographical Perspectives.* Oxbow pp. 133-147

Zeng J, Nojiri Y, Fujinuma Y, Murphy P, Wong CS (2002) Distribution of delta-$pCO_2$ and $CO_2$ fluxes in the northern North Pacific: results from a commercial vessel in 1996-1999. *Deep-Sea Research*

Zeng N, Neelin J, Lau K-M, Compton J (1999) Enhancement of interdecadal climate variability in the Sahel by vegetation interaction. *Science* 286:1537-1540

Zhang DE (1984) Synoptic climatic studies of dust fall in China since the historic time. *Quaternary Science Reviews* Ser.B27:825-836

Zhang H, Henderson-Sellers A, McAvaney B, Pitman A (1997) Uncertainties in GCM evaluations of tropical deforestation: A comparison of two model simulations. In: Howe W and Henderson-Sellers A (eds.) *Assessing Climate Change: Results from the Model Evaluation Consortium for Climate Assessment.* Gordon and Breach Science Publisher, Sydney pp. 418

Zhang R-H, Rothstein LM, Busalacchi AJ (1998) Origin of upper-ocean warming and El Niño change on decadal scales in the tropical Pacific Ocean. *Nature* 391:879-883

Zhang Y, Wallace JM, Battisti DS (1997) ENSO-like interdecadal variability: 1900-93. *Journal of Climate* 10:1004-1020

Zhou W, Donahue DJ, Porter SC, Jull TA, Li XS, Stuiver M, An Z, Matsumoto E, Dong G (1996) Variability of monsoon climate in East Asia at the end of the last glaciation. *Quaternary Research* 46:219-229

Zhu ZR, Wyrwoll KH, Collins LB, Chen JH, Wasserburg GJ, Eisenhauer A (1993) High-precision U-series dating of Last Interglacial events by mass spectrometry: Houtman Abrolhos Islands, Western Australia. *Earth and Planetary Science Letters* 118:281-293

Zielinski G (1995) Stratospheric loading and optical depth estimates of explosive volcanism over the last 2100 years as derived from the GISP2 Greenland ice core. *Journal of Geophysical Research* 100D:20937-20955

Zielinski G, Mayewski PA, Meeker LD, Whitlow S, Twickler MS, Morrison M, Meese DA, Gow AJ, Alley RB (1994) Record of explosive volcanism since 7000 B.C. from the GISP2 Greenland ice core and implications for the volcano-climate system. *Science* 267:256-258

Zielinski GA (2000) Use of paleo-records in determining variability within the volcanism-climate system. *Quaternary Science Reviews* 19:417-438

Zolitschka B (1998) A 14000 year sediment yield record from western Germany based on annually laminated sediments. *Geomorphology* 22:1-17

# Subject Index

# Acknowledgements

The research on which this book is based has been supported by many organizations. The PAGES program acknowledges longstanding and continuing financial support from the Swiss National Science Foundation (SNF), United States National Science Foundation (NSF), United States National Oceanic and Atmospheric Administration (NOAA) and the International Geosphere Biosphere Programme (IGBP).

The editors thank all of the scientists who provided reviews of various chapters. These include: José Boninsegna (Argentine Institute for Snow and Glacier Research, Mendoza, Argentina), Carole Crumley (University of North Carolina, USA), Peter Fawcett (University of New Mexico, USA), Isabelle Larocque (PAGES International Project Office, Switzerland), Jürg Luterbacher (University of Bern, Switzerland), Vera Markgraf (University of Colorado, USA), Frank Oldfield (University of Liverpool, UK), Margit Schwikowski (Paul Scherrer Institut (Switzerland), Ryuji Tada (University of Tokyo, Japan), Ricardo Villaba (Argentine Institute for Snow and Glacier Research, Mendoza, Argentina), Robert Wasson (Australian National University) as well as those who chose to remain anonymous.

The authors and publisher gratefully acknowledge permission to reprint the following material: *Figure 1.7: Reprinted from: Quaternary Science Reviews* 19: Knox, J.C., Sensitivity of modern and Holocene floods to climate change, 439-457, 2000; Copyright 2002, with permission from Elsevier Science. *Figure 3.9: Reprinted from: Science, 291:* Tudhope, A.W., C.P. Chilcott, M.T. McCulloch, E.R. Cook, J. Chappell, R.M. Ellam, D.W. Lea, J.M. Lough, and G.B. Shimmield, Variability in the El Nino-Southern Oscillation through a Glacial-Interglacial cycle, 1511-1517, 2001; Copyright 2002, American Association for the Advancement of Science. *Figure 5.11: Reprinted from: Ecosystems* 4: Shafer, S., Bartlein, P.J. and R.S. Thompson, Potential changes in the distributions of western North America tree and shrub taxa under future climate scenarios, 200-215, 2001, Copyright 2002, Springer-Verlag GmbH & Co. *Figure 5.12: Reprinted from: Conservation Biology* 11(3): Bartlein, P.J., Whitlock, C. and S.L. Shafer, Future Climate in the Yellowstone National Park Region and its Potential Impact on Vegetation, 782-792, 1997; Copyright 2002, Blackwell Science, Inc. *Figure 6.4: Reprinted from: Geophysical Research Letters* 26: Mann, M.E., Bradley, R.S. and M.K. Hughes, Northern hemisphere temperatures during the past millennium: inferences, uncertainties, and limitations, 759-762, 1999; Copyright 2002, American Geophysical Union, Reproduced by permission of AGU. *Figure 6.8: Reprinted from: Earth Interactions* 4-4, Mann, M.E., Gille, E., Bradley, R.S., Hughes, M.K., Overpeck, J.T., Webb, R.S. and F.T. Keimig, Annual temperature patterns in past centuries: an interactive presentation, 2000; Copyright 2002, American Geophysical Union, Reproduced by permission of AGU. *Figure 6.18: Reprinted from: Science* 262, Swetnam, T.W., Fire history and climate change in Giant Sequoia groves, 885-889, 1993; Copyright 2002, American Association for the Advancement of Science.

Finally, the authors are very grateful to Christoph Kull and Isabelle Laroque for their hard work in producing camera-ready copy from drafts covered with scribbles.